Universitext

Series Editors

Nathanaël Berestycki, Universität Wien, Vienna, Austria

Carles Casacuberta, Universitat de Barcelona, Barcelona, Spain

John Greenlees, University of Warwick, Coventry, UK

Angus MacIntyre, Queen Mary University of London, London, UK

Claude Sabbah, École Polytechnique, CNRS, Université Paris-Saclay, Palaiseau, France

Endre Süli, University of Oxford, Oxford, UK

Universitext is a series of textbooks that presents material from a wide variety of mathematical disciplines at master's level and beyond. The books, often well class-tested by their author, may have an informal, personal, or even experimental approach to their subject matter. Some of the most successful and established books in the series have evolved through several editions, always following the evolution of teaching curricula, into very polished texts.

Thus as research topics trickle down into graduate-level teaching, first textbooks written for new, cutting-edge courses may find their way into *Universitext*

Frank-Olaf Schreyer

An Introduction to Algebraic Geometry

A Computational Approach

Frank-Olaf Schreyer
Fakultät für Mathematik und Informatik
Universität des Saarlandes
Saarbrücken, Germany

ISSN 0172-5939   ISSN 2191-6675 (electronic)
Universitext
ISBN 978-3-031-84833-9   ISBN 978-3-031-84834-6 (eBook)
https://doi.org/10.1007/978-3-031-84834-6

© The Editor(s) (if applicable) and The Author(s), under exclusive license to Springer Nature Switzerland AG 2025

This work is subject to copyright. All rights are solely and exclusively licensed by the Publisher, whether the whole or part of the material is concerned, specifically the rights of translation, reprinting, reuse of illustrations, recitation, broadcasting, reproduction on microfilms or in any other physical way, and transmission or information storage and retrieval, electronic adaptation, computer software, or by similar or dissimilar methodology now known or hereafter developed.
The use of general descriptive names, registered names, trademarks, service marks, etc. in this publication does not imply, even in the absence of a specific statement, that such names are exempt from the relevant protective laws and regulations and therefore free for general use.
The publisher, the authors and the editors are safe to assume that the advice and information in this book are believed to be true and accurate at the date of publication. Neither the publisher nor the authors or the editors give a warranty, expressed or implied, with respect to the material contained herein or for any errors or omissions that may have been made. The publisher remains neutral with regard to jurisdictional claims in published maps and institutional affiliations.

This Springer imprint is published by the registered company Springer Nature Switzerland AG
The registered company address is: Gewerbestrasse 11, 6330 Cham, Switzerland

If disposing of this product, please recycle the paper.

For Christine, Nora and Sarah

Preface

Algebraic Geometry is a huge area of mathematics which has gone through several phases: Hilbert's fundamental paper from 1890, sheaves and cohomology introduced by Serre in the 1950s, Grothendieck's theory of schemes in the 1960s, and so on.

This book covers the basic material known before Serre's introduction of sheaves to the subject. We will prove most of Hilbert's important theorems from his paper [44], with an emphasis on computational methods. In particular, we will use Gröbner bases systematically.

The highlights are the Nullstellensatz, Gröbner bases, Hilbert's syzygy theorem and the Hilbert function, Bézout's theorem, Mora division, semi-continuity of the fiber dimension, Bertini's theorem, Cremona resolution of plane curves and parametrization of rational curves. We will also take a glance at the Hilbert scheme and interpret the ideals of leading terms as limits under one-parameter subgroups in $\mathrm{PGL}(n+1)$. Finally, we discuss the Riemann-Roch theorem and its basic applications. Our proof of the Riemann-Roch theorem is incomplete, because it quotes three results, which have a much more transparent proof using coherent sheaves and their cohomology.

To make the book accessible to computer science students, the book has very few prerequisites. It is only assumed that the students are familiar with basic linear algebra. For students with a background in algebra, e.g. primary decomposition and localization, some chapters are superfluous.

The goal of the book is to give a self-motivating introduction to the material, and not to be a final reference book. Thus in many cases we repeat a result in a more general setting, instead of giving the most general result from the start. For example, the proof of Hilbert's Nullstellensatz in Chapter 1 uses integral ring extensions implicitly. However, the proper definition is only given in Chapter 6, where it is also used to prove basic results on the Krull dimension of a ring.

Information for the experts

The first 6 chapters are devoted to affine geometry. In Chapter 1 we give an algorithm which decides whether an algebraic system of equations in n variables $x_1, \ldots, x_n$ with rational coefficients has a solution in $\mathbb{C}^n$ by combining Hilbert's

Nullstellensatz with a Gröbner basis computation in $\mathbb{Q}[x_1, \ldots, x_n]$. This result makes it clear that we should distinguish between a field of definition k and an algebraically closed extension field K, which we do throughout the book. We denote by $\mathbb{A}^n$ the affine space over K, while $\mathbb{A}^n(k)$ denotes its subset of k-rational points.

The same Gröbner basis computation decides whether there are only finitely many solutions and computes the dimension of the solution space in case there are infinitely many solutions. However to prove that the dimension notion coming out of the proof is independent from coordinate choices takes some time until Chapter 6.

Chapter 2 establishes the strong Nullstellensatz and the algebra-geometry dictionary. In particular, we introduce the coordinate ring $K[A]$ of an algebraic variety A and morphisms.

In Chapter 3 we show that an algebraic set decomposes into irreducible components and give more generally Emmy Noether's proof of the existence of a primary decomposition of an ideal in Noetherian ring. The concept of associated primes of a finitely generated module over a Noetherian ring gives a structure result for modules, which later on in Chapter 9 is used to define the intersection multiplicity of a projective algebraic variety with a hypersurface along a subvariety.

Chapter 4 is devoted to localization and its exactness properties.

In Chapter 5 we discuss dominant rational maps between affine algebraic varieties and define the dimension of an affine variety A as the transcendence degree $\operatorname{trdeg}_K K(A)$ of its function field.

In Chapter 6 we formulate and prove a Gröbner basis criterion for the dimension of an algebraic set and for its unmixedness. We define integral ring extensions and the Krull dimension of a ring. Our proof that $k[x_1, \ldots, x_n]$ has Krull dimension n uses the dimension concept of algebraic varieties $A \subset \mathbb{A}^n$.

Chapter 7 is about constructive ideal and module theory. We give algorithms which compute

- the intersection of ideals $I, J \subset R = k[x_1, \ldots, x_n]$,
- the kernel of a ring homomorphism between affine k-algebras,
- the kernel, cokernel and image of an R-module homomorphism,
- a presentation of $\operatorname{Hom}(M, N)$ for finitely presented R-modules M and N.

In Chapter 8 we start studying projective geometry. We define $\mathbb{P}^n$, its standard charts, projective algebraic sets, homogeneous ideals and their homogeneous coordinate rings. We define the degree of a projective algebraic set via its Hilbert polynomial. The polynomial nature of the Hilbert function is deduced from Hilbert's syzygy theorem, for which we give a constructive proof with Gröbner basis.

Chapter 9 contains two versions of Bézout's theorem. The proof for the intersection of plane curves via resultants is moved to the exercise section. We illustrate its use by computing rational parametrization of plane curves with enough singularities in a few examples, a topic to which we will return in Chapter 15. The second version is Bézout's theorem for the intersection of a projective variety with a hypersurface, which we prove via the computation of a Hilbert function in two ways.

In Chapter 10 we study power series and local rings. We introduce Grauert's division theorem and Gröbner bases for the ring of formal power series $k[[x_1, \ldots, x_n]]$. A first application is the proof of the lower bound on the intersection multiplicity of two plane curves at a point p by the product of their multiplicities at p. To compute the Krull dimension of $k[[x_1,\ldots,x_n]]$ we use the Weierstrass preparation theorem. Since Grauert's division theorem gives only a convergent procedure we introduce Mora's division algorithm for applications in the local ring $\mathcal{O}_{\mathbb{A}^n,p}$. The techniques of this chapter allow us to compute the completion $\widehat{\mathcal{O}}_{A,p}$, the tangent cone $gr_\mathfrak{m}\mathcal{O}_{A,p}$ and the tangent space T_pA of an affine algebraic set A at a point p. Finally we speak about discrete valuation rings.

Chapter 11 contains basic constructions of projective geometry. We define the Segre product, morphisms and prove some dimension bounds for projective varieties and their intersections. The study of the Veronese embeddings allows us to prove that every quasi-projective algebraic set is covered by affine algebraic sets. This fact is used to prove that the image of an algebraic set under a projective morphism is closed in the target. It is also needed in the proof of the semi-continuity of the fiber dimension for projective morphisms. Finally we prove that the general fiber of a surjective projective morphism between varieties has dimension the difference of the dimensions of the source and target. We prove these with a Gröbner basis computation over a function field. The proof actually gives the flattening stratification, though we do not introduce the notion of flatness. By the same method we prove that for a homogeneous ideal $I \subset \mathbb{Z}[x_0,\ldots,x_n]$ the Hilbert function of $I_\mathbb{Q} \subset \mathbb{Q}[x_0,\ldots,x_n]$ and $I_p \subset \mathbb{Z}/p[x_0,\ldots,x_n]$ coincide for almost all primes p. This fact is fundamental for computer algebra experiments in algebraic geometry because Gröbner basis computations over finite fields avoid the coefficient explosion which frequently happens for Gröbner basis computations over $\mathbb{Q}$.

Chapter 12 contains the blow-up and the quadratic transformation. We prove that resolution of plane curve singularities follows from the existence of a Cremona resolution, that is, the existence of a sequence of quadratic transformations which transform an irreducible curve into a curve with only ordinary singularities. The existence of a Cremona resolution is established later in Chapter 15.

Chapter 13 concerns families of varieties. The main examples for the proof of the Riemann-Roch formula are linear systems of plane curves with assigned base points of given multiplicities. We also take a look at the Grassmannian and the Hilbert scheme, for which we take an ad hoc definition of subschemes of $\mathbb{A}^n$ and $\mathbb{P}^n$. We simply identify subschemes of $\mathbb{A}^n$ with arbitrary ideals in $K[x_1,\ldots,x_n]$ and subschemes of $\mathbb{P}^n$ with saturated homogeneous ideals of $K[x_0,\ldots,x_n]$. The construction of the Hilbert scheme is of course incomplete. To give these ideas some concrete content we study the Hilbert scheme $\text{Hilb}_{3t+1}(\mathbb{P}^3)$ and prove that it has (at least) two rational components $H_{12} \cup H_{15}$ corresponding to twisted cubic curves or to the union of a plane cubic with a point respectively. Using a Gröbner basis we compute the strata of the Hilbert scheme of ideals I whose ideals of leading forms coincide with $\text{Lt}(I) = (x_0^2, x_0x_1, x_0x_2, x_1^3)$ following Piene and Schlessinger [71]. Finally we interpret the association $I \rightsquigarrow \text{Lt}(I)$ as a limit of one-parameter families in $\text{PGL}(n+1,K)$ within the Hilbert scheme.

In Chapter 14 we prove Bertini's theorem and introduce the dual variety of a projective variety. Applications of Bertini's theorem include the interpretation of the degree of a projective variety as the number of intersection points with a general linear subspace of complementary dimension. We then prove the existence of a Cremona resolution, and show that intersection multiplicities of two plane curves over $\mathbb{C}$ have a dynamical interpretation.

Chapter 15 bounds the number of singular points of an irreducible plane curve C of degree d:

$$\binom{d-1}{2} \geq \sum_{p \in C} \binom{r_p}{2},$$

where r_p denotes the multiplicity of C in p. We prove that curves where equality holds have a rational parametrization, and define the geometric genus $g \geq 0$ of a curve with only ordinary singularities as the difference.

Chapter 16 concerns the famous Riemann-Roch theorem. We introduce divisors D on curves, the Riemann-Roch spaces $L(D)$ and prove Riemann's inequality for smooth irreducible projective curves via double liaison with hypersurfaces of sufficiently large degree. Next we introduce rational differential forms and compute the degree of a (canonical) divisor W of a rational differential form ω on an irreducible plane curve with only ordinary singularities. It has degree $\deg W = 2g - 2$. This proves that the geometric genus of a curve does not depend on the choice of a plane model with only ordinary singularities. It also allows us to prove the Riemann-Hurwitz formula for separable morphisms between smooth projective curves. In the case when $K = \mathbb{C}$ we prove that the geometric genus is a topological invariant of the underlying Riemann surface, by comparing it with the Riemann-Hurwitz formula for triangulated branched coverings.

Section 16.3 contains the proof of the Riemann-Roch theorem. We take the approach of Brill and Noether, but only quote the completeness of the adjoint systems. Max Noether got this result from his famous AF+BG theorem A.2.5. His approach can be found in Fulton's book [26]. I think that the proof of the completeness of the adjoint systems and also the AF+BG theorem, is much more transparent if one uses some basic results from coherent sheaves and their cohomology. In Appendix A I explain how this follows from general results on cohomology of coherent sheaves.

Section 16.4 contains Riemann's count that curves of genus g depend on $3g - 3$ moduli. The final section of Chapter 16 concerns the Clifford index and a discussion of the famous conjecture of Green [35] on the graded Betti numbers of canonical curves.

Appendix B contains partial solutions of exercises which require Macaulay2 [34] computations and the computation for $\text{Hilb}_{3t+1}(\mathbb{P}^3)$ from Section 13.5. Complete Macaulay2 code is provided for convenience.

Acknowledgment I thank the students of various classes in Saarbrücken and Perugia for their helpful comments during the lectures. Special thanks goes to Sabrina Hinrichs for pointing out many typos in the manuscript. I thank Wolfram Decker for various discussions on the course material and Nikos Tsakanikas for suggestions.

Saarbrücken, November 2024, *Frank-Olaf Schreyer*

Contents

1 Hilbert's Nullstellensatz 1
  1.1 Basic questions 1
  1.2 Gröbner bases 7
  1.3 Buchberger's criterion 14
  1.4 Proof of Hilbert's Nullstellensatz 24

2 The algebra-geometry dictionary 31
  2.1 Hilbert's strong Nullstellensatz 31
  2.2 Coordinate rings and morphisms 36

3 Noetherian rings and primary decomposition 39
  3.1 The ascending chain condition 39
  3.2 Primary decomposition 43
  3.3 Associated primes 46
  3.4 Filtration of modules with primes 49

4 Localization 51
  4.1 Fractions 51
  4.2 Some local properties 54
  4.3 Extended and contracted ideals 56
  4.4 Ideal theory of localizations 56

5 Rational functions and dimension 61
  5.1 The rational function field of a variety 61
  5.2 Dominant rational maps 64
  5.3 Appendix: The transcendence degree 66

6 Integral ring extensions and Krull dimension 69
  6.1 A Gröbner basis criterion for the dimension 69
  6.2 Integral ring extensions 70
  6.3 The lying-over theorem 73
  6.4 Krull dimension 75

7	**Constructive ideal and module theory**	79
	7.1 Syzygies and applications	79
	7.2 Elimination and the kernel of a ring homomorphism	82
	7.3 Homomorphisms between modules	84
	7.4 Cokernel, image and kernel of an R-module homomorphism	86

8	**Projective algebraic geometry**	91
	8.1 The projective space	91
	8.2 The algebra-geometry dictionary in the projective case	97
	8.3 Hilbert's syzygy theorem	100
	8.4 The Hilbert function	105

9	**Bézout's theorem**	111
	9.1 Rational functions and regular functions on projective varieties	111
	9.2 Intersection multiplicities for plane curves	112
	9.3 Bézout's theorem for plane curves	116
	9.4 Bézout's theorem for the intersection with a hypersurface	120

10	**Local rings and power series**	125
	10.1 Local rings	125
	10.2 Formal power series and completions	128
	10.3 A lower bound on intersection multiplicities	133
	10.4 Mora division	135
	10.5 Differentiation and the tangent space	137
	10.6 Discrete valuation rings	143

11	**Products and morphisms of projective varieties**	147
	11.1 The Segre product	147
	11.2 Morphisms	150
	11.3 Dimension bounds	153
	11.4 The Veronese embeddings	154
	11.5 Morphisms from projective algebraic sets	158
	11.6 Semi-continuity of the fiber dimension	160

12	**Resolution of curve singularities**	165
	12.1 The blow-up	165
	12.2 Blow-up of smooth projective surfaces	169

13	**Families of varieties**	175
	13.1 The family of hypersurfaces	175
	13.2 Linear systems of plane curves	177
	13.3 The Grassmannian	179
	13.4 A glance at the Hilbert scheme	183
	13.5 $\text{Hilb}_{3t+1}(\mathbb{P}^3)$	186
	13.6 Ideals of leading terms from a Hilbert scheme point of view	187

14 Bertini's theorem and applications 191
- 14.1 The dual variety 191
- 14.2 The Riemann-Hurwitz formula 195
- 14.3 Dynamical interpretation of intersection numbers 197

15 The geometric genus of a plane curve 199
- 15.1 A bound on the number of singular points 199
- 15.2 Existence of a Cremona resolution 201
- 15.3 Rational curves 203

16 Riemann-Roch 207
- 16.1 Divisors 208
- 16.2 Rational differentials and canonical divisors 215
- 16.3 Proof of the Riemann-Roch theorem 222
- 16.4 Riemann's count 232
- 16.5 The Clifford index and syzygies of canonical curves 236

A A glimpse of sheaves and cohomology 249
- A.1 Sheaves 249
- A.2 Cohomology 255
- A.3 Differentials and the adjunction sequence 260
- A.4 Intersection theory of curves on smooth projective surfaces 263

B Code for Macaulay2 computations 271
- B.1 Solutions to Exercises in Chapter 1, 2, 3, 4 and 6 271
- B.2 Solutions to Exercises in Chapter 8, 9, 10, 11 and 12 273
- B.3 The Computation in Section 13.5 283
- B.4 Solutions to Exercises in Chapter 14 and 16 285

References 291

Glossary 295

Index 297

Chapter 1
Hilbert's Nullstellensatz

In this chapter we develop an algorithm which decides, for an algebraic system of equations given by finitely many polynomials in $\mathbb{Q}[x_1, \ldots, x_n]$, whether this system has a solution in $\mathbb{C}^n$. This is based on Hilbert's Nullstellensatz and the solution of the ideal membership problem via Gröbner bases. The main theorems proved here are Buchberger's Gröbner basis criterion and Hilbert's Nullstellensatz. Buchberger's algorithm is a common generalization of Gaussian elimination for linear systems of equations and the euclidean algorithm in the univariate case to the case of multivariate polynomials.

Our proof of Hilbert's Nullstellensatz is effective. It allows us to compute effectively a solution of an algebraic system of equations, if such solution exists. Moreover the proof suggests a dimension notation for the solution set. However, the proof that this notation is well-defined will only be completed in Chapter 6.

1.1 Basic questions

One of the basic tasks in mathematics is to solve algebraic systems of equations.

Example 1.1.1.

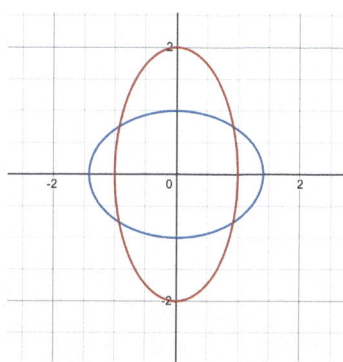

The equations

$$\frac{x^2}{2} + y^2 = 1, \quad x^2 + \frac{y^2}{4} = 1$$

define two ellipses which intersect in four points.

Definition 1.1.2. Let k be a field, for example $\mathbb{Q}, \mathbb{R}$ or $\mathbb{C}$. The **vanishing locus** in k of a polynomial
$$f = f(x_1,\ldots,x_n) \in k[x_1,\ldots,x_n]$$
in n variables $x_1,\ldots,x_n$ with coefficients in k is the set
$$V(f) = \{a = (a_1,\ldots,a_n) \in k^n \mid f(a_1,\ldots,a_n) = 0\} \subset k^n =: \mathbb{A}^n(k)$$
in the **affine n-space**. Given finitely many polynomials $f_1,\ldots,f_r \in k[x_1,\ldots,x_n]$ we denote the common solution space of the system of equations $f_1 = 0, \ldots, f_r = 0$ by
$$V(f_1,\ldots,f_r) = \bigcap_{j=1}^{r} V(f_j).$$

Question 1.1.3 (Most basic questions). Given $f_1,\ldots,f_r \in k[x_1,\ldots,x_n]$ we may ask:

1. Does the corresponding system of equations have a solution? Is the vanishing locus $V(f_1,\ldots,f_r)$ non-empty?
2. If so, how many solutions are there?
3. If there are infinitely many solutions, what is the dimension of the solution space?
4. If there are infinitely many solutions, can we parametrize the solution space?

Examples 1.1.4.

a)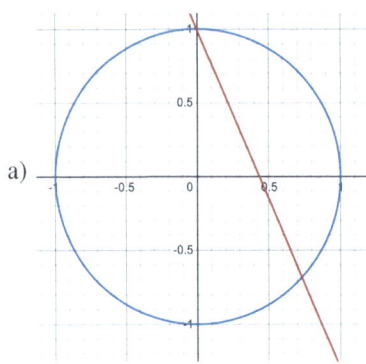

The circle $x^2 + y^2 = 1$ can be parametrized by rational functions:
$$y = tx + 1 \text{ and } x^2 + y^2 = 1$$
$$\Rightarrow x^2 + (tx+1)^2 = 1$$
$$\Rightarrow x + t^2 x + 2t = 0$$
$$\Rightarrow x = \frac{-2t}{1+t^2}, \ y = \frac{1-t^2}{1+t^2}.$$

1.1 Basic questions

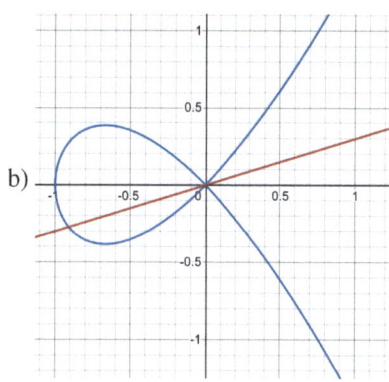

The nodal cubic $y^2 = x^3 + x^2$ has a parametrization with polynomial functions:

$$y = tx \text{ and } y^2 = x^3 + x^2$$
$$\Rightarrow t^2 x^2 = x^3 + x^2$$
$$\Rightarrow x = t^2 - 1,\ y = t(t^2 - 1).$$

The answer to the first question in 1.1.3 depends very much on the nature of the field.

a) In the case of $\mathbb{C}$, solvability can be decided with Hilbert's Nullstellensatz (1899).
b) In the case of $\mathbb{R}$, quantifier elimination (Tarski 1948) leads to an answer.
   Example. $\exists x \in \mathbb{R} : x^2 + px + q = 0 \iff p^2 - 4q \geq 0$.
c) In case of $\mathbb{Q}$, there exists no general algorithm which decides whether a system of algebraic equations has an integer solution. (Matiyasevich's solution (1970) of Hilbert's 10th problem, [58]). It is not known whether there exists an algorithm which decides whether there exists a solution in $\mathbb{Q}$, see [72] for some recent results.

Hilbert's Nullstellensatz uses the concept of ideals, which we discuss next.

Definition 1.1.5. Let R be a (commutative) ring (with 1). A non-empty subset $I \subset R$ is an **ideal** if

1) $a, b \in I \Rightarrow a + b \in I$, and
2) $r \in R, a \in I \Rightarrow ra \in I$

holds.

Examples 1.1.6. 1) Let $\varphi \colon R \to S$ be a ring homomorphism. Then the kernel of the ring homomorphism $\ker \varphi = \{a \in R \mid \varphi(a) = 0\}$ is an ideal.
2) Let $f_1, \ldots, f_r \in R$ be elements of a ring. Then

$$(f_1, \ldots, f_r) = \{f \mid \exists g_1, \ldots, g_r \in R : f = g_1 f_1 + \cdots + g_r f_r\}$$

is an ideal, **the ideal generated by** $f_1, \ldots, f_r$.

Definition 1.1.7 (Residue rings). Let R be a ring, $I \subset R$ an ideal. Then

$$a \equiv b \mod I \iff a - b \in I$$

is an equivalence relation on R. We denote the **residue class** of a by

$$\bar{a} = \{b \in R \mid b \equiv a\} = a + I \subset R.$$

The set of residue classes

$$R/I = \{\bar{a} \mid a \in R\} \subset 2^R$$

carries the structure of a ring defined by

$$\bar{a} + \bar{b} := \overline{a+b}, \bar{a} \cdot \bar{b} := \overline{ab}.$$

This is the unique ring structure on R/I which makes the map

$$\pi: R \to R/I, a \mapsto \bar{a}$$

into a ring homomorphism. Note that $\ker \pi$ equals I.

Examples 1.1.8. 1) For $n \in \mathbb{Z}$ an integer, the residue ring $\mathbb{Z}/(n)$ has n elements, namely $\bar{0}, \bar{1}, \ldots, \overline{n-1}$. The residue ring $\mathbb{Z}/(p)$ is a field if and only if p is a prime number. We denote the field with p elements by

$$\mathbb{F}_p := \mathbb{Z}/(p).$$

2) The polynomial $f = x^2 + x + 1 \in \mathbb{F}_2[x]$ has no zero in $\mathbb{F}_2$. The ring

$$\mathbb{F}_4 = \mathbb{F}_2[x]/(x^2 + x + 1)$$

is a field with 4 elements.

3) All finite fields $\mathbb{F}_q$ can be constructed similarly. The number of elements $q = p^r$ is necessarily a prime power, and

$$\mathbb{F}_q \cong \mathbb{F}_p[x]/(f)$$

for f a monic irreducible polynomial of degree r in $\mathbb{F}_p[x]$.

Theorem 1.1.9 (Division with remainder). *Let k be a field and $f \in k[x]$ a univariate polynomial which is not the zero polynomial. For all $g \in k[x]$ there exist unique polynomials $q, r \in k[x]$ such that*

$$g = qf + r \text{ and } \deg r < \deg f.$$

*The polynomial r is called the **remainder** of g divided by f.*

Proof. Existence follows by induction on $\deg g$. For $\deg g < \deg f$ we may take $q = 0$ and $r = g$. If $m = \deg g \geq n = \deg f$ with leading coefficients b_m and a_n, we consider the difference

$$g^{(1)} = g - \frac{b_m}{a_n} x^{m-n} f.$$

In this difference the leading coefficient cancels, hence $\deg g^{(1)} < \deg g$. By the induction hypothesis, we have an expression $g^{(1)} = q^{(1)} f + r$. Then

1.1 Basic questions

$$g = \left(\frac{b_m}{a_n}x^{m-n} + q^{(1)}\right)f + r$$

is the desired expression for g.

Uniqueness: Suppose $g = qf + r = q'f + r'$ are two expressions. Then

$$0 = (q - q')f + (r - r').$$

If $q \neq q'$, then $\deg(q - q')f \geq \deg f > \deg(r - r')$ and $(q - q')f + (r - r') \neq 0$, a contradiction. Thus $q = q'$ and $r = r'$. □

How to compute in $k[x]/(f)$? Let k be a field and $f \in k[x] \setminus \{0\}$ a univariate polynomial. Suppose f is monic of degree $d = \deg f > 0$, i.e.,

$$f = x^d + a_{d-1}x^{d-1} + \cdots + a_1 x^1 + a_0.$$

Then every element $\overline{g} \in k[X]/(f)$ has a unique representative $r \in k[x]$ by a polynomial of degree $\leq d - 1$. The elements $1, x, \ldots, x^{d-1}$ represent a k-vector space basis of $k[x]/(f)$.

Given two elements $\overline{g}, \overline{h} \in k[x]/(f)$, we compute their product by taking representatives g, h and the remainder r of gh divided by f.

Example. Consider $\overline{x} \in \mathbb{F}_4 = \mathbb{F}_2[x]/(x^2 + x + 1)$. Then

$$\overline{x}^2 = -\overline{x} - 1 = \overline{x} + 1$$

and

$$\overline{x}^3 = \overline{x}^2 \overline{x} = (\overline{x} + 1)\overline{x} = \overline{x}^2 + \overline{x} = 1.$$

Hence the multiplicative group $(\mathbb{F}_4^*, \cdot)$ is cyclic of order 3.

Definition 1.1.10. Let k be a field. An **affine k-algebra** is a ring of the form

$$R = k[x_1, \ldots, x_n]/(f_1, \ldots, f_r).$$

One of the goals of this chapter is to learn how to compute in such rings. In particular, we want to decide whether an element $\overline{f}$ is zero in this ring.

Ideal membership problem. Given a field k, an ideal $(f_1, \ldots, f_r) \subset k[x_1, \ldots, x_n]$ and an element $f \in k[x_1, \ldots, x_n]$ decide whether $f \in (f_1, \ldots, f_r)$.

Theorem 1.1.11 (Hilbert's Nullstellensatz, weak version). *Let K be an algebraically closed field. Let $f_1, \ldots, f_r \in K[x_1, \ldots, x_n]$ be polynomials. Then*

$$V(f_1, \ldots, f_r) = \emptyset \iff 1 \in (f_1, \ldots, f_r).$$

Thus combined with an algorithm for the membership problem, we can decide whether an algebraic system of equations has a solution. One direction in Hilbert's Nullstellensatz is easy. Suppose $1 \in (f_1, \ldots, f_r)$, say $1 = g_1 f_1 + \cdots + g_r f_r$. If $a \in V(f_1, \ldots, f_r)$, then

$$1 = g_1(a)f_1(a) + \cdots + g_r(a)f_r(a) = 0,$$

a contradiction. Thus $V(f_1, \ldots, f_r) = \emptyset$.

Definition 1.1.12. A field K is **algebraically closed** if every non-constant univariate polynomial $f \in K[X]$ has a root in K.

The condition K algebraically closed is clearly a necessary assumption in Hilbert's Nullstellensatz:
If $f \in k[x]$ is a univariate polynomial of positive degree which has no root in k, then $V(f)$ is the empty subset of $\mathbb{A}^1(k)$. But $1 \notin (f)$, since non-zero elements of (f) have degree $\geq \deg f$.

Fundamental theorem of algebra. *The field of complex numbers $\mathbb{C}$ is algebraically closed.* □

Theorem (Steinitz, [82]). *Every field k is contained in an algebraically closed extension field $K \supset k$.* □

Solvability with Computer Algebra. For $f_1, \ldots, f_r \in \mathbb{Q}[x_1, \ldots, x_n]$ we consider the vanishing locus

$$V(f_1, \ldots, f_r) := \{a \in \mathbb{C}^n \mid f_1(a) = 0, \ldots, f_r(a) = 0\} \subset \mathbb{A}^n(\mathbb{C})$$

over $\mathbb{C}$. Due to the Nullstellensatz we can decide $V(f_1, \ldots, f_r) = \emptyset$ with a computation over $\mathbb{Q}$:

The condition $1 = g_1 f_1 + \cdots + g_r f_r$ can be viewed as a linear system of equations for unknown coefficients of $g_1, \ldots, g_r$. If this system has a solution over $\mathbb{C}$, it also has a solution over $\mathbb{Q}$. Thus

$$V(f_1, \ldots, f_r) = \emptyset \subset \mathbb{A}^n(\mathbb{C}) \iff 1 \in (f_1, \ldots, f_r) \subset \mathbb{Q}[x_1, \ldots, x_n].$$

Implementing $\mathbb{C}$ into a computer requires numerical methods. But $\mathbb{Q}$ is accessible to exact computer algebra methods.

Definition 1.1.13. The set $\mathbb{A}^n = K^n$ will always denote the **affine n-space** over an algebraically closed field K. An **algebraic subset** $X \subset \mathbb{A}^n$ is a set of the form

$$X = V(f_1, \ldots, f_r) \subset \mathbb{A}^n$$

for polynomials $f_1, \ldots, f_r \in K[x_1, \ldots, x_n]$.

If $f_1, \ldots, f_r \in k[x_1, \ldots, x_n]$ for a subfield $k \subset K$, then we call k **a field of definition** of X. In this case

$$X(k) = X \cap \mathbb{A}^n(k) \subset \mathbb{A}^n = \mathbb{A}^n(K)$$

denotes the set of k-**rational points of** X.

Diophantine equations. Let $f_1, \ldots, f_r \in \mathbb{Z}[x_1, \ldots, x_n]$ be polynomials with integral coefficients and let $X = V(f_1, \ldots, f_r)$ be their vanishing locus. Then for

1.2 Gröbner bases

any prime number p we can reduce the coefficients mod p to obtain equations in $\mathbb{F}_p[x_1,\ldots,x_n]$. Thus $X(\mathbb{F}_p)$ makes sense, and the numbers $N_r = |X(\mathbb{F}_{p^r})|$ of $\mathbb{F}_{p^r}$-rational points are defined.

We will see that for almost all prime numbers p, the growth of N_r determines the dimension of X over $\mathbb{C}$:

$$N_r = O(p^{rd}) \iff \dim_{\mathbb{C}} X = d.$$

If we want to study $X(\mathbb{Q})$, then the study of $X(\mathbb{F}_{p^r})$ and $X(\mathbb{R})$ gives some partial information. There is a huge branch of mathematics devoted to this approach to diophantine equations.

Exercise 1.1.14. Let k be an infinite field and $f \in k[x_1,\ldots,x_n]$ a non-zero polynomial. Show that there exists a point $a \in \mathbb{A}^n(k)$ such that $f(a) \neq 0$.

Exercise 1.1.15. Let $k[x]$ be a polynomial ring in one variable over a field. Prove that $k[x]$ is a principal ideal domain, that is, every ideal $I \subset k[x]$ is generated by a single polynomial.

Exercise 1.1.16. One of the following two systems of three equations in two variables has a solution in $\mathbb{A}^2(\mathbb{C})$, the other one not.

a) 1) $28x^3 + 63x^2y + 14xy^2 + 7y^3 + 63xy + 28y^2 + 28x + 49y + 58 = 0$,
   2) $109x^3 + 99x^2y + 41xy^2 + 16y^3 + 27x^2 + 108xy + 46y^2 + 28x + 103y + 88 = 0$,
   3) $35x^3 + 30x^2y + 13xy^2 + 5y^3 + 9x^2 + 33xy + 14y^2 + 8x + 32y + 32 = 0$.
b) 1) $5x^3 + 8x^2y + 7xy^2 + 7y^3 - 7x^2 + xy - 6y^2 + x - 2y - 14 = 0$,
   2) $3x^3 + 15x^2y + 4xy^2 + y^3 - 3x^2 - 10xy + 3y^2 - 4x - 3y - 6 = 0$,
   3) $8x^3 + 14x^2y + 8xy^2 + 4y^3 - 9x^2 - 6x - 4y - 15 = 0$.

Which system has no solution?

1.2 Gröbner bases

Gröbner bases are our tools to solve the ideal membership problem. They are based on monomial orders.

Definition 1.2.1. A **monomial** in $k[x_1,\ldots,x_n]$ is an element of the form

$$x^\alpha = x_1^{\alpha_1} \cdot \ldots \cdot x_n^{\alpha_n},$$

where $\alpha = (\alpha_1,\ldots,\alpha_n) \in \mathbb{N}^n = \mathbb{Z}_{\geq 0}^n$ is a multi-exponent. Thus $x^\alpha x^\beta = x^{\alpha+\beta}$. A **term** in $k[x_1,\ldots,x_n]$ is an element of the form ax^α with $a \in k$. Every element $f \in k[x_1,\ldots,x_n]$ is a finite sum of terms, i.e.,

$$f = \sum f_\alpha x^\alpha.$$

Example 1.2.2 (A motivating example). Consider the ideal $I = (x^2 + xy, y^2 + xy)$ in the polynomial ring $k[x, y]$. Applying division with remainder we can use $x^2 + xy$ to remove from an $f \in k[x, y]$ any multiple of x^2:

$$f = q(x^2 + xy) + r \text{ with } r \in k[y] + xk[y].$$

Likewise, we can use $y^2 + xy$ to remove multiples of y^2.

Can we use both generators to remove multiples of x^2 or y^2 simultaneously? The answer is no! If the answer were yes, then $\overline{1}, \overline{x}, \overline{y}, \overline{xy}$ would generate $k[x, y]/I$ as a k-vector space. But this is an infinite-dimensional k-vector space, since we have a surjection

$$k[x, y]/I \twoheadrightarrow k[x, y]/(x + y) \cong k[y].$$

What went wrong? We did not choose the leading terms x^2 and y^2 in a compatible way!

Definition 1.2.3. A **monomial order** $>$ on $k[x_1, \ldots, x_n]$ is a total order of the monomials in $k[x_1, \ldots, x_n]$ satisfying

$$x^\alpha > x^\beta \implies x^\alpha x^\gamma > x^\beta x^\gamma$$

for any triple of monomials. For $f = \sum f_\alpha x^\alpha$ we define the **leading term** with respect to the relation $>$ as

$$\mathrm{Lt}(f) = f_\alpha x^\alpha \text{ where } x^\alpha = \max\{x^\beta \mid f_\beta \neq 0\} \text{ and } \mathrm{Lt}(0) = 0.$$

Note that since a monomial order is a total order we actually have

$$x^\alpha > x^\beta \iff x^\alpha x^\gamma > x^\beta x^\gamma$$

because $x^\alpha x^\gamma < x^\beta x^\gamma \implies x^\alpha < x^\beta$ holds.

In our example above the choice $\mathrm{Lt}(x^2 + xy) = x^2$ implies $x^2 > xy \iff x > y \iff xy > y^2 \implies \mathrm{Lt}(y^2 + xy) = xy$. So our choice above was not compatible with a monomial order.

Abusing notation we write for non-zero terms

$$ax^\alpha \geq bx^\beta \text{ if } x^\alpha \geq x^\beta \quad (:\iff x^\alpha > x^\beta \text{ or } x^\alpha = x^\beta).$$

Note that $\geq$ is not an order on the set of non-zero terms since

$$ax^\alpha \geq bx^\beta \text{ and } bx^\beta \geq ax^\alpha \implies x^\alpha = x^\beta,$$

but $a \neq b$ is possible.

Proposition 1.2.4. *Let $>$ be a monomial order. Then*

1.2 Gröbner bases

1. $\mathrm{Lt}(fg) = \mathrm{Lt}(f)\mathrm{Lt}(g)$,
2. $\mathrm{Lt}(f+g) \leq \max(\mathrm{Lt}(f), \mathrm{Lt}(g))$ *and equality holds unless* $\mathrm{Lt}(f) + \mathrm{Lt}(g) = 0$.

□

Definition 1.2.5. A **global** monomial order on $k[x_1, \ldots, x_n]$ is a monomial order satisfying
$$x_j > 1 \text{ for } j = 1, \ldots, n.$$
In contrast, a **local** monomial order on $k[x_1, \ldots, x_n]$ is a monomial order satisfying
$$x_j < 1 \text{ for } j = 1, \ldots, n.$$

The key property of global monomial orders is that there are **no** infinite descending sequences $m_1 > m_2 > \ldots$ of monomials, which will follow from Dixon's lemma below.

In contrast, for a local monomial order
$$1 > x_1 > x_1^2 > \ldots > x_1^k > \ldots$$
is an infinite descending sequence.

Local orders are useful for computations in power series rings $k[[x_1, \ldots, x_n]]$. We will consider those later in Chapter 10.

Definition 1.2.6. For a monomial x^α the **degree** is defined by
$$\deg x^\alpha = \sum_{j=1}^n \alpha_j = |\alpha|.$$
For a non-zero polynomial $f = \sum f_\alpha x^\alpha$ the degree is
$$\deg f = \max\{\deg x^\alpha \mid f_\alpha \neq 0\}.$$

There are plenty of different global monomial orders. The most important ones are the following:

1) The **lexicographic** monomial order is defined by
$$x^\alpha >_{\mathrm{lex}} x^\beta$$
if the first non-zero entry of $\alpha - \beta \in \mathbb{Z}^n$ is positive. Thus
$$x_1 x_3 >_{\mathrm{lex}} x_1 >_{\mathrm{lex}} x_2^k >_{\mathrm{lex}} x_2^2.$$

2) The **reversed degree lexicographic** order is defined as follows:
$$x^\alpha >_{\mathrm{rdlex}} x^\beta$$

if $\deg x^\alpha > \deg x^\beta$ or $\deg x^\alpha = \deg x^\beta$ and the last non-zero entry of $\alpha - \beta \in \mathbb{Z}^n$ is negative. Thus

$$x_3^3 >_{\text{rdlex}} x_1^2 >_{\text{rdlex}} x_2^2 >_{\text{rdlex}} x_1 x_3.$$

3) **Weight orders**. Let $w = (w_1, \ldots, w_n) \in \mathbb{R}_{>0}^n$ be a weight vector. Consider the linear form $w(\alpha) = \sum_{j=1}^n w_j \alpha_j$. We define

$$x^\alpha >_w x^\beta \text{ if } w(\alpha) > w(\beta) \text{ or } w(\alpha) = w(\beta) \text{ and } x^\alpha >_{\text{tb}} x^\beta$$

where $>_{\text{tb}}$ denotes a tiebreak order, for example $>_{\text{lex}}$. If the weights $w_1, \ldots, w_n$ are $\mathbb{Q}$-linearly independent, then the condition $x^\alpha >_{\text{tb}} x^\beta$ is superfluous.

Definition 1.2.7. Let J be an arbitrary set of polynomials. The **ideal generated by** J is

$$I = (J) = \{f \mid \exists r \in \mathbb{N}, f_1, \ldots, f_r \in J \text{ and } g_1, \ldots, g_r \in k[x_1, \ldots, x_n]$$
$$\text{such that } f = g_1 f_1 + \cdots + g_r f_r\}.$$

Definition 1.2.8. A **monomial ideal** $I \subset k[x_1, \ldots, x_n]$ is an ideal satisfying

$$f = \sum f_\alpha x^\alpha \in I \implies x^\alpha \in I \, \forall \alpha \text{ with } f_\alpha \neq 0.$$

In other words, I is generated by monomials.

Lemma 1.2.9 (Dixon's Lemma). *Every monomial ideal I is finitely generated, i.e., there exists a finite set J of monomials such that $I = (J)$.*

Proof. Induction on n. Let $I \subset k[x_1, \ldots, x_n]$ be a non-zero monomial ideal, $x^\alpha \in I$ and $\alpha = (\alpha_1, \ldots, \alpha_n)$. For $j = 1, \ldots, n$ and $\gamma = 0, \ldots, \alpha_j - 1$ consider the monomial ideal $I_{j,\gamma}$ generated by

$$\{x^\beta \subset K[x_1, \ldots, x_{j-1}, x_{j+1}, \ldots, x_n] \mid x_j^\gamma x^\beta \in I\}$$

in a polynomial ring with $n-1$ variables. By induction hypothesis all $I_{j,\gamma}$ are finitely generated, say by a set of monomials $J_{j,\gamma}$. Then

$$J = \{x^\alpha\} \cup \bigcup_{j,\gamma} \{x_j^\gamma x^\beta \mid x^\beta \in J_{j,\gamma}\}$$

is a finite set of generators of I. □

Definition 1.2.10. A monomial x^α is a **minimal generator** of a monomial ideal I if it is not divisible by any other monomial in I. In other words, x^α cannot be dropped from a generating set by monomials. The minimal generators of a monomial ideal form a generating set.

Corollary 1.2.11. *Let $>$ be a global monomial order and $m_1 \geq m_2 \geq \ldots \geq m_j \geq \ldots$ a descending chain of monomials. Then there exists an $N \in \mathbb{N}$ such that*

1.2 Gröbner bases

$$m_j = m_N \; \forall j \geq N.$$

Proof. A global monomial order > refines divisibility in $k[x_1, \ldots, x_n]$:

$$x^\alpha | x^\beta \iff \beta - \alpha \in \mathbb{Z}_{\geq 0}^n \implies x^{\beta-\alpha} \geq 1 \implies x^\beta \geq x^\alpha.$$

Consider the ideal $I = (\{m_j \mid j \in \mathbb{N}\})$. By Dixon's lemma, I is generated by a finite set J of monomials. Set $N = \min\{\ell \mid J \subset \{m_1, \ldots, m_\ell\}\}$. For $j \geq N$ every monomial m_j is divisible by a generator $m_\nu \in J$. Thus we have $m_j \geq m_\nu \geq m_N \geq m_j$ and equality holds: $m_j = m_N$. $\square$

Theorem 1.2.12 (Division with remainder). *Let > be a fixed global monomial order on $k[x_1, \ldots, x_n]$, and let $f_1, \ldots, f_r \in k[x_1, \ldots, x_n]$ be non-zero polynomials. For every $f \in k[x_1, \ldots, x_n]$ there exist uniquely determined $g_1, \ldots, g_r \in k[x_1, \ldots, x_n]$ and a unique remainder $h \in k[x_1, \ldots, x_n]$ satisfying*

1) $f = g_1 f_1 + \cdots + g_r f_r + h$,
2) a) *No term of $g_j \operatorname{Lt}(f_j)$ is divisible by a leading term $\operatorname{Lt}(f_i)$ for some $i < j$,*
   b) *No term of h is divisible by a leading term $\operatorname{Lt}(f_j)$ for any $j = 1, \ldots, r$.*

Proof. Uniqueness: Taking differences it suffices to prove that

$$0 = g_1 f_1 + \cdots + g_r f_r + h \implies g_1 = 0, \ldots, g_r = 0, h = 0.$$

Since non-zero leading terms of the form $\operatorname{Lt}(g_j f_j) = \operatorname{Lt}(g_j) \operatorname{Lt}(f_j)$ and $\operatorname{Lt}(h)$ belong to different monomials by condition 2), they cannot cancel in the sum. So all are zero, hence all g_j and h are zero.

Existence: The theorem is trivially true for monomial ideals. Thus we can write

$$f = g_1^{(0)} \operatorname{Lt}(f_1) + \cdots + g_r^{(0)} \operatorname{Lt}(f_r) + h^{(0)}$$

satisfying 2a) and 2b). Consider

$$f^{(1)} = f - (g_1^{(0)} f_1 + \cdots + g_r^{(0)} f_r + h^{(0)}).$$

In the difference on the right-hand side the leading term cancels. Hence either $f^{(1)} = 0$ and we are done, or

$$\operatorname{Lt}(f^{(1)}) < \operatorname{Lt}(f).$$

Continuing with $f^{(1)}$ we obtain a sequence of polynomials

$$f^{(\nu+1)} = f^{(\nu)} - (g_1^{(\nu)} f_1 + \cdots + g_r^{(\nu)} f_r + h^{(\nu)}),$$

where

$$f^{(\nu)} = g_1^{(\nu)} \operatorname{Lt}(f_1) + \cdots + g_r^{(\nu)} \operatorname{Lt}(f_r) + h^{(\nu)}$$

whose leading terms form a descending sequence

$$\text{Lt}(f) > \text{Lt}(f^{(1)}) > \text{Lt}(f^{(2)}) > \ldots.$$

So after a finite number of steps we arrive at $f^{(N+1)} = 0$, and then

$$\text{the } g_j = \sum_{\nu=0}^{N} g_j^{(\nu)} \text{ and } h = \sum_{\nu=0}^{N} h^{(\nu)}$$

are the desired coefficients and remainder. □

Definition 1.2.13. Let $>$ be a global monomial order and $I \subset k[x_1,\ldots,x_n]$ an ideal. The **ideal of leading terms** of I is the ideal generated by the leading terms of elements of I:

$$\text{Lt}(I) = (\{\text{Lt}(f) \mid f \in I\}).$$

Elements $f_1,\ldots,f_r \in I$ form a **Gröbner basis** of I (with respect to $>$) if

$$\text{Lt}(I) = (\text{Lt}(f_1),\ldots,\text{Lt}(f_r)).$$

Theorem 1.2.14 (Hilbert's, 1899). *Every ideal in $k[x_1,\ldots,x_n]$ is finitely generated.*

Gordon's proof of Hilbert's basis theorem. Let $I \subset k[x_1,\ldots,x_n]$ be an ideal. Consider the leading term ideal $\text{Lt}(I)$. This is a monomial ideal, hence it is finitely generated by Dixon's lemma.

Let $f_1,\ldots,f_r \in I$ be elements whose leading terms generate $\text{Lt}(I)$. We claim

$$I = (f_1,\ldots,f_r).$$

Indeed $(f_1,\ldots,f_r) \subset I$ is clear, since $f_1,\ldots,f_r \in I$. For the other inclusion, let $f \in I$ be an arbitrary element. Consider the remainder h of f divided by $f_1,\ldots,f_r$,

$$h = f - (g_1 f_1 + \cdots + g_r f_r).$$

Then on the one hand, we have $h \in I$ and on the other hand no non-zero term of h lies in $\text{Lt}(I) = (\text{Lt}(f_1),\ldots,\text{Lt}(f_r))$ by condition 2b). Thus $\text{Lt}(h) = 0$. Hence $h = 0$ and $f \in (f_1,\ldots,f_r)$. □

Corollary 1.2.15. *Let $I \subset k[x_1,\ldots,x_n]$ be an ideal, $f \in k[x_1,\ldots,x_n]$ and $f_1,\ldots,f_r$ a Gröbner basis for I. Then $f \in I$ if and only if the remainder h of f divided by $f_1,\ldots,f_r$ is zero.* □

Theorem 1.2.16 (Macaulay). *Let $f_1,\ldots,f_r$ be a Gröbner basis of an ideal $I \subset k[x_1,\ldots,x_n]$ with respect to a global monomial order. Then the monomials $\{x^\alpha \mid x^\alpha \notin \text{Lt}(I)\}$ represent a k-vector space basis for $k[x_1,\ldots,x_n]/I$.*

1.2 Gröbner bases

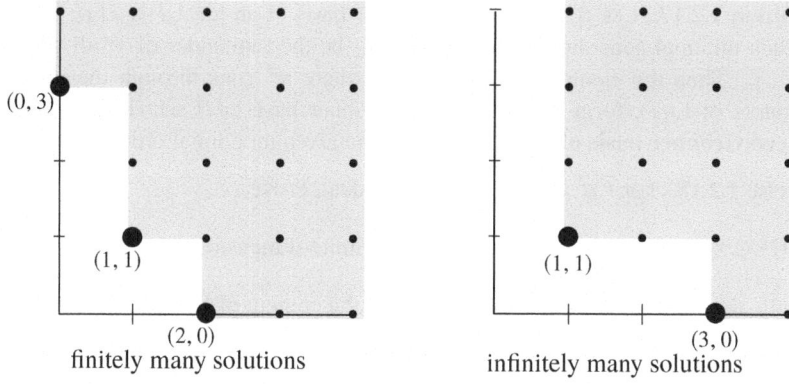

finitely many solutions   infinitely many solutions

By Exercise 2.1.14 $V(I) \subset \mathbb{A}^n$ is finite if and only if $k[x_1,\ldots,x_n]/I$ is a finite-dimensional k-vector space.

Proof. Let $\overline{f}$ be an element of $k[x_1,\ldots,x_n]/I$ and $f \in k[x_1,\ldots,x_n]$ a representative. Then the remainder h of f divided by $f_1,\ldots,f_r$ represents the same element: $\overline{f} = \overline{h}$. Since $\mathrm{Lt}(I) = (\mathrm{Lt}(f_1),\ldots,\mathrm{Lt}(f_r))$, the remainder h is a linear combination of the $x^\alpha \notin \mathrm{Lt}(I)$ by condition 2b). So the $\overline{x}^\alpha$ with $x^\alpha \notin \mathrm{Lt}(I)$ span $k[x_1,\ldots,x_n]/I$ as an k-vector space. They are linearly independent by Corollary 1.2.15. □

Example of a division. Consider $f_1 = x^2y - y^3$, $f_2 = x^3 \in k[x,y]$ and $>_{\mathrm{lex}}$. Then

$$\mathrm{Lt}(f_1) = x^2y \text{ and } \mathrm{Lt}(f_2) = x^3.$$

We divide $f = x^3y$ by f_1, f_2:

$$f = x\,\mathrm{Lt}(f_1) + 0\,\mathrm{Lt}(f_2) + 0, \text{ hence}$$
$$f^{(1)} = f - (xf_1 + 0f_2 + 0) = xy^3.$$

In the second step we obtain

$$xy^3 = 0\,\mathrm{Lt}(f_1) + 0\,\mathrm{Lt}(f_2) + xy^3, \text{ hence}$$
$$f^{(2)} = f^{(1)} - (0f_1 + 0f_2 + xy^3) = 0.$$

The final result is $f = xf_1 + 0f_2 + xy^3$.
On the other hand if we divide $f = x^3y$ by $x^3, x^2y - y^3$ we obtain

$$f = y\,\mathrm{Lt}(x^3) + 0\,\mathrm{Lt}(x^2y - y^3)) + 0, \text{ hence}$$
$$f^{(0)} = x^3y - (y(x^3) + 0(x^2y - y^3) + 0) = 0$$

and the final result is $f = yf_2 + 0f_1 + 0$. Thus

Warning: The remainder of the division by polynomials $f_1,\ldots,f_r$ can depend on the order of $f_1,\ldots,f_r$!
This does not happen if $f_1,\ldots,f_r$ is a Gröbner basis.

Definition 1.2.17. Let $f_1,\ldots,f_r$ be a Gröbner basis of an ideal $I \subset k[x_1,\ldots,x_n]$. For each minimal generator x^α of $\mathrm{Lt}(I)$ let h_α be the remainder of x^α divided by $f_1,\ldots,f_r$. Then the elements $x^\alpha - h_\alpha \in I$, where x^α runs through the minimal generators of $\mathrm{Lt}(I)$, form a distinguished Gröbner basis of I which is called the **reduced Gröbner basis of** I with respect to the given monomial order $>$.

Exercise 1.2.18. Let $I \subset k[x_1,\ldots,x_n]$ be an ideal. Prove

$$V(I) \subset \mathbb{A}^n \text{ is finite} \Leftarrow k[x_1,\ldots,x_n]/I \text{ is a finite-dimensional } k\text{-vector space.}$$

Actually the conditions are equivalent. This is the content of Exercise 2.1.14.

Exercise 1.2.19. Let $>$ be a monomial order and let $M \subset k[x_1,\ldots,x_n]$ be a finite set of monomials. Prove that there exists a weight order $>_w$ (with $\mathbb{Q}$-linearly independent weights) which induces the same order on the monomials of M as $>$.

Hint: Consider the convex hull C of the set

$$\{\alpha - \beta \mid x^\alpha, x^\beta \in M \text{ with } x^\alpha \geq x^\beta\} \subset \mathbb{R}^n$$

and prove that $0 \in C$ is a vertex of C, i.e., 0 is not a linear combination of other points in C with strictly positive coefficients.

1.3 Buchberger's criterion

Let $f_1,\ldots,f_r \in k[x_1,\ldots,x_n]$ be polynomials. How to compute a Gröbner basis for $I = (f_1,\ldots,f_r)$?

The easiest way to discover a new leading term of $(f_1,\ldots,f_r)$ is to consider a difference where the leading terms cancel. Consider the monomial $m_{ij} = \gcd(\mathrm{Lt}(f_i), \mathrm{Lt}(f_j))$ and the **S-polynomial**

$$S(f_i, f_j) := \frac{\mathrm{Lt}(f_i)}{m_{ij}} f_j - \frac{\mathrm{Lt}(f_j)}{m_{ij}} f_i.$$

The leading term in this difference cancels, so we might discover a new leading term of I.

Theorem 1.3.1 (Buchberger's criterion). *Let $f_1,\ldots,f_r \in k[x_1,\ldots,x_n]$ be polynomials and $>$ be a global monomial order. $f_1,\ldots,f_r$ is a Gröbner basis for $(f_1,\ldots,f_r)$ if and only if for each pair i,j the remainder of $S(f_i, f_j)$ divided by $f_1,\ldots,f_r$ is zero.*

Algorithm 1.3.2 (Buchberger).
Input. A global monomial order and polynomials $f_1,\ldots,f_r$.
Output. A Gröbner basis $f_1,\ldots,f_s$ for $(f_1,\ldots,f_r)$.

1. Initialize $s = r$ and $L = \{f_1,\ldots,f_r\}$;

1.3 Buchberger's criterion

2. for all i, j with $1 \leq i < j \leq s$ do
   compute the remainder h of $S(f_i, f_j)$ divided by $f_1, \ldots, f_s$;
   if $h \neq 0$ then
   $$f_{s+1} = h; L = L \cup \{f_{s+1}\}; s = s + 1;$$
3. return L.

The algorithm terminates, since monomial ideals are finitely generated.

Example 1.3.3. Consider $f_1 = x^3$, $f_2 = x^2y - y^3 \in K[x, y]$ and $>_{\text{lex}}$. Then

$$\text{Lt}(f_1) = x^3, \text{Lt}(f_2) = x^2y$$

$m_{12} = x^2$ and $S(f_1, f_2) = xf_2 - yf_1 = -xy^3 = 0f_1 + 0f_2 - xy^3$ has a non-zero remainder. Thus

$$f_3 = -xy^3.$$

$m_{13} = x$ and $S(f_1, f_3) = x^2 f_3 - (-y^3) f_1 = 0$.
$m_{23} = xy$ and $S(f_2, f_3) = xf_3 - (-y^2) f_2 = -y^5$. Thus

$$f_4 = -y^5.$$

The S-polynomials $S(f_1, f_4)$ and $S(f_3, f_4)$ are zero. Finally, $m_{24} = y$ and $S(f_2, f_4) = x^2 f_4 - (-y^4) f_2 = -y^7 = 0f_1 + 0f_2 + 0f_3 + y^2 f_4 + 0$.
So $f_1, \ldots, f_4$ is a Gröbner basis.

Example 1.3.4 (3×3-minors of a 3×5-matrix). Consider the ideal $I \subset k[x_1, \ldots, z_5]$ generated by the 3×3 minors of the matrix

$$\begin{pmatrix} x_1 & x_2 & x_3 & x_4 & x_5 \\ y_1 & y_2 & y_3 & y_4 & y_5 \\ z_1 & z_2 & z_3 & z_4 & z_5 \end{pmatrix}$$

and $>_{\text{lex}}$. There are $10 = \binom{5}{3}$ minors. To check that they form a Gröbner basis we have to check $45 = \binom{10}{2}$ S-pairs. Slightly changing the focus in Buchberger's criterion one can get away with 15 tests only. We are going to explain how this works next.

Definition 1.3.5. Let $I, J \subset R$ be ideals in a ring. Then the **colon ideal** is

$$I : J = \{r \in R \mid rJ \subset I\}.$$

Notation. Let $f_1, \ldots, f_r \in k[x_1, \ldots, x_n]$ be polynomials. We define $r-1$ monomial ideals as follows

$$M_j = (\text{Lt}(f_1), \ldots, \text{Lt}(f_{j-1})) : \text{Lt}(f_j)$$

for $j = 2, \ldots, r$.

For each minimal generator $x^\alpha \in M_j$ the multiple $x^\alpha f_j$ is an expression not allowed in the division theorem by condition 2a).

Theorem 1.3.6 (Buchberger's criterion, second version). *With notation as above, $f_1, \ldots, f_r$ is a Gröbner basis for $(f_1, \ldots, f_r)$ if and only if for each $j = 2, \ldots, r$ and each minimal generator x^α of M_j the remainder of $x^\alpha f_j$ divided by $f_1, \ldots, f_r$ is zero.*

Example 1.3.4 continued. The minors of

$$\begin{pmatrix} x_1 & x_2 & x_3 & x_4 & x_5 \\ y_1 & y_2 & y_3 & y_4 & y_5 \\ z_1 & z_2 & z_3 & z_4 & z_5 \end{pmatrix}$$

suitably ordered have the following leading terms:

j	$\mathrm{Lt}(f_j)$	M_j
1	$x_1 y_2 z_3$	0
2	$x_1 y_2 z_4$	(z_3)
3	$x_1 y_3 z_4$	(y_2)
4	$x_2 y_3 z_4$	(x_1)
5	$x_1 y_2 z_5$	(z_3, z_4)
6	$x_1 y_3 z_5$	(y_2, z_4)
7	$x_2 y_3 z_5$	(x_1, z_4)
8	$x_1 y_4 z_5$	(y_2, y_3)
9	$x_2 y_4 z_5$	(x_1, y_3)
10	$x_3 y_4 z_5$	(x_1, x_2)

$$0 = \det \begin{pmatrix} x_1 & x_2 & x_3 & x_4 \\ y_1 & y_2 & y_3 & y_4 \\ z_1 & z_2 & z_3 & z_4 \\ z_1 & z_2 & z_3 & z_4 \end{pmatrix}$$

$$\implies z_3 f_2 = z_4 f_1 + z_2 f_3 - z_1 f_4 + 0.$$

Similarly, all other remainders are zero.

Hence $f_1, \ldots, f_{10}$ is a Gröbner basis.

For our proof of Buchberger's criterion we need the concept of modules and division with remainder in free modules.

Definition 1.3.7. Let R be a ring. An **R-module** M is an abelian group together with an operation

$$R \times M \to M, (a, m) \mapsto am$$

satisfying the usual associativity and distributivity laws: For all $a, b \in R$ and all $m, n \in M$ we have

$$a(bm) = (ab)m,$$
$$1m = m,$$
$$(a + b)m = am + bm,$$
$$a(m + n) = am + an.$$

For a field k a k-module is simply a k-vector space.

Examples 1.3.8. 1) The ring R is an R-module.
2) A free module is a module of the form $F = R^r$. It has basis vectors

$$e_j = (0, \ldots, 1, \ldots, 0)^t$$

with 1 in the j-th position. An element of F is simply a column vector

$$(a_1, \ldots, a_r)^t = \sum a_j e_j$$

with entries in R.

3) A **submodule** $N \subset M$ of a module M is a subgroup N satisfying

$$n \in N \Rightarrow an \in N \quad \forall a \in R \; \forall n \in N.$$

Thus an ideal I is a submodule of R.

If $f_1, \ldots f_r \in M$, then

$$(f_1 \ldots, f_r) = \{g_1 f_1 + \cdots + g_r f_r \mid g_j \in R\}$$

is a submodule of M.

Definition 1.3.9. An **R-module homomorphism** $\varphi \colon M \to N$ is a group homomorphism satisfying additionally $\varphi(am) = a\varphi(m)$.

The kernel $\ker \varphi$ is a submodule of M and the image $\operatorname{im}(\varphi)$ is a submodule of N.

To say that a module is generated by elements $f_1, \ldots, f_r \in M$ is equivalent to saying that

$$\varphi \colon F = R^r \to M, e_j \mapsto f_j$$

defines a surjective R-module homomorphism.

Definition 1.3.10. A **syzygy** between elements $f_1, \ldots, f_r \in M$ is an element $(g_1, \ldots, g_r)^t$ in $F = R^r$ satisfying $\sum g_j f_j = 0$.

In other words, it is an element of $\ker \varphi$, where $\varphi \colon F = R^r \to M$ is defined by $e_j \mapsto f_j$.

Definition 1.3.11 (Quotient modules). Let $N \subset M$ be a submodule. Then

$$f \equiv g \mod N :\Leftrightarrow f - g \in N$$

defines an equivalence relation on M with equivalence classes

$$f + N = \{f + h \mid h \in N\}.$$

The set of equivalence classes $M/N = \{f + N \mid f \in M\} \subset 2^M$ carries a unique R-module structure such that the projection

$$\pi \colon M \to M/N, f \mapsto f + N$$

becomes an R-module homomorphism.

Theorem 1.3.12 (Homomorphism theorem). *Let $\varphi \colon M \to N$ be an R-module homomorphism. Then there is a canonical R-module isomorphism*

$$M/\ker(\varphi) \cong \operatorname{im}(\varphi).$$

Proof. The map $f + \ker(\varphi) \mapsto \varphi(f)$ is a well-defined isomorphism: Indeed, if $f_1 + \ker \varphi = f_2 + \ker \varphi$ then $f_1 - f_2 \in \ker \varphi$. Thus

$$\varphi(f_2) = \varphi(f_2) + \varphi(f_1 - f_2) = \varphi(f_2 + f_1 - f_2) = \varphi(f_1).$$

This homomorphism is clearly surjective. It is injective, because $\varphi(f) = 0$ holds if and only if $f \in \ker\varphi \Leftrightarrow f + \ker\varphi = \ker\varphi$. □

For $\varphi\colon M \to N$ we define the **cokernel** of φ as

$$\operatorname{coker}(\varphi) = N/\operatorname{im}(\varphi).$$

Definition 1.3.13 (Finitely presented modules). An R-module M is **finitely generated** if there exists a surjection $\varphi\colon R^r \to M$. The module M is **finitely presentable** if one can choose the surjection $\varphi\colon R^r \to M$ such that the **syzygy module** $\ker(\varphi)$ is finitely generated as well. In that case we obtain a sequence

$$R^s \xrightarrow{\varphi_1} R^r \xrightarrow{\varphi} M \longrightarrow 0$$

with $\operatorname{im}(\varphi_1) = \ker(\varphi)$ and $M \cong \operatorname{coker}(\varphi_1)$. Such a sequence is called a **finite presentation** of M.

A homomorphism $R^s \to R^r$ between free modules can be described by an $r \times s$-matrix with entries in R. We can specify a finitely presented module via a matrix φ_1.

Tasks of constructive module theory. Not so easy are the following tasks: Given two finitely presented modules

$$R^s \xrightarrow{\varphi_1} R^r \longrightarrow M \longrightarrow 0$$

and

$$R^\ell \xrightarrow{\psi_1} R^k \longrightarrow N \longrightarrow 0,$$

1) compute the R-module $\operatorname{Hom}(M,N)$ of all R-module homomorphisms,
2) decide whether M and N are isomorphic.

We will approach these questions for $R = k[x_1,\ldots,x_n]$ using Gröbner bases for submodules of free modules in Section 7.

Notation. We denote the polynomial ring by $S = k[x_1,\ldots,x_n]$ and use

$$F = S^m$$

to denote the free S-module with m basis elements $e_j = (0,\ldots 1,\ldots 0)^t$.

Definition 1.3.14. A **monomial** in F is an element of the form $x^\alpha e_j$, a **term** in F is an element of the form $ax^\alpha e_j$ with $a \in k$.

A **monomial order** on F is a complete order $>$ of all the monomials in F satisfying

$$x^\alpha e_j > x^\beta e_i \implies x^\gamma x^\alpha e_j > x^\gamma x^\beta e_i$$

1.3 Buchberger's criterion

for any two monomials in F and any monomial x^γ in S.

Every element $f \in F$ is a finite sum of terms, and we can define the **leading term** of f as before: If $f = \sum_{\alpha,j} f_{\alpha,j} x^\alpha e_j$, then $\mathrm{Lt}(f) = f_{\beta,i} x^\beta e_i$ where

$$x^\beta e_i = \max\{x^\alpha e_j \mid f_{\alpha,j} \neq 0\}.$$

A monomial order $>$ on F is **global** if

$$x_i e_j > e_j \text{ holds for } i = 1, \ldots, n \text{ and } j = 1, \ldots, m.$$

Examples 1.3.15. Let $>$ be a global monomial order on S. We can define a monomial order on F in two ways:

1) $x^\alpha e_j >_1 x^\beta e_i$ iff $x^\alpha > x^\beta$ or ($x^\alpha = x^\beta$ and $j > i$),
2) $x^\alpha e_j >_2 x^\beta e_i$ iff $j > i$ or ($j = i$ and $x^\alpha > x^\beta$),

which we call the **monomial before component order** and **component before monomial order**, respectively.

There are many more ways to define global monomial orders on F, for example, weight orders, where also the e_j get some weights.

A monomial order on $F = S^m$ gives m monomial orders on S using the isomorphism

$$S \cong Se_j.$$

These might not coincide, but in all examples we will consider, they do.

Theorem 1.3.16 (Division with remainder). *Let $>$ be a global monomial order on $F = S^m$ and let $f_1, \ldots, f_r \in F$ be non-zero polynomial vectors. For every $f \in F$ there exist uniquely determined $g_1, \ldots, g_r \in S$ and a unique remainder $h \in F$ satisfying*

1) $f = g_1 f_1 + \cdots + g_r f_r + h$,
2) a) *No term of $g_j \mathrm{Lt}(f_j)$ is a multiple of a leading term $\mathrm{Lt}(f_i)$ for some $i < j$,*
   b) *No term of h is a multiple of a leading term $\mathrm{Lt}(f_j)$.*

Proof. As before we write

$$f = g_1^{(0)} \mathrm{Lt}(f_1) + \cdots + g_r^{(0)} \mathrm{Lt}(f_r) + h^{(0)}$$

satisfying 2a) and 2b). Consider

$$f^{(1)} = f - (g_1^{(0)} f_1 + \cdots + g_r^{(0)} f_r + h^{(0)}).$$

Then $\mathrm{Lt}(f^{(1)}) < \mathrm{Lt}(f)$ and we can iterate until $f^{(N)} = 0$. □

Remarks 1.3.17. 1) Notice that to perform the division algorithm we do not need to know the monomial order precisely. We only need to know the leading terms $\mathrm{Lt}(f_j)$.
2) The role of the global monomial order is to guarantee that the algorithm terminates.
3) This in turn is based on the fact that monomial submodules of F are finitely

generated.
4) We deduce the descending chain condition:
Every strictly decreasing chain $m_1 > m_2 > \ldots$ of monomials in F with respect to a global monomial order is finite.

Let $I \subset F$ be a submodule. Then $\mathrm{Lt}(I) = (\{\mathrm{Lt}(f) \mid f \in I\})$ is the **module of leading terms** of I. Elements $f_1, \ldots, f_r \in I$ form a **Gröbner basis** for I if and only if $\mathrm{Lt}(I) = (\mathrm{Lt}(f_1), \ldots, \mathrm{Lt}(f_r))$.

- Since every monomial module is finitely generated, every submodule of F has a Gröbner basis.
- The remainder of $f \in F$ by a Gröbner basis $f_1, \ldots, f_r$ is zero if and only if $f \in (f_1, \ldots, f_r)$.
- In particular, a Gröbner basis of I is a generating set of I.
- The monomials $x^\alpha e_j \in F$ with $x^\alpha e_j \notin \mathrm{Lt}(I)$ represent a k-vector space basis of the quotient module $M = F/I$.

For submodules $N_1, N_2 \subset M$ of an R-module M the colon ideal is defined as

$$N_1 : N_2 = \{a \in R \mid aN_2 \subset N_1\}.$$

Notation. Let $f_1, \ldots, f_r \in F$ be polynomial vectors. We define monomial ideals as follows

$$M_j = (\mathrm{Lt}(f_1), \ldots, \mathrm{Lt}(f_{j-1})) : \mathrm{Lt}(f_j)$$

for $j = 2, \ldots, r$.

For each minimal generator $x^\alpha \in M_j$ the multiple $x^\alpha f_j$ is an expression not allowed in the division theorem by condition 2a).

Theorem 1.3.18 (Buchberger's criterion). *With notation as above, $f_1, \ldots, f_r$ is a Gröbner basis for $(f_1, \ldots, f_r)$ if and only if for each $j = 2, \ldots, r$ and each minimal generator x^α of M_j the remainder of $x^\alpha f_j$ divided by $f_1, \ldots, f_r$ is zero.*

Proof. If $f_1, \ldots, f_r$ is a Gröbner basis, then the remainder of $x^\alpha f_j$ is zero, because $x^\alpha f_j$ is in $(f_1, \ldots, f_r)$. For the converse assume that the condition of the criterion is satisfied. Then for each minimal generator $x^\alpha \in M_j$ we have a division expression with remainder zero:

$$x^\alpha f_j = \sum_{i=1}^{r} g_i^{(j,\alpha)} f_i$$

satisfying condition 2a). Consider $F_1 = S^r$ and the S-module homomorphism

$$\varphi : F_1 \to F, \; e_i \mapsto f_i.$$

Then

$$G^{(j,\alpha)} = x^\alpha e_j - \sum_{i=1}^{r} g_i^{(j,\alpha)} e_i$$

1.3 Buchberger's criterion

is a syzygy of $f_1, \ldots, f_r$, in other words, it is an element of $\ker(\varphi)$. Our proof of Buchberger's criterion uses division with remainder in F_1.

Definition 1.3.19. The **induced monomial order** on F_1 is defined by

$$x^\alpha e_j > x^\beta e_i \iff x^\alpha \operatorname{Lt}(f_j) > x^\beta \operatorname{Lt}(f_i) \text{ or}$$
$$x^\alpha \operatorname{Lt}(f_j) = x^\beta \operatorname{Lt}(f_i) \text{ up to a non-zero factor in } k$$
$$\text{and } j > i.$$

We could avoid the phrase "up to a non-zero factor in k" if we assume that the f_j are monic, i.e., that they have leading coefficients 1.

Lemma 1.3.20. *With respect to the induced monomial orders the syzygies $G^{(j,\alpha)} \in F_1$ have the leading terms*

$$\operatorname{Lt}(G^{(j,\alpha)}) = x^\alpha e_j.$$

Proof. Since $x^\alpha f_j = \sum_{i=1}^r g_i^{(j,\alpha)} f_i$ satisfies condition 2a), we have

$$\operatorname{Lt}(x^\alpha f_j) = \max\{\operatorname{Lt}(g_i^{(j,\alpha)} f_i)\},$$

and equality is achieved for $\ell = \min\{i \mid x^\alpha \operatorname{Lt}(f_j) \in (\operatorname{Lt}(f_i))\}$:

$$x^\alpha \operatorname{Lt}(f_j) = \operatorname{Lt}(g_\ell^{(j,\alpha)}) \operatorname{Lt}(f_\ell).$$

All other terms of any $g_i^{(j,\alpha)} \operatorname{Lt}(f_i)$ are strictly smaller than $x^\alpha \operatorname{Lt}(f_j)$. Since $\ell < j$, we obtain

$$\operatorname{Lt}(G^{(j,\alpha)}) = x^\alpha e_j$$

from the definition of the induced order. □

We are now ready to complete the proof of Buchberger's criterion. Consider an arbitrary element $f = a_1 f_1 + \cdots + a_r f_r \in (f_1, \ldots, f_r)$. We consider

$$A = \sum_{i=1}^r a_i e_i \in F_1$$

and the remainder $H = \sum_{i=1}^r g_i e_i$ of A divided by the $G^{(j,\alpha)}$'s. Since the $G^{(j,\alpha)}$'s are syzygies of $f_1, \ldots, f_r$, we have

$$f = a_1 f_1 + \cdots + a_r f_r = g_1 f_1 + \cdots + g_r f_r.$$

Indeed

$$A = \sum_{(j,\alpha)} g_{j,\alpha} G^{(j,\alpha)} + H \in F_1 \implies \varphi(A) = \varphi(H).$$

From the definition of the monomial ideals M_j and the $G^{(j,\alpha)}$'s we see that the coefficients g_i of the remainder $H = \sum_{i=1}^r g_i e_i$ have no term t such that $t \operatorname{Lt}(f_i) \in$

($Lt(f_1),\ldots,Lt(f_{i-1})$). In other words, the coefficients $g_1,\ldots g_r$ satisfy the condition 2a) for division by $f_1,\ldots,f_r$ in F. Thus

$$Lt(f) = \max\{Lt(g_j f_j)\} \in (Lt(f_1),\ldots,Lt(f_r)).$$

Since $f \in (f_1,\ldots,f_r)$ was an arbitrary element we conclude

$$Lt((f_1,\ldots,f_r)) = (Lt(f_1),\ldots,Lt(f_r)).$$

Hence $f_1,\ldots,f_r$ is a Gröbner basis of $(f_1,\ldots,f_r)$. □

Corollary 1.3.21 (Schreyer). *If $f_1,\ldots,f_r \in F$ is a Gröbner basis, then the $G^{(j,\alpha)}$'s in F_1 form a Gröbner basis of the syzygy module $\ker(\varphi)$, where*

$$\varphi \colon F_1 \to F, \ e_j \mapsto f_j.$$

Proof. Let G be an element of $\ker(\varphi)$. Consider the remainder $H = (g_1,\ldots,g_r)^t$ of the division of G by the $G^{(j,\alpha)}$. The coefficients g_j satisfy condition 2a) for the division by $f_1,\ldots,f_r$. Thus

$$Lt(g_1 f_1 + \cdots + g_r f_r) = \max\{Lt(g_j f_j)\}.$$

On the other hand, $g_1 f_1 + \cdots + g_r f_r = \varphi(H) = \varphi(G) = 0$. Thus all $g_j = 0$ and hence H is zero.

Thus every $G \in \ker(\varphi)$ has remainder zero under the division by the $G^{(j,\alpha)}$'s. Applying the condition 2a) for the division by the $G^{(j,\alpha)}$'s, we see that

$$Lt(G) \in (\{Lt(G^{(j,\alpha)})\}).$$

□

Example 1.3.22. We compute a Gröbner basis of $I = (y - x^2, z - x^3) \subset k[x,y,z]$ with respect to the lexicographic order $>_{\text{lex}}$.

$x^2 - y$	$-x$	$-y$	$-z$		
$x^3 - z$	1				
$xy - z$	-1	x	y	$-z$	$-y^2$
$xz - y^2$		1	x	y	z
$y^3 - z^2$				1	x
			z	$-y$	x	-1

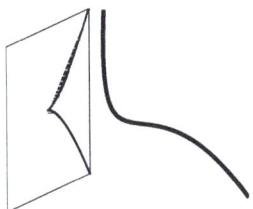

Note that $y^3 - z^2 \in (y - x^2, z - x^3) \cap k[y,z]$.

Exercise 1.3.24 establishes that computing a Gröbner basis of $I \subset k[x_1,\ldots,x_n]$ with respect to $>_{\text{lex}}$ allows us to compute the **j-th elimination ideal**

1.3 Buchberger's criterion

$$I_j = I \cap k[x_{j+1}, \ldots, x_n]$$

obtained from I by eliminating the first j variables.

Algorithm 1.3.23 (Submodule membership).
Input. $f_1, \ldots, f_r \in F$ and a further polynomial vector $f \in F$.
Output. The boolean value t of "$f \in (f_1, \ldots, f_r)$?" and if $t = true$ coefficients $g_1, \ldots, g_r \in S$ such that $f = g_1 f_1 + \cdots + g_r f_r$.

1. Choose a global monomial order $>$ on F.
2. Compute a Gröbner basis $f_1, \ldots, f_s$ of $(f_1, \ldots, f_r)$ with Buchberger's algorithm.
3. Divide f by $f_1, \ldots, f_s$ with remainder:

$$f = \tilde{g}_1 f_1 + \cdots + \tilde{g}_s f_s + h.$$

4. If $h \neq 0$, then return $t = false$ else $t = true$ and recursively substitute f_j by a linear combination of $f_1, \ldots, f_{j-1}$ for $j = s, \ldots, r+1$ to obtain an expression of the form $f = g_1 f_1 + \cdots + g_r f_r$.
5. Return t and $g_1, \ldots, g_r$.

Exercise 1.3.24 (Key property of $>_{\text{lex}}$). 1) Suppose $f \in k[x_1, \ldots, x_n]$ and $1 \leq j \leq n-1$. Then

$$\text{Lt}_{\text{lex}}(f) \in k[x_{j+1}, \ldots, x_n] \iff f \in k[x_{j+1}, \ldots, x_n].$$

2) Let $f_1, \ldots, f_r$ be a Gröbner basis of $I \subset k[x_1, \ldots, x_n]$ with respect to $>_{\text{lex}}$. Then

$$\{f_s \mid \text{Lt}_{\text{lex}}(f_s) \in k[x_{j+1}, \ldots, x_n]\}$$

is a Gröbner basis of $I_j = I \cap k[x_{j+1}, \ldots, x_n]$.

Exercise 1.3.25. Let $S = k[x_1, \ldots, x_n]$ denote the polynomial ring and $F = S^m$. Prove the correctness of the following variant of the division algorithm.

Algorithm 1.3.26 (Division with remainder).
Input. A global monomial order $>$ on F and non-zero polynomial vectors $f_1, \ldots, f_r$ in F and a further polynomial vector $f \in F$.
Output. Coefficients $g_1, \ldots, g_r \in S$ and a remainder $h \in F$ satisfying

$$f = g_1 f_1 + \cdots + g_r f_r + h$$

and the conditions 2a) and 2b) from Theorem 1.3.16.

1. Initialize $f' = f$ and $g_1 = 0, \ldots, g_r = 0$ and $h = 0$.
2. While $f' \neq 0$ do
   a. Determine the leading term $\text{Lt}(f')$ with respect to $>$.
   b. Determine $j = \min\{i \mid \text{Lt}(f_i) \mid \text{Lt}(f')\}$ if this set is non-empty. Write $\text{Lt}(f') = t \, \text{Lt}(f_j)$ for a term $t \in S$. Replace g_j by $g_j + t$ and f' by $f' - t g_j$.
   c. If the set above is empty then replace h by $h + \text{Lt}(f')$ and replace f' by $f' - \text{Lt}(f')$.

3. Return $g_1, \ldots, g_r$ and h.

What are the advantages/disadvantages of this version compared to the version explained in the proof of Theorem 1.3.16? How can one combine this algorithm with Buchberger's algorithm to detect new Gröbner basis elements faster?

Exercise 1.3.27. A binomial $f \in k[x_1, \ldots, x_n]$ is a polynomial which has exactly two terms
$$f = ax^\alpha - bx^\beta.$$
A binomial ideal is an ideal generated by binomials and monomials. Prove: Binomial ideals have a Gröbner basis consisting of binomials and monomials.

Exercise 1.3.28. Implement the computer algebra system Macaulay2
https://macaulay2.com
on your machine. Do Exercise 1.1.16 again.

1.4 Proof of Hilbert's Nullstellensatz

Remember that $\mathbb{A}^n = K^n$ always denotes the affine space over an algebraically closed extension field K of our ground field k.

Definition 1.4.1. Let $I \subset k[x_1, \ldots, x_n]$ be an ideal. We define the **vanishing locus** of I as
$$V(I) = \{a \in \mathbb{A}^n \mid f(a) = 0 \,\forall f \in I\}$$
Since I is finitely generated, say $I = (f_1, \ldots, f_r)$, we have
$$V(I) = V(f_1, \ldots, f_r).$$

With this notation we may rephrase Hilbert's Nullstellensatz as follows.

Theorem 1.4.2. Let $I \subset k[x_1, \ldots, x_n]$ be an ideal. Then $V(I) \subset \mathbb{A}^n$ is empty if and only if $1 \in I \subset k[x_1, \ldots, x_n]$.

Notice that the left-hand side concerns solutions over the algebraically closed extension field K while the condition on the right-hand side can be answered with a Gröbner basis computation over k. A typical example is $K = \mathbb{C}$ and $k = \mathbb{Q}$.

The basic approach of the proof of the Nullstellensatz is an induction on the number of variables. If $n = 1$, the theorem holds, because $k[x]$ is a principal ideal domain. So every ideal
$$(0) \subsetneq I \subsetneq k[x]$$
is generated by a monic polynomial of positive degree, $I = (f)$, and f has a zero in K because K is algebraically closed.

Basic approach of the induction step. Let $I \subset k[x_1, \ldots, x_n]$. Consider the projection
$$\mathbb{A}^n \to \mathbb{A}^{n-1}, (a_1, \ldots, a_n) \mapsto (a_2, \ldots, a_n)$$

1.4 Proof of Hilbert's Nullstellensatz

and the ideal
$$I_1 = I \cap k[x_2, \ldots, x_n]$$
obtained by eliminating x_1. If $I \neq (1)$, then $I_1 \neq (1)$, so by the induction hypothesis $V(I_1) \subset \mathbb{A}^{n-1}$ is non-empty.

Let
$$a' = (a_2, \ldots, a_n) \in V(I_1) \subset \mathbb{A}^{n-1}$$
be a point and consider the ideal
$$(\{f(x_1, a') \mid f \in I\}) \subset K[x_1].$$

This is a principal ideal, and a root a_1 of its generator would give a solution
$$a = (a_1, a') \in V(I) \subset \mathbb{A}^n.$$

So, we have a diagram

$$\begin{array}{ccc} V(I) & \subset & \mathbb{A}^n \\ \downarrow & & \downarrow \pi \\ V(I_1) & \subset & \mathbb{A}^{n-1}. \end{array}$$

Since $I_1 \subset I$, we have $\pi(V(I)) \subset V(I_1)$. However, the map is not necessarily surjective.

Example 1.4.3. For $I = (xy - 1)$ we have $I_1 = (0) \subset K[y]$. The origin $a' = 0$ in $V(I_1) = \mathbb{A}^1$ has no preimage, because $(\{f(x, a') \mid f \in I\}) = (x0 - 1) = (-1)$ has no zero.

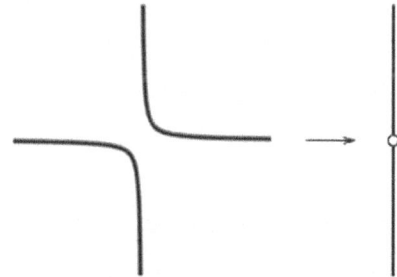

In a certain sense the solution $(1/t, t)$ approaches $(\infty, 0)$ for $t \to 0$.
See Example 11.2.5 for a proper explanation.

Theorem 1.4.4 (Projection theorem). *Let $I \subset k[x_1, \ldots, x_n]$ be an ideal, and consider $I_1 = I \cap k[x_2, \ldots, x_n]$. Suppose I contains an element*
$$f = x_1^d + c_1(x_2, \ldots, x_n)x_1^{d-1} + \cdots + c_d(x_2, \ldots, x_n)$$
which is monic in x_1. Then the projection
$$\pi \colon \mathbb{A}^n \to \mathbb{A}^{n-1}, (a_1, \ldots, a_n) \mapsto (a_2, \ldots, a_n)$$

onto the last $n-1$ components satisfies $\pi(V(I)) = V(I_1)$.

Remark 1.4.5. Since $f \in I$, we can have at most d points $a \in V(I)$ over any given point $a' \in V(I_1)$.

Proof of the projection theorem. We already know that $\pi(V(I)) \subset V(I_1)$, because $I_1 \subset I$. For the converse inclusion we have to find, for every $a' \in \mathbb{A}^{n-1}$ which is not contained in $\pi(V(I))$, a polynomial $h \in I_1$ with $h(a') \neq 0$. We will do this in three steps. The first two handle the case of an algebraically closed field $\overline{K}$. Consider the ideal
$$J = (I) \subset \overline{K}[x_1, \ldots, x_n]$$
generated by I in the polynomial ring over the extension field $\overline{K}$.

Step 1. *For every polynomial $g \in \overline{K}[x_1, \ldots, x_n]$ there exists a polynomial $\tilde{g}$ in $\overline{K}[x_1, \ldots, x_n]$ of degree $< d$ in x_1 such that*
$$\tilde{g}(x_1, a') = 0 \text{ and } g \equiv \tilde{g} \mod J.$$

Consider the surjective ring homomorphism
$$\varphi \colon \overline{K}[x_1, \ldots, x_n] \to \overline{K}[x_1], g \mapsto g(x_1, a').$$

Since $a' \notin \pi(V(I))$, the Nullstellensatz in one variable implies $\varphi(J) = \overline{K}[x_1]$. Thus for every $g \in \overline{K}[x_1, \ldots, x_n]$ there exists a $g_1 \in J$ with $\varphi(g) = \varphi(g_1)$. Consider $g_2 = g - g_1$. Since f is monic in x_1, division of g_2 by f gives an expression $g_2 = qf + \tilde{g}$. The remainder $\tilde{g}$ has degree smaller d in x_1. Applying φ to this equation yields
$$0 = q(x_1, a')f(x_1, a') + \tilde{g}(x_1, a').$$

Thus $\tilde{g}(x_1, a')$ is the unique remainder of 0 under the division by $f(x_1, a')$. Hence $\tilde{g}(x_1, a')$ is the zero polynomial in $\overline{K}[x_1]$ and
$$\tilde{g} - g = g_2 - qf - g = g - g_1 - qf - g = -g_1 - qf \in J.$$

Thus $g \equiv \tilde{g} \mod J$. This establishes the assertion of Step 1.

Step 2. Applying step 1 to the polynomials $1, x_1, x_1^2, \ldots, x_1^{d-1}$ we find expressions
$$\begin{aligned}
1 &\equiv g_{00} + g_{01}x_1 + \cdots + g_{0,d-1}x_1^{d-1} & \mod J \\
x_1 &\equiv g_{10} + g_{11}x_1 + \cdots + g_{1,d-1}x_1^{d-1} & \mod J \\
&\vdots \\
x_1^{d-1} &\equiv g_{d-1,0} + g_{d-1,1}x_1 + \cdots + g_{d-1,d-1}x_1^{d-1} & \mod J
\end{aligned}$$

with $g_{ij} \in \overline{K}[x_2, \ldots, x_n]$ and $g_{ij}(a') = 0$. In matrix form we have

1.4 Proof of Hilbert's Nullstellensatz

$$(E_d - B)\begin{pmatrix} 1 \\ x_1 \\ \vdots \\ x_1^{d-1} \end{pmatrix} \equiv 0 \mod J,$$

where $B = (g_{ij})$ and E_d is the $d \times d$ identity matrix.

Multiplying the last equation by the cofactor matrix of $(E_d - B)$ we arrive at

$$\det(E_d - B)\begin{pmatrix} 1 \\ x_1 \\ \vdots \\ x_1^{d-1} \end{pmatrix} \equiv 0 \mod J.$$

In particular, $h = \det(E_d - B) \in J \cap K[x_2, \ldots, x_n] = J_1$. Since $h(a') = \det E_d = 1 \neq 0$ we have found a polynomial in J_1 which does not vanish at a'. This completes the proof of the projection theorem in the case of an algebraically closed field.

Step 3. The ideal $J \cap K[x_2, \ldots, x_n]$ can be computed via a Gröbner basis computation with respect to $<_{\text{lex}}$ from generators of I by Exercise 1.3.24. Thus h above is a linear combination of the generators of I_1 and at least one generator cannot vanish in a'. This completes the proof of the projection theorem. □

To complete the proof of the Nullstellensatz we prove that we can achieve the assumptions of the Projection theorem 1.4.4 after a change of coordinates.

Lemma 1.4.6. *Let* $f \in k[x_1, \ldots, x_n]$ *be a non-constant polynomial.*

1. *If k is an infinite field and $a_2, \ldots, a_n \in k$ are sufficiently general elements, then substituting*

$$x_j = \tilde{x}_j + a_j x_1$$

for $j = 2, \ldots, n$ into f gives a polynomial

$$\tilde{f} = a x_1^d + c_1(\tilde{x}_2, \ldots, \tilde{x}_n) x_1^{d-1} + \cdots + c_d(\tilde{x}_2, \ldots, \tilde{x}_n)$$

with $d \geq 1$, $a \in k \setminus \{0\}$ and $c_j \in k[\tilde{x}_2, \ldots, \tilde{x}_n]$.

2. *If k is an arbitrary field, then a substitution of the form*

$$x_j = \tilde{x}_j + x_1^{(r^{j-1})}$$

for $j = 2, \ldots, n$ and $r \in \mathbb{N}$ sufficiently large yields a polynomial $\tilde{f}$ of the same shape.

Proof. Let $d = \deg f$ denote the degree of f and let

$$f = f_d + \cdots + f_1 + f_0 \text{ with } f_j = \sum_{|\alpha|=j} f_\alpha x^\alpha$$

be the decomposition of $f = \sum f_\alpha x^\alpha$ into homogeneous parts. Then $f_d(1, x_2, \ldots, x_n)$ is not the zero polynomial. If k is infinite, then there exists $(a_2, \ldots, a_n) \in \mathbb{A}^{n-1}(k)$ with $a = f_d(1, a_2, \ldots a_n) \neq 0$ by Exercise 1.1.14. The substitution $x_j = \tilde{x}_j + a_j x_1$ yields

$$f_d(x_1, \tilde{x}_2 + a_2 x_1, \ldots, \tilde{x}_n + a_n x_1) = a x_1^d + \text{terms of lower degree in } x_1.$$

Thus $f(x_1, \tilde{x}_2 + a_2 x_1, \ldots, \tilde{x}_n + a_n x_1)$ has the desired shape.

In case 2, we take

$$r > \max\{e \mid \exists \alpha \; \exists j \text{ with } f_\alpha \neq 0 \text{ and } \alpha_j = e\}$$

larger than any exponent occurring in a term of f. Then the monomials

$$x_1^{\sum_{j=1}^{n} \alpha_j r^{j-1}} \text{ for } \alpha \text{ with } f_\alpha \neq 0$$

are all distinct, and the largest one will give the desired leading term after the substitution

$$x_j = \tilde{x}_j + x_1^{(r^{j-1})}.$$

$\square$

Example 1.4.3 continued. For $f = xy - 1$ every substitution $y = \tilde{y} + a_2 x$ for $a_2 \neq 0$ has the desired effect:

$$\tilde{f} = a_2 x^2 + x\tilde{y} - 1.$$

Proof of the Nullstellensatz. Let $I \subsetneq k[x_1, \ldots, x_n]$ be a proper ideal. We have to prove that $V(I) \neq \emptyset$. If $I = (0)$, then $V(I) = \mathbb{A}^n$. Otherwise, there exists a non-constant polynomial $f \in I$. After a change of coordinates as in Lemma 1.4.6 we may assume that f is monic in x_1. Thus the projection $V(I) \to V(I_1)$ is surjective. Since $1 \notin I_1 \subset I$, we obtain $V(I_1) \neq \emptyset$ from the induction hypothesis. Hence $V(I) \neq \emptyset$ holds as well. $\square$

Remark 1.4.7. Notice that we can perform the change of coordinates over the field of definition of I. Thus for example if $I \subset \mathbb{C}[x_1, \ldots, x_n]$ is generated by polynomials in $\mathbb{Q}[x_1, \ldots, x_n]$, we can take a linear change of coordinates defined over $\mathbb{Q}$.

1.4 Proof of Hilbert's Nullstellensatz

Theorem 1.4.8 (Tower of projections). *Suppose that $I \subsetneq k[x_1,\ldots,x_n]$ is a proper ideal. Consider the elimination ideals $I_j = I \cap k[x_{j+1},\ldots,x_n]$. Set*

$$c = \min\{j \mid I_j = (0)\}$$

and suppose that for each j with $0 \leq j \leq c - 1$ the ideal I_j contains an x_{j+1}-monic polynomial of some degree d_j. Then the projection $\pi_c \colon V(I) \to \mathbb{A}^{n-c}$ onto the last $n - c$ components is surjective and each fiber

$$\pi_c^{-1}(a_{c+1},\ldots,a_n)$$

is finite of cardinality at most $\prod_{j=0}^{c-1} d_j$. □

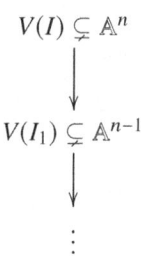

Remarks 1.4.9. 1) If I has an infinite field of definition $L \subset k$, we can reach the assumption of the tower theorem by a triangular change of coordinates defined over L:

$$\begin{pmatrix} x_1 \\ x_2 \\ x_3 \\ \vdots \\ x_n \end{pmatrix} = \begin{pmatrix} 1 & & & & 0 \\ a_{21} & 1 & & & \\ a_{31} & a_{32} & 1 & & \\ \vdots & & & \ddots & \\ a_{n1} & a_{n2} & \cdots & & 1 \end{pmatrix} \begin{pmatrix} \tilde{x}_1 \\ \tilde{x}_2 \\ \tilde{x}_3 \\ \vdots \\ \tilde{x}_n \end{pmatrix},$$

with $a_{ij} \in L$.

2) In the situation of the tower theorem it is tempting to define

$$\dim V(I) = n - c \text{ and } \operatorname{codim} V(I) = c,$$

because the projection $\pi_c \colon V(I) \to \mathbb{A}^{n-c}$ is surjective with finite fibers. A problem with this definition is that it is not clear that this is independent from the choice of coordinates. We will prove that this assertion holds in Corollary 6.1.3.

Exercise 1.4.10. Consider the ideal $I = (xy(x + y) + 1) \subset \mathbb{F}_2[x, y]$. Determine coordinates in which I satisfies the extra hypothesis of the projection theorem. Show that the extra hypothesis cannot be achieved by means of a linear change of coordinates.

Exercise 1.4.11. Consider the curve $C = V(f_1, f_2) \subset \mathbb{A}^3(\mathbb{C})$, where

$$f_1 = x^2 - yz + z, \quad f_2 = y^3 - 2y^2 + y - z^2.$$

Prove that the f_1, f_2 is the reduced lexicographic Gröbner basis for the ideal (f_1, f_2) with variables ordered as $x > y > z$,
If we reorder the variables as $y > z > x$, the reduced lexicographic Gröbner basis of (f_1, f_2) consists of five polynomials. Prove that they are given by

$$y^3 - 2y^2 + y - z^2, \quad y^2x^2 - yx^2 - z^3, \quad yz - z - x^2,$$
$$yx^4 - z^4, \quad z^5 - zx^4 - x^6.$$

Consider $C_1 = V(y^3 - 2y^2 + y - z^2) \subset \mathbb{A}^2$ and $C_2 = V(z^5 - zx^4 - x^6) \subset \mathbb{A}^2$ and the projections $C \to C_1$ and $C \to C_2$ onto the curve in the yz- and xz-plane. How many preimage points do points in C_1 and C_2 have?

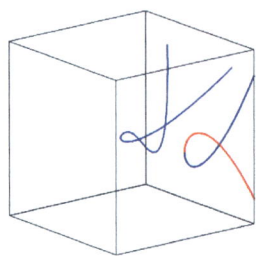 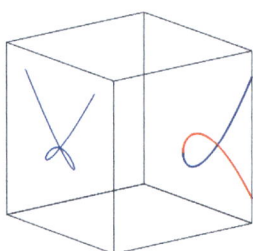

Note that the preimages in C of a point on the red part of C_1 consists of a pair of points whose x-coordinates form a pair of complex conjugate numbers.

Chapter 2
The algebra-geometry dictionary

In this chapter we develop the algebra-geometry dictionary. We will see that the algebraic subsets of $\mathbb{A}^n$ satisfy the axioms of closed subsets of a topology, the Zariski topology. Moreover, the concepts of vanishing loci and vanishing ideals induce for an algebraically closed field K bijections between the algebraic subsets of the affine n-space $\mathbb{A}^n$ and radical ideals in $K[x_1, \ldots, x_n]$. Under this bijection irreducible algebraic sets correspond to prime ideals and points to maximal ideals.

We will define the coordinate ring $K[A]$ of an algebraic subset $A \subset \mathbb{A}^n$, and define morphisms between (affine) algebraic sets. It turns out that a morphism $\varphi \colon A \to B$ between affine algebraic sets corresponds to a K-algebra homomorphism $\varphi^* \colon K[B] \to K[A]$ between their coordinate rings.

2.1 Hilbert's strong Nullstellensatz

Recall our convention that K always denotes an algebraically closed field.

Definition 2.1.1. For any ideal $J \subset K[x_1, \ldots, x_n]$ we have defined its **vanishing locus** as
$$V(J) = \{a \in \mathbb{A}^n \mid f(a) = 0 \ \forall f \in J\}.$$
Conversely for $A \subset \mathbb{A}^n$ an arbitrary subset we define the **vanishing ideal** as
$$I(A) = \{f \in K[x_1, \ldots, x_n] \mid f(a) = 0 \ \forall a \in A\}.$$

Example 2.1.2. Consider the set $C = \{(t, t^2, t^3) \in \mathbb{A}^3 \mid t \in \mathbb{A}^1\}$. The vanishing ideal of C is the kernel of the ring homomorphism
$$\varphi \colon K[x, y, z] \to K[t], x \mapsto t, y \mapsto t^2, z \mapsto t^3.$$
Claim. $I(C) = \ker \varphi = (y - x^2, z - x^3)$.

Proof of the Claim. The inclusion $(y - x^2, z - x^3) \subset I(C)$ is clear. For the converse pick a global monomial order such that $\text{Lt}(y - x^2) = y$ and $\text{Lt}(z - x^3) = z$. Let $f \in \ker \varphi$. Division with remainder gives

$$f = g_1(y - x^2) + g_2(z - x^3) + h$$

with no term of h divisible by y or z, i.e., $h \in K[x] \subset K[x, y, z]$. Substituting gives $0 = f(t, t^2, t^3) = h(t)$. Hence h is the zero polynomial and $f \in (y - x^2, z - x^3)$. □

$$C = \{(t, t^2, t^3) \in \mathbb{A}^3 \mid t \in \mathbb{A}^1\}$$

is called the **twisted cubic curve**.

Thus we have correspondences

$$\{\text{ideals of } K[x_1, \ldots, x_n]\} \xrightleftharpoons[I]{V} \{\text{subsets of } \mathbb{A}^n\}$$

$$J \mapsto V(J), \qquad I(A) \mapsfrom A.$$

Proposition 2.1.3 (Basic properties of the correspondence V). *Let* $S = K[x_1, \ldots, x_n]$ *and let* $I, J, I_\lambda \subset S$ *be ideals.*

1) $V(0) = \mathbb{A}^n$ *and* $V(1) = \emptyset$.
2) $I \subset J \implies V(I) \supset V(J)$.
3) $V(I) \cup V(J) = V(I \cap J) = V(I \cdot J)$.
4) $\bigcap_\lambda V(I_\lambda) = V(\sum_\lambda I_\lambda)$.
5) $V(x_1 - a_1, \ldots, x_n - a_n) = \{(a_1, \ldots, a_n)\}$.

Proof. Only 3) needs an argument. Since $I \cdot J \subset I \cap J \subset J$, the inclusions

$$V(J) \subset V(I \cap J) \subset V(I \cdot J)$$

follow by property 2. For the converse let $a \in V(I \cdot J)$ be a point not contained in $V(J)$. By assumption $\exists g \in J$ with $g(a) \neq 0$. Let $f \in I$ be arbitrary. Since $f \cdot g \in I \cdot J$, we have $f(a)g(a) = 0$. Since $g(a) \neq 0$, we deduce $f(a) = 0$. Hence $a \in V(I)$. □

Definition 2.1.4. An **algebraic subset** $A \subset \mathbb{A}^n$ is a subset of the form $A = V(J)$.

Conditions 1), 3) and 4) of Proposition 2.1.3 can be rephrased by saying that the collection of algebraic subsets of $\mathbb{A}^n$ form the closed sets of a topology on $\mathbb{A}^n$. We call this topology the **Zariski topology** on $\mathbb{A}^n$. The complement $U = \mathbb{A}^n \setminus A$ of an algebraic set A is called **Zariski open**.

Recall, a **topology** on a set X is a subset $\mathcal{T} \subset 2^X$ satisfying

i) $\emptyset \in \mathcal{T}, X \in \mathcal{T}$,
ii) $U_1, U_2 \in \mathcal{T} \implies U_1 \cap U_2 \in \mathcal{T}$, and
iii) $U_\lambda \in \mathcal{T} \implies \bigcup_\lambda U_\lambda \in \mathcal{T}$.

2.1 Hilbert's strong Nullstellensatz

The elements $U \in \mathcal{T}$ are called the open sets of the topology, and their complements $A = X \setminus U$ are called the closed sets of the topology. The closure of an arbitrary subset $Y \subset X$ is

$$\overline{Y} = \bigcap_{\substack{A \supset Y \\ \text{closed}}} A.$$

This is the smallest closed set containing Y.

Proposition 2.1.5 (Basic properties of the correspondence I). *Let $S = K[x_1, \ldots, x_n]$ and let $A, B \subset \mathbb{A}^n$.*

1) $I(\emptyset) = (1)$ and $I(\mathbb{A}^n) = (0)$.
2) $A \subset B \implies I(A) \supset I(B)$.
3) $I(A \cup B) = I(A) \cap I(B)$.
4) $V(I(A)) \supset A$ and equality holds if A is an algebraic subset. The equality

$$V(I(A)) = \overline{A}$$

always holds.
5) $I(\{(a_1, \ldots, a_n)\}) = (x_1 - a_1, \ldots, x_n - a_n)$.

$\square$

Remark 2.1.6. If $\mathbb{F}_q \subset K$ is a finite subfield, then the set of $\mathbb{F}_q$-rational points $\mathbb{A}^n(\mathbb{F}_q)$ is algebraic, since it is a finite union of q^n points.

$$I(\mathbb{A}^n(\mathbb{F}_q)) = (x_1^q - x_1, \ldots, x_n^q - x_n)$$

is defined over the prime field $\mathbb{F}_p$, where $p = \text{char } K$ and $q = p^r$.

Our next goal is to describe $I(V(J))$.

Definition 2.1.7. Let R be a ring and $J \subset R$ an ideal. The **radical** of J is the ideal

$$\text{rad}(J) = \{f \in R \mid \exists n \in \mathbb{N} \text{ such that } f^n \in J\}.$$

To see that this is indeed an ideal we use

$$f^n \in J \text{ and } g^m \in J \implies (f + g)^{n+m-1} \in J.$$

Theorem 2.1.8 (Hilbert's Nullstellensatz, strong version). *Let K be an algebraically closed field and let $J \subset K[x_1, \ldots, x_n]$ be an ideal. Then*

$$I(V(J)) = \text{rad}(J).$$

Proof. The inclusion $\text{rad}(J) \subset I(V(J))$ is elementary:

$$f \in \mathrm{rad}(J) \implies f^N \in J \text{ for some } N \in \mathbb{N}$$
$$\implies 0 = f^N(a) = (f(a))^N \ \forall a \in V(J)$$
$$\implies f(a) = 0 \ \forall a \in V(J)$$
$$\implies f \in \mathrm{I}(V(J)).$$

Let $J = (f_1, \ldots, f_r)$ and $f \in \mathrm{I}(V(J))$. We have to show that

$$f^m \in (f_1, \ldots, f_r)$$

for a suitable $m \in \mathbb{N}$. We use the trick of Rabinowitch: Consider an additional variable y and the ideal

$$(f_1, \ldots, f_r, yf - 1) \subset K[x_1, \ldots, x_n, y].$$

If $(a, b) \in \mathbb{A}^n \times \mathbb{A}^1 = \mathbb{A}^{n+1}$ lies in $V(f_1, \ldots, f_r, yf - 1)$, then $f_1(a) = 0, \ldots, f_r(a) = 0$. Thus $a \in V(J)$. Hence $f(a) = 0$ and the last polynomial does not vanish, because $(fy - 1)(a, b) = f(a)b - 1 = -1 \neq 0$. Thus $V(f_1, \ldots, f_r, yf - 1) = \emptyset$ and the weak version of the Nullstellensatz, Theorem 1.1.11, implies

$$1 = g_1 f_1 + \cdots + g_r f_r + g_{r+1}(yf - 1)$$

for suitable polynomials $g_1, \ldots, g_{r+1} \in K[x_1, \ldots, x_n, y]$.

Let m be the maximal power in which y occurs in $g_1, \ldots, g_r$. Then

$$f^m \equiv \tilde{g}_1 f_1 + \cdots + \tilde{g}_r f_r \mod (yf - 1)$$

for polynomials $\tilde{g}_1, \ldots, \tilde{g}_r \in K[x_1, \ldots, x_n]$, since we can remove the appearance of y in $f^m g_i$ using $fy \equiv 1 \mod (yf - 1)$.

Since $K[x_1, \ldots, x_n]$ is a subring of $K[x_1, \ldots, x_n, y]/(yf - 1)$, we obtain

$$f^m = \tilde{g}_1 f_1 + \cdots + \tilde{g}_r f_r \in K[x_1, \ldots, x_n].$$

Hence $f \in \mathrm{rad}(J)$. □

Thus for an algebraically closed field K the correspondences V and I induce bijections

$$\{ \text{radical ideals of } K[x_1, \ldots, x_n]\} \underset{\mathrm{I}}{\overset{V}{\longleftrightarrow}} \{\text{algebraic subsets of } \mathbb{A}^n\}$$

$$J \mapsto V(J), \quad \mathrm{I}(A) \mapsfrom A.$$

A **radical ideal** in a ring R is an ideal J satisfying $\mathrm{rad}(J) = J$. The radical of an ideal is always a radical ideal, because $\mathrm{rad}(\mathrm{rad}(I)) = \mathrm{rad}(I)$ holds.

Definition 2.1.9. An ideal $\mathfrak{p} \subsetneq R$ in a ring R is called a **prime ideal** if

$$ab \in \mathfrak{p} \implies a \in \mathfrak{p} \text{ or } b \in \mathfrak{p}$$

holds for all $a, b \in R$, equivalently, $R/\mathfrak{p}$ is an integral domain.

A **maximal ideal** $\mathfrak{m} \subsetneq R$ is an ideal which is maximal with respect to inclusion for proper ideals, i.e.,

$$\mathfrak{m} \subset I \subsetneq R \implies \mathfrak{m} = I$$

holds for all proper ideals $I \subsetneq R$. An equivalent condition is that $R/\mathfrak{m}$ is a field.

In the ring $K[x_1, \ldots, x_n]$ these types of ideals have a geometric interpretation.

Definition 2.1.10. An algebraic set $A \subset \mathbb{A}^n$ satisfying

$$A = A_1 \cup A_2 \implies A = A_1 \text{ or } A = A_2$$

for all **algebraic** subsets A_1, A_2 is called **irreducible**. Irreducible algebraic sets are also called **varieties**.

Example 2.1.11.

$$V(xy, yz) = V(y) \cup V(x, z)$$

is a reducible algebraic set.

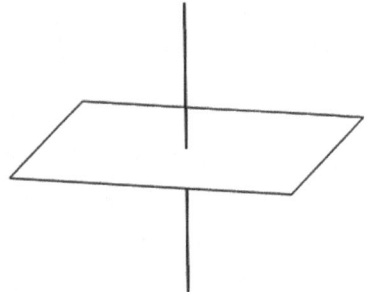

Proposition 2.1.12. *An algebraic subset $A \subset \mathbb{A}^n$ is irreducible if and only if $I(A) \subset K[x_1, \ldots, x_n]$ is a prime ideal.*

Proof. Suppose $A = A_1 \cup A_2$ with $A \supsetneq A_j$ for $j = 1, 2$. Consider $f_j \in I(A_j) \setminus I(A)$. Then $f_1 f_2 \in I(A)$ with both factors not in $I(A)$. So $I(A)$ is not prime. Conversely if $I(A)$ is not prime and $fg \in I(A)$ a product whose factors are not in $I(A)$, then

$$A = V(I(A)) = V((fg) + I(A)) = V((f) + I(A)) \cup V((g) + I(A))$$

shows that A is not irreducible. □

Example 2.1.11 continued. $V(y)$ and $V(x, z)$ are irreducible, because $K[x, y, z]/(y) \cong K[x, z]$ and $K[x, y, z]/(x, z) \cong K[y]$ are integral domains. Thus

$$V(xy, yz) = V(y) \cup V(x, z)$$

is a decomposition into irreducible algebraic sets.

Theorem 2.1.13. *Let K be an algebraically closed field. The correspondences V and I induce bijections*

$$\{\text{radical ideals of } K[x_1,\ldots,x_n]\} \leftrightarrow \{\text{algebraic subsets of } \mathbb{A}^n\}$$
$$\cup \qquad\qquad\qquad\qquad \cup$$
$$\{\text{prime ideals of } K[x_1,\ldots,x_n]\} \leftrightarrow \{\text{irreducible alg. subsets of } \mathbb{A}^n\}$$
$$\cup \qquad\qquad\qquad\qquad \cup$$
$$\{\text{maximal ideals of } K[x_1,\ldots,x_n]\} \leftrightarrow \{\text{points of } \mathbb{A}^n\}$$

Proof. Only the last bijection still needs a proof. If $\mathfrak{m} \subset K[x_1,\ldots,x_n]$ is a maximal ideal, then $V(\mathfrak{m}) \neq \emptyset$ by the Nullstellensatz. If $a = (a_1,\ldots,a_n) \in V(\mathfrak{m})$, then

$$\mathfrak{m} \subset (x_1 - a_1, \ldots x_n - a_n)$$

and the maximality of $\mathfrak{m}$ implies that equality holds. □

Exercise 2.1.14. Let $K \supset k$ be an algebraically closed extension field. Let $I \subset k[x_1,\ldots,x_n]$ be an ideal. Prove

$$V(I) \subset \mathbb{A}^n \text{ is finite} \iff k[x_1,\ldots,x_n]/I \text{ is a finite-dimensional } k\text{-vector space.}$$

Moreover

$$|V(I)| = \dim_K K[x_1,\ldots,x_n]/\mathrm{rad}(I) \leq \dim_k k[x_1,\ldots,x_n]/I.$$

2.2 Coordinate rings and morphisms

Definition 2.2.1. The **coordinate ring** of an algebraic set $A \subset \mathbb{A}^n$ is the residue ring

$$K[A] = K[x_1,\ldots,x_n]/I(A).$$

This can be regarded as a subring of the ring $K^A = \{f\colon A \to K\}$ of K-valued functions on A. It is the K-subalgebra generated by the coordinate functions $x_j|_A$, the restriction of x_j to A.

Definition 2.2.2. Let $A \subset \mathbb{A}^n$ and $B \subset \mathbb{A}^m$ be algebraic sets. A **morphism**

$$\Phi\colon A \to \mathbb{A}^m, a \mapsto \Phi(a) = (\overline{f}_1(a),\ldots,\overline{f}_m(a))$$

is a map given by an m-tuple of functions $\overline{f}_1,\ldots,\overline{f}_m \in K[A]$. A **morphism**

$$\varphi\colon A \to B$$

is given by a morphism $\Phi\colon A \to \mathbb{A}^m$ such that $\Phi(a) \in B\ \forall a \in A$. Thus for a morphism $\varphi\colon A \to B$ we always have a diagram

2.2 Coordinate rings and morphisms

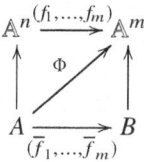

where $f_1,\ldots,f_m$ are representatives of $\overline{f}_1,\ldots,\overline{f}_m$. A morphism $\Phi\colon A \to \mathbb{A}^m$ specifies a **ring homomorphism**

$$\Phi^*\colon K[y_1,\ldots,y_m] \to K[A],\ y_j \mapsto \overline{f}_j,$$

and conversely any *K*-**algebra homomorphism** Φ^* induces a morphism $\Phi\colon A \to \mathbb{A}^m$. A morphism $\varphi\colon A \to B$ corresponds to a *K*-algebra homomorphism

$$\varphi^*\colon K[B] \to K[A].$$

While Φ is easy to specify, morphisms $\varphi\colon A \to B$ are difficult to find: The tuple $(\overline{f}_1,\ldots,\overline{f}_m)$ has to satisfy

$$F(\overline{f}_1,\ldots,\overline{f}_m) = 0 \in K[A]$$

for all equations $F(y_1,\ldots,y_m) \in I(B) \subset K[y_1,\ldots,y_m]$.

Thus we have

$$\operatorname{Mor}(A,B) \cong \operatorname{Hom}_{K\text{-algebra}}(K[B],K[A]),\quad \varphi \mapsto \varphi^*.$$

Definition 2.2.3. A morphism $\varphi\colon A \to B$ is an **isomorphism** if there exists a morphism $\psi\colon B \to A$ with

$$\psi \circ \varphi = \operatorname{id}_A \text{ and } \varphi \circ \psi = \operatorname{id}_B.$$

Proposition 2.2.4. *Algebraic sets A and B are isomorphic if and only if their coordinate rings are isomorphic K-algebras, i.e., $K[A] \cong K[B]$.* □

Examples 2.2.5. 1) $A = V(y - x^2) \subset \mathbb{A}^2$ and $\mathbb{A}^1$ are isomorphic because

$$K[x] \to K[x,y]/(y - x^2)$$

is a *K*-algebra isomorphism.

2) The inclusion $K[x] \hookrightarrow K[x,x^{-1}] \cong K[x,y]/(xy - 1)$ defines a morphism of the hyperbola $A = V(xy - 1)$ to $\mathbb{A}^1$. This corresponds to the projection onto the *x*-axis. In particular, we see that the image of a morphism is not necessarily again an algebraic set.

3) The map

$$\mathbb{A}^1 \to B = V(z^2 - y^3) \subset \mathbb{A}^2,\ x \mapsto (x^2,x^3)$$

is a morphism because $(x^3)^2 - (x^2)^3 = 0$. Although this is a bijection as a map of sets, it is not an isomorphism, because

$$K[y,z]/(z^2 - y^3) \cong K[x^2, x^3] \hookrightarrow K[x]$$

is not surjective.

4) An isomorphism of an algebraic set $A \subset \mathbb{A}^n$ with an algebraic subset $B \subset \mathbb{A}^m$ is given by an m-tuple $(\overline{f}_1, \ldots, \overline{f}_m)$ of K-algebra generators of $K[A]$: $I(B) = \ker \Phi^*$, where

$$\Phi^* : K[y_1, \ldots, y_m] \to K[A], \quad y_i \mapsto \overline{f}_i.$$

Indeed Φ^* is surjective if and only if $\overline{f}_1, \ldots, \overline{f}_m$ are K-algebra generators of $K[A]$. In that case $K[B] = K[y_1, \ldots, y_m]/I(B) \cong K[A]$ follows from the Homomorphism theorem 1.3.12. In particular, we see that any algebraic set is isomorphic to many different algebraic sets $B \subset \mathbb{A}^m$.

Exercise 2.2.6. Prove that the algebraic sets $V(y - x^2)$ and $V(xy - 1)$ in $\mathbb{A}^2$ are not isomorphic.

Exercise 2.2.7. Prove that an algebraically closed field has infinitely many elements.

Exercise 2.2.8 (Real cubic with an isolated point).

Consider the set $A \subset \mathbb{A}^2(\mathbb{R})$ parametrized by

$$A = \{(t^2 + 1, t^3 + t) \mid t \in \mathbb{R}\}.$$

Prove that $I(A) = (y^2 - x^3 + x^2)$ and conclude that the point $o = (0,0)$ lies in the algebraic closure $\overline{A}$ of A.

Chapter 3
Noetherian rings and primary decomposition

In this chapter (and the following two chapters) we introduce some concepts from Commutative Algebra that are needed in Algebraic Geometry. We introduce Noetherian rings and give Emmy Noether's proof of the existence of a primary decomposition of an ideal in Noetherian ring. In Algebraic Geometry this is well motivated since the primary decomposition of a radical ideal corresponds to the decomposition of an algebraic set into its irreducible components.

The concept of associated primes of a module leads to a first structure result for finitely generated modules M over Noetherian rings R: They have a filtration

$$0 = M_0 \subset M_1 \subset \ldots \subset M_N = M$$

such that the quotients $M_i/M_{i-1} \cong R/\mathfrak{p}_i$ for prime ideals $\mathfrak{p}_i \subset R$. This allows us to deduce the first uniqueness theorem for primary decomposition.

3.1 The ascending chain condition

In Example 2.1.11 we saw that

$$V(xy, yz) = V(y) \cup V(x, z)$$

is a decomposition into irreducible algebraic subsets.

Theorem 3.1.1 (Component decomposition). *Let K be an algebraically closed field, and let $A \subset \mathbb{A}^n$ be an algebraic subset. Then there exist finitely many irreducible algebraic subsets $C_j \subset \mathbb{A}^n$ such that*

$$A = C_1 \cup C_2 \cup \ldots \cup C_r.$$

Definition 3.1.2. A component decomposition $A = C_1 \cup C_2 \cup \ldots \cup C_r$ is called **irredundant** if $C_i \not\subset C_j$ for $i \neq j$.

By deleting C_i which are contained in a C_j, one can pass from a component decomposition to an irredundant one.

Theorem 3.1.3 (Uniqueness of the component decomposition). *An irredundant decomposition $A = C_1 \cup C_2 \cup \ldots \cup C_r$ into irreducible algebraic sets C_j is unique up to order.*

Proof. Suppose $A = C_1 \cup C_2 \cup \ldots \cup C_r = C'_1 \cup C'_2 \cup \ldots \cup C'_s$ are two irredundant decompositions. Then each C'_ℓ is contained in some C_j because

$$C'_\ell = (C'_\ell \cap C_1) \cup (C'_\ell \cap C_2) \cup \ldots \cup (C'_\ell \cap C_r)$$

implies $C'_\ell = (C'_\ell \cap C_j)$ for some j since C'_ℓ is irreducible.

Similarly, each C_j is contained in C'_k for some k. So $C'_\ell \subset C_j \subset C'_k$ and we have equality because $C'_1 \cup C'_2 \cup \ldots \cup C'_s$ is irredundant. Thus each C'_ℓ coincides with a unique C_j and vice versa. In particular, r equals s. □

Towards the existence of a component decomposition. Suppose $A \subset \mathbb{A}^n$ is an algebraic set. If A is irreducible, we are done. Otherwise, we can decompose

$$A = A_1 \cup A_2$$

into proper algebraic subsets. If these are irreducible, we are done. Otherwise, we decompose each reducible subset again. Thus we get a tree of smaller and smaller algebraic subsets. The problem is to show that this process terminates. If we translate this with our algebra-geometry dictionary, we get a tree of larger and larger radical ideals in the polynomial ring $K[x_1, \ldots, x_n]$.

Theorem 3.1.4 (Noetherian rings). *Let R be a ring. The following conditions are equivalent*

1) *Each ideal $I \subset R$ is finitely generated.*
2) *Every ascending chain of ideals*

$$I_1 \subset I_2 \subset \ldots \subset I_j \subset \ldots$$

   becomes stationary, i.e., there exists an N such that $I_N = I_{N+1} = I_{N+2} = \ldots$.
3) *Every nonempty set $\mathcal{M}$ of ideals contains maximal elements with respect to inclusion, i.e.,*

$$\exists I \in \mathcal{M} \text{ such that } I \subset J \Rightarrow I = J \; \forall J \in \mathcal{M}.$$

It was Emmy Noether who noticed the importance of these conditions. To honor her we call rings which satisfy the equivalent conditions **Noetherian**. The polynomial ring $k[x_1, \ldots, x_n]$ is Noetherian by Hilbert's basis theorem 1.2.14.

Proof of the theorem. 1) $\Rightarrow$ 2): Let $I_1 \subset I_2 \subset \ldots$ be a chain of ideals. Then

3.1 The ascending chain condition

$$J = \bigcup_{j=1}^{\infty} I_j$$

is an ideal as well:

$$f, g \in J \Rightarrow f \in I_i, g \in I_j \text{ for indices } i, j \in \mathbb{N}$$
$$\Rightarrow f + g \in I_{\max(i,j)} \subset J.$$

By 1) the ideal J is finitely generated, say $J = (f_1, \ldots, f_r)$. Each f_ν lies in an $I_{j(\nu)}$ for some $j(\nu)$. If we take $N = \max\{j(\nu) \mid \nu = 1, \ldots, r\}$, then

$$J = (f_1, \ldots, f_r) \subset I_N \subset I_{N+1} \subset \ldots \subset J$$

and we have equality $I_j = I_{j+1}$ for all j greater than or equal to N.

2) $\Rightarrow$ 3): Let $\mathcal{M}$ be a non-empty set of ideals. Suppose there are no maximal elements in $\mathcal{M}$. Then for each $I \in \mathcal{M}$ we find a $J \in \mathcal{M}$ with $I \subsetneq J$. Inductively, we find a chain

$$I_1 \subsetneq I_2 \subsetneq \ldots$$

which does not become stationary, in contradiction to 2).

3) $\Rightarrow$ 1): Let J be an ideal and consider the set

$$\mathcal{M} = \{I \subset J \mid I \text{ is finitely generated}\}.$$

The set $\mathcal{M}$ is not empty because (0) lies in $\mathcal{M}$. Let $I = (f_1, \ldots, f_r) \in \mathcal{M}$ be a maximal element. We have to prove $I = J$. Let $f \in J$ be an arbitrary element. Then the ideal $(f_1, \ldots, f_r, f)$ is also finitely generated and $I \subset (f_1, \ldots, f_r, f) \subset J$. By the maximality of I in $\mathcal{M}$ we get $I = (f_1, \ldots, f_r, f)$, i.e., $f \in I$. This proves $J \subset I$ and equality holds. □

Using our dictionary we obtain

Corollary 3.1.5.

2') Every descending chain of algebraic subsets

$$A_1 \supset A_2 \supset \ldots$$

becomes stationary.

3') Every nonempty set $\mathcal{M}$ of algebraic subsets of $\mathbb{A}^n$ has a minimal element with respect to inclusion. □

Proof of the existence of a component composition. This is a typical proof by the so-called Noetherian induction. Consider the set

$$\mathcal{M} = \{A \subset \mathbb{A}^n \mid A \text{ is a non-empty algebraic set which is}$$
$$\text{not a finite union of irreducible algebraic subsets}\}.$$

We have to prove that $\mathcal{M}$ is empty.

Suppose $\mathcal{M} \neq \emptyset$. Then we can consider a minimal element $A \in \mathcal{M}$. The algebraic set A is not irreducible by the definition of $\mathcal{M}$. Thus there exists a decomposition

$$A = A_1 \cup A_2$$

into strictly smaller algebraic sets. By the minimality of A both A_1 and A_2 are finite unions of irreducible algebraic sets. But then so is A, a contradiction. We must have $\mathcal{M} = \emptyset$. □

Using the algebra-geometry dictionary we obtain:

Theorem 3.1.6. *Let $R = K[x_1, \ldots, x_n]$ be the polynomial ring over an algebraically closed field K. Every radical ideal $I \subset R$ is a finite intersection of prime ideals*

$$I = \mathfrak{p}_1 \cap \mathfrak{p}_2 \cap \ldots \cap \mathfrak{p}_r.$$

□

Remark 3.1.7. Note that a finite intersection of prime ideals in an arbitrary ring is always a radical ideal.

Example 3.1.8 (A natural appearance of non-radical ideals in geometry).
Consider the intersection

$$V(xy, yz) \cap V(y - x - t) = V(xy, yz, y - x - t)$$

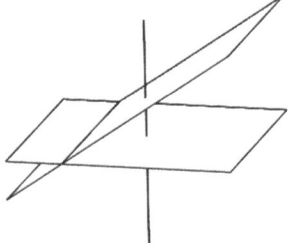

with a moving plane $H_t = V(y - x - t)$. For $t \neq 0$ the intersection has two components, the line $L_t = V(y, x + t)$ and the point $p_t = V(x, z, y - t)$. For $t = 0$ the intersection is defined by the ideal

$$(xy, yz, y - x) = (x^2, xz, y - x).$$

This is not a radical ideal.

Definition 3.1.9. Let M be an R-module and $m \in M$. The **annihilator** of m is the ideal

$$\text{ann}(m) = \{a \in R \mid am = 0 \in M\}.$$

Example 3.1.8 continued. The annihilator of the element $\overline{x} \in K[x, y, z]/(x^2, xz, y-x)$ as a $K[x, y, z]$-module is

$$\text{ann}(\overline{x}) = (x, z, y - x) = (x, y, z).$$

The corresponding point $V(x, y, z)$ is the limit of the point $p_t = (0, t, 0)$ for $t \to 0$ which lies on the limit line $L_0 = V(y, x)$ of the L_t's.

Exercise 3.1.10. Let R be a Noetherian ring. Prove that $R[x]$ is Noetherian as well. Deduce that any finitely generated R-algebra S is also Noetherian.

3.2 Primary decomposition

Remark. This gives another proof that $k[x_1,\ldots,x_n]$ is Noetherian. Hilbert used this approach. As in all such proofs the proof uses division with remainder in one way or another. The proof using a Gröbner basis has the advantage that it separates the induction done in Dixon's lemma from the division. This is the reason for its strength in applications.

Exercise 3.1.11. An R-module M is called Noetherian if it satisfies the analogous equivalent conditions for submodules instead of ideals. Prove:

1. Let
$$0 \longrightarrow M' \xrightarrow{\psi} M \xrightarrow{\varphi} M'' \longrightarrow 0$$
be a short exact sequence of R-modules, i.e., the homomorphism ψ is injective, the homomorphism φ is surjective and $\ker \varphi = \operatorname{im} \psi$. Then M is Noetherian if and only if M' and M'' are Noetherian.
2. An R-module M over a Noetherian ring R is Noetherian if and only if M is finitely generated.

Exercise 3.1.12. Consider $R = C^0[0,1] = \{f : [0,1] \to \mathbb{R} \mid f \text{ is continuous}\}$, the ring of continuous functions on the interval $[0,1]$. Prove that R is neither an integral domain nor Noetherian.

3.2 Primary decomposition

Definition 3.2.1. A **primary ideal** $\mathfrak{q}$ in a ring R is a proper ideal satisfying
$$fg \in \mathfrak{q} \Rightarrow f \in \mathfrak{q} \text{ or } g^n \in \mathfrak{q} \text{ for some } n \in \mathbb{N}$$
for all $f, g \in R$.

Proposition 3.2.2. *The radical* $\mathfrak{p} = \operatorname{rad}(\mathfrak{q})$ *of a primary ideal* $\mathfrak{q}$ *is a prime ideal.*

In this situation, $\mathfrak{q}$ is called a $\mathfrak{p}$-**primary ideal**.

Proof of the proposition. Suppose $fg \in \operatorname{rad}(\mathfrak{q})$ and $g \notin \operatorname{rad}(\mathfrak{q})$. Then for a power we obtain $(fg)^n = f^n g^n \in \mathfrak{q}$. Since no power g^m of g lies in $\mathfrak{q}$ we have $f^n \in \mathfrak{q}$ by the defining property of primary ideals. Thus $f \in \operatorname{rad}(\mathfrak{q})$. □

Theorem 3.2.3 (Primary decomposition). *Let* $I \subsetneq R$ *be a proper ideal in a Noetherian ring* R. *Then* I *is a finite intersection*
$$I = \mathfrak{q}_1 \cap \mathfrak{q}_2 \cap \ldots \cap \mathfrak{q}_r$$
of primary ideals $\mathfrak{q}_j$.

Corollary 3.2.4. *Let* $I \subset R$ *be a radical ideal in a Noetherian ring. Then*
$$I = \mathfrak{p}_1 \cap \ldots \cap \mathfrak{p}_r$$

is the intersection of finitely many prime ideals $\mathfrak{p}_i$.

Remark 3.2.5. The corollary generalizes Theorem 3.1.6 to arbitrary Noetherian rings. Emanuel Lasker [54] established primary decomposition for polynomial rings in 1905. Emmy Noether [82] gave a simplified proof for arbitrary Noetherian rings in 1921. Notice however that primary decomposition is not stable under field extensions: The ideal
$$(x^2 + y^2) \subset \mathbb{R}[x, y]$$
is a prime ideal, while in $\mathbb{C}[x, y]$ this ideal is no longer prime:
$$(x^2 + y^2) = (x + iy) \cap (x - iy) \subset \mathbb{C}[x, y].$$
Emmy Noether's proof is another case of Noetherian induction.

Proof. We proceed in two steps.

Definition 3.2.6. An ideal $I \subsetneq R$ satisfying
$$I = I_1 \cap I_2 \implies I_1 = I \text{ or } I_2 = I$$
for all ideals $I_1, I_2 \subset R$ is called **irreducible**.

Step 1. By property 3) in the definition of Noetherian rings, the set of ideals which are not the intersection of finitely many irreducible ideals is empty by the same argument as in the component decomposition.

Step 2. *Irreducible ideals are primary ideals.*

Let $I \subsetneq R$ be an irreducible ideal, $fg \in I$ and $f \notin I$. We have to prove that some power g^m lies in I. Consider the ascending chain of ideals
$$I : g \subset I : g^2 \subset \dots.$$
By property 2) in the definition of Noetherian rings there exists an $m \in \mathbb{N}$ such that
$$I : g^m = I : g^{m+1}.$$
We first claim
$$I : g^m = I : g^{m+1} \implies (I : g^m) \cap ((I + (g^m)) = I.$$
Let $a + bg^m$ with $a \in I$ and $b \in R$ be an arbitrary element of the intersection. So $(a + bg^m)g^m \in I$ and hence $bg^{2m} \in I$. Writing $bg^{2m} = bg^{m-1}g^{m+1}$ the assumption gives $bg^{2m-1} \in I$. By the same argument
$$bg^k g^{m+1} \in I \implies bg^k g^m \in I$$
holds for every $k \geq 0$. Finally, we obtain $bg^m \in I$ and hence $a + bg^m \in I$. This proves

3.2 Primary decomposition

$$(I : g^m) \cap ((I + (g^m)) \subset I.$$

Since the other inclusion holds trivially, we arrive at the claim. Now we use the irreducibility of I. Since $f \in I : g^m$, but $f \notin I$, we conclude

$$I + (g^m) = I, \text{ i.e., } g^m \in I.$$

This completes the proof of Theorem 3.2.3. □

Lemma 3.2.7. *If q_1 and q_2 are $\mathfrak{p}$-primary, then $q_1 \cap q_2$ is $\mathfrak{p}$-primary as well.*

Proof. If $fg \in q_1 \cap q_2$ and $g \notin \mathfrak{p}$, we have to show $f \in q_1 \cap q_2$. This is clear because $f \in q_j$ for all $j = 1, 2$ by the defining condition of primary ideals. □

A primary decomposition $I = q_1 \cap q_2 \cap \ldots \cap q_r$ is called **minimal** if

1. $q_j \not\supset \bigcap_{i \neq j} q_i$ for any j,
2. The prime ideals $\mathfrak{p}_j = \mathrm{rad}(q_j)$ are pairwise distinct.

By dropping superfluous terms and by collecting primary ideals with the same radical into a single primary ideal one can always pass from an arbitrary primary decomposition to a minimal one.

Example 3.2.8 (Embedded Point). The ideal (x^2, xy) has many different minimal primary decompositions. Indeed,

$$(x^2, xy) = (x) \cap (x^2, y) = (x) \cap (x^2, xy, y^n)$$

are different primary decompositions with associated primes (x) and (x, y). Thus minimal primary decompositions are not necessarily unique.

Theorem 3.2.9 (First Uniqueness Theorem). *The associated primes $\{\mathfrak{p}_1, \ldots, \mathfrak{p}_r\}$ of a minimal primary decomposition of*

$$I = q_1 \cap q_2 \cap \ldots \cap q_r$$

are uniquely determined by I.

The minimal elements in the set of associated primes $\{\mathfrak{p}_1, \ldots, \mathfrak{p}_r\}$ are called **isolated primes** or **minimal primes** of I. The non-isolated primes are called **embedded primes**. The last notation is motivated by geometry: If $\mathfrak{p}_i \subset \mathfrak{p}_j$ then $V(\mathfrak{p}_j)$ is embedded into $V(\mathfrak{p}_i)$ in the case of $R = K[x_1, \ldots, x_n]$.

Theorem 3.2.10 (Second Uniqueness Theorem). *The primary ideal q_i corresponding to the isolated primes $\mathfrak{p}_i$ of a minimal primary decomposition*

$$I = q_1 \cap q_2 \cap \ldots \cap q_r$$

are uniquely determined by I.

Corollary 3.2.11. *Proper ideals I in a Noetherian ring which have no embedded primes have a unique primary decomposition.* □

Example 3.2.8 continued. In our example above, i.e.,
$$(x^2, xy) = (x) \cap (x^2, y) = (x) \cap (x^2, xy, y^n),$$
the ideal $\mathfrak{p}_1 = (x)$ is an isolated prime and $\mathfrak{p}_2 = (x, y)$ an embedded prime. The primary ideal $\mathfrak{q}_1 = (x)$ is the same for each minimal decomposition.

Exercise 3.2.12. Let R be a Noetherian ring, let $\mathfrak{m}$ be a maximal ideal of R, and let I be any ideal of R. Show that the following are equivalent:

1. The ideal I is $\mathfrak{m}$-primary.
2. $\mathrm{rad}(I) = \mathfrak{m}$.
3. $\mathfrak{m} \supset I \supset \mathfrak{m}^N$ for some $N \geq 1$.

Exercise 3.2.13. 1. When is a monomial ideal a prime ideal?
2. Characterize monomial primary ideals.

Exercise 3.2.14. Consider the monomial ideal $I = (xy, xz, yz) \subset \mathbb{Q}[x, y, z]$. Compute a primary decomposition of I and I^2.

3.3 Associated primes

Definition 3.3.1. Let M be an R-module. An **associated prime** of M is a prime ideal $\mathfrak{p}$ of the form
$$\mathfrak{p} = \mathrm{ann}(m) = \{r \in R \mid rm = 0\}$$
for some non-zero element $m \in M$.

Proposition 3.3.2. *The maximal elements with respect to inclusion of the set*
$$\mathcal{M} = \{\mathrm{ann}(m) \mid m \in M, m \neq 0\}$$
are associated primes of M.

Proof. Let $\mathrm{ann}(m) \in \mathcal{M}$ be maximal and $f, g \in R$ elements with
$$fg \in \mathrm{ann}(m).$$
Suppose $g \notin \mathrm{ann}(m)$. Then $gm \neq 0$ and $\mathrm{ann}(m) \subset \mathrm{ann}(gm)$. Since $\mathrm{ann}(m) \in \mathcal{M}$ is maximal, we have $\mathrm{ann}(m) = \mathrm{ann}(gm)$ and $f \in \mathrm{ann}(gm) = \mathrm{ann}(m)$. Thus $\mathrm{ann}(m)$ is a prime ideal. □

Definition 3.3.3. Let M be an R-module. Then
$$\mathrm{Ass}(M) = \{\mathfrak{p} \mid \mathfrak{p} \text{ is an associated prime of } M\}$$

3.3 Associated primes

denotes the **set of associated primes of** M. For a non-zero module M over a Noetherian ring $\mathrm{Ass}(M)$ is non-empty, since the set $\mathcal{M}$ above is non-empty.

Definition 3.3.4. A **short exact sequence** of R-modules is a sequence

$$0 \longrightarrow M' \xrightarrow{\psi} M \xrightarrow{\varphi} M'' \longrightarrow 0$$

which consists of an injective R-module homomorphism ψ and a surjective R-module homomorphism φ such that

$$\ker(\varphi) = \mathrm{im}(\psi).$$

If we identify M' with a submodule of M via ψ, then M'' is isomorphic to the quotient module M/M':

$$M'' \cong M/\ker(\varphi) = M/M'.$$

Proposition 3.3.5. *Let*

$$0 \longrightarrow M' \xrightarrow{\psi} M \xrightarrow{\varphi} M'' \longrightarrow 0$$

be a short exact sequence of R-modules. Then

$$\mathrm{Ass}(M') \subset \mathrm{Ass}(M) \subset \mathrm{Ass}(M') \cup \mathrm{Ass}(M'').$$

Proof. The first inclusion is clear. For the second consider a $\mathfrak{p} \in \mathrm{Ass}(M) \setminus \mathrm{Ass}(M')$ and an element $m \in M$ such that $\mathfrak{p} = \mathrm{ann}(m)$. Then

$$Rm \cong R/\mathfrak{p}.$$

Since $\mathfrak{p}$ is prime, every non-zero element of $gm \in Rm$ has annihilator $\mathrm{ann}(gm) = \mathfrak{p}$ as well: $f \in \mathrm{ann}(gm) \neq 0 \Rightarrow fg \in \mathrm{ann}(m) = \mathfrak{p} \Rightarrow f \in \mathfrak{p}$, since $g \notin \mathrm{ann}(m)$. Since $\mathfrak{p} \notin \mathrm{Ass}(M')$, it follows that $Rm \cap M' = 0$. Thus Rm is isomorphic to its image $\varphi(Rm)$ in M'' and we obtain $\mathfrak{p} = \mathrm{ann}(\varphi(m)) \in \mathrm{Ass}(M'')$. □

Corollary 3.3.6. $\mathrm{Ass}(M' \oplus M'') = \mathrm{Ass}(M') \cup \mathrm{Ass}(M'')$.

Proof. For $M = M' \oplus M''$ we have two short exact sequences

$$0 \longrightarrow M' \longrightarrow M \longrightarrow M'' \longrightarrow 0$$

and

$$0 \longrightarrow M'' \longrightarrow M \longrightarrow M' \longrightarrow 0.$$

Hence

$$\mathrm{Ass}(M') \cup \mathrm{Ass}(M'') \subset \mathrm{Ass}(M' \oplus M'') \subset \mathrm{Ass}(M') \cup \mathrm{Ass}(M'')$$

follows from Proposition 3.3.5. □

Theorem 3.3.7. *Let $I = \mathfrak{q}_1 \cap \ldots \cap \mathfrak{q}_r$ be a minimal primary decomposition of an ideal in a Noetherian ring R. Then the collection of associated primes of R/I as an R-module is precisely the set*

$$\mathrm{Ass}(R/I) = \{\mathfrak{p}_1, \ldots, \mathfrak{p}_r\},$$

where $\mathfrak{p}_i = \mathrm{rad}(\mathfrak{q}_i)$.

Proof. We first establish the special case when $I = \mathfrak{q}$ is a $\mathfrak{p}$-primary ideal: We claim

$$\mathrm{Ass}(R/\mathfrak{q}) = \{\mathfrak{p}\}.$$

Indeed, suppose $g \in \mathrm{ann}(\overline{f}) = \mathfrak{p}'$ for some $\overline{f} \in R/\mathfrak{q}$ lies in an associated prime. Then $gf \in \mathfrak{q}$. Since $f \notin \mathfrak{q}$, we obtain $g^n \in \mathfrak{q}$ for some $n \in \mathbb{N}$, i.e., $\mathfrak{p}' \subset \mathrm{rad}(\mathfrak{q})$. Since $\mathfrak{q} \subset \mathfrak{p}'$, we deduce

$$\mathrm{rad}(\mathfrak{q}) \subset \mathrm{rad}(\mathfrak{p}') = \mathfrak{p}' \subset \mathrm{rad}(\mathfrak{q})$$

and equality holds.

Now consider the R-module homomorphism

$$\psi: R \to R/\mathfrak{q}_1 \oplus \ldots \oplus R/\mathfrak{q}_r, \; f \mapsto (f + \mathfrak{q}_1, \ldots, f + \mathfrak{q}_r).$$

Since $\ker(\psi) = I$, we get an inclusion

$$R/I \hookrightarrow R/\mathfrak{q}_1 \oplus \ldots \oplus R/\mathfrak{q}_r.$$

Hence we obtain $\mathrm{Ass}(R/I) \subset \{\mathfrak{p}_1, \ldots, \mathfrak{p}_r\}$ from Proposition 3.3.5 and Corollary 3.3.6.

To show equality we use that the primary decomposition is irredundant. Thus for each i in $\{1, \ldots, r\}$ we have

$$\bigcap_{j \neq i} \mathfrak{q}_j \supsetneq \bigcap_{j=1}^{r} \mathfrak{q}_j.$$

Consider an element f_i in the complement and the residue class $\overline{f}_i \in R/I$. The R-module homomorphism ψ maps the submodule $R\overline{f}_i \subset R/I$ into the summand $R/\mathfrak{q}_i$. Thus

$$\mathrm{Ass}(R\overline{f}_i) \subset \mathrm{Ass}(R/\mathfrak{q}_i) = \{\mathfrak{p}_i\}$$

and equality holds. Hence $\{\mathfrak{p}_i\} = \mathrm{Ass}(R\overline{f}_i) \subset \mathrm{Ass}(R/I)$. □

Abusing notation the **associated primes of an ideal** I refer to $\mathrm{Ass}(R/I)$, where we regard R/I as an R-module.

Notice that $\mathrm{Ass}(I)$ where we regard I as an R-module is not so interesting. For example, if R is an integral domain, then $\mathrm{Ass}(I) = \mathrm{Ass}(R) = \{(0)\}$.

Thus the associated primes of I are precisely the prime ideals which occur in a minimal primary decomposition of I. This completes the proof of the first uniqueness theorem 3.2.9.

3.4 Filtration of modules with primes

The concept of associated primes gives a structure result for modules.

Theorem 3.4.1 (Filtration with prime ideal quotients). *Let M be a finitely generated non-zero module over a Noetherian ring R. Then there exists a filtration*

$$0 = M_0 \subset M_1 \subset \ldots \subset M_N = M$$

by submodules such that all quotients

$$M_i/M_{i-1} \cong R/\mathfrak{p}_i$$

for some prime ideals $\mathfrak{p}_i$ of R.

Proof. Since R is Noetherian, the set of proper ideals

$$\mathcal{M} = \{\text{ann}(m) \mid m \in M, m \neq 0\}$$

is not empty, and a maximal element of this set is a prime ideal $\mathfrak{p}_1 = \text{ann}(m_1)$ such that

$$Rm_1 \cong R/\mathfrak{p}_1.$$

We take $M_1 = Rm_1$. Now suppose that $M_0 \subset \ldots \subset M_{\nu-1}$ are already constructed. If $M_{\nu-1} \subsetneq M$, then we consider an associated prime $\mathfrak{p}_\nu = \text{ann}(\overline{m}_\nu) \in \text{Ass}(M/M_{\nu-1})$, and define

$$M_\nu = \pi^{-1}(R\overline{m}_\nu) = Rm_\nu + M_{\nu-1},$$

where $\pi \colon M \to M/M_{\nu-1}$ is the natural projection and $\pi(m_\nu) = \overline{m}_\nu$. By Noether's isomorphism theorem, Exercise 3.4.2 b) below, we deduce

$$M_\nu/M_{\nu-1} \cong Rm_\nu/Rm_\nu \cap M_{\nu-1} \cong Rm_\nu/\mathfrak{p}_\nu m_\nu \cong R\overline{m}_\nu \cong R/\mathfrak{p}_\nu.$$

By Exercise 3.1.11, M is Noetherian. Hence any ascending chain of submodules of M becomes stationary and the process stops for some $N \in \mathbb{N}$ with $M_N = M$. □

Exercise 3.4.2 (Noether's Isomorphism Theorems). Prove:
a) Let $L \subset N \subset M$ be submodules of an R-module M. Then N/L is a submodule of M/L and

$$(M/L)/(N/L) \cong M/N.$$

b) Let $L, N \subset M$ be submodules of an R-module M. Then the intersection $L \cap N$ and the sum $L + N = \{\ell + n \in M \mid \ell \in L \text{ and } n \in N\}$ are submodules of M and

$$N/(L \cap N) \cong (N + L)/L.$$

Hint: Apply the homomorphism theorem 1.3.12 to suitable homomorphisms.

Exercise 3.4.3. 1) Two ideals I and J of a ring R are called **coprime** if $I + J = R$. Prove: The ideals I and J are coprime if and only if $\text{rad}(I)$ and $\text{rad}(J)$ are coprime.

2) Let $I = \mathfrak{q}_1 \cap \ldots \cap \mathfrak{q}_r \subset R$ be a primary decomposition with pairwise coprime associated primes $\mathfrak{p}_i = \operatorname{rad} \mathfrak{q}_i$. Then

$$R/I \cong \bigoplus_{i=1}^{r} R/\mathfrak{q}_i$$

as rings.

Chapter 4
Localization

In this chapter we introduce localization of a ring in a multiplicative subset. In the case of an integral domain R the localization in the multiplicative subset of all non-zero elements is the familiar construction of its quotient field $Q(R)$. Localization in general takes a more restricted set of denominators. The localization of the coordinate ring $K[A]$ at a maximal ideal $\mathfrak{m}_p$ corresponding to a point $p \in A$ leads to the ring $\mathcal{O}_{A,p}$, the local ring of A at p, whose properties depend only on an arbitrary small Zariski open neighborhood of $p \in A$. This explains the name.

We establish the functorial properties of localizations, and prove the second uniqueness theorem for primary decomposition.

4.1 Fractions

If we want to add or multiply two fractions, we have to be able to multiply the denominators:
$$\frac{a}{s} + \frac{b}{t} = \frac{at + bs}{st}.$$

Definition 4.1.1. A **multiplicative** subset $U \subset R$ of a ring R is a subset which satisfies

a) $1 \in U$
b) $s, t \in U \implies st \in U$.

Examples 4.1.2. The most important multiplicative sets are:

1) $U = \{f^n \mid n \in \mathbb{N}\}$, the powers of an element $f \in R$,
2) $U = R \setminus \mathfrak{p}$, the complement of a prime ideal,
3) $U = \{r \in R \mid rs \neq 0 \ \forall s \in R \setminus \{0\}\}$, the set of non-zero divisors.

If R is an integral domain, then (0) is a prime ideal and the set of non-zero divisors coincides with the complement of (0).

Localization in U. Let $U \subset R$ be a multiplicative subset of a ring. We will define a ring of fractions
$$R[U^{-1}] = \{\frac{a}{s} \mid a \in R \text{ and } s \in U\}$$
as follows: Consider the following equivalence relation on $R \times U$:

$(a_1, s_1) \sim (a_2, s_2)$ if and only if there exists a $u \in U$ such that $u(s_2 a_1 - s_1 a_2) = 0 \in R$.

The factor u is needed for the transitivity, since R might not be an integral domain.

$(a_1, s_1) \sim (a_2, s_2)$ and $(a_2, s_2) \sim (a_3, s_3)$
$\Rightarrow \exists u, v \in U$ such that $u(s_2 a_1 - s_1 a_2) = 0$ and $v(s_3 a_2 - s_2 a_3) = 0$
$\Rightarrow 0 = v s_3 u(s_2 a_1 - s_1 a_2) - u s_1 v(s_3 a_2 - s_2 a_3) = u v s_2 (s_3 a_1 - s_1 a_3)$
$\Rightarrow (a_1, s_1) \sim (a_3, s_3)$ since $u v s_2 \in U$.

The fraction
$$\frac{a}{s} = \{(b, t) \in R \times U \mid (a, s) \sim (b, t)\}$$
denotes the equivalence class of (a, s). Then
$$R[U^{-1}] = (R \times U)/\sim$$
defines the localization as a set. It is a subset of $2^{R \times U}$. The usual formulas give $R[U^{-1}]$ the structure of a commutative ring with $1 = \frac{1}{1}$. Of course, one has to verify that addition and multiplication are well-defined. For example, if $(a_1, s_1) \sim (a_2, s_2)$, then
$$\frac{a_1}{s_1} + \frac{b}{t} = \frac{a_1 t + s_1 b}{s_1 t} = \frac{a_2 t + s_2 b}{s_2 t} = \frac{a_2}{s_2} + \frac{b}{t}$$
because $u(s_2 a_1 - s_1 a_2) = 0$ gives $u(s_2 t(t a_1 + s_1 b) - s_1 t(a_2 t + s_2 b)) = t^2 u(s_2 a_1 - s_1 a_2) = 0$. The map
$$\iota: R \to R[U^{-1}], r \mapsto \frac{r}{1}$$
is a ring homomorphism which might not be injective:
$$\ker(\iota) = \{r \in R \mid \exists u \in U \text{ with } ur = 0\}.$$
Notice that the elements $\iota(u)$ for $u \in U$ are units in $R[U^{-1}]$: $\frac{u}{1}\frac{1}{u} = 1$.

Localization of modules. Let M be an R-module and $U \subset R$ a multiplicative subset. Then we can define $M[U^{-1}]$ similarly:
$$(m_1, s_1) \sim (m_2, s_2) \text{ iff } \exists u \in U \text{ such that } u(s_2 m_1 - s_1 m_2) = 0 \in M$$
is an equivalence relation on $M \times U$, and the set of equivalence classes
$$M[U^{-1}] = \{\frac{m}{s} \mid m \in M, s \in U\}$$

4.1 Fractions

becomes an $R[U^{-1}]$-module by

$$\frac{a}{s} \cdot \frac{m}{t} = \frac{am}{st}.$$

Functoriality of localization. Let $\varphi: M \to N$ be an R-module homomorphism. Then

$$\varphi[U^{-1}]: M[U^{-1}] \to N[U^{-1}]$$

defined by

$$\varphi[U^{-1}](\frac{m}{s}) = \frac{\varphi(m)}{s}$$

is a well-defined $R[U^{-1}]$-module homomorphism. If $M' \xrightarrow{\psi} M \xrightarrow{\varphi} M''$ are two composable morphisms with $\varphi \circ \psi = 0$, then the same holds for the localizations. More is true.

Definition 4.1.3. A sequence

$$M' \xrightarrow{\psi} M \xrightarrow{\varphi} M''$$

of R-module homomorphisms is **exact at** M if $\ker(\varphi) = \operatorname{im}(\psi)$.

Proposition 4.1.4 (Exactness of localization). *Let* $M' \xrightarrow{\psi} M \xrightarrow{\varphi} M''$ *be exact at* M. *Let* U *be a multiplicative subset. Then the induced sequence*

$$M'[U^{-1}] \xrightarrow{\psi[U^{-1}]} M[U^{-1}] \xrightarrow{\varphi[U^{-1}]} M''[U^{-1}]$$

is exact at $M[U^{-1}]$.

Proof. The inclusion $\operatorname{im}(\psi[U^{-1}]) \subset \ker(\varphi[U^{-1}])$ is clear because $\varphi \circ \psi = 0$. To prove the converse inclusion let $m/s \in \ker(\varphi[U^{-1}])$. Then $\varphi(m)/s = 0 \in M''[U^{-1}]$, i.e., there exists a $u \in U$ such that $u\varphi(m) = 0 \in M''$. But $u\varphi(m) = \varphi(um)$ since φ is R-linear. Hence $um \in \ker(\varphi) = \operatorname{im}(\psi)$. So there exists an $m' \in M'$ such that $\psi(m') = um$. Thus

$$\frac{m'}{us} \mapsto \frac{um}{us} = \frac{m}{s}.$$

$\square$

If $N \subset M$ is a submodule, then by the proposition applied to the exact sequence

$$0 \to N \to M$$

we may regard $N[U^{-1}]$ as a submodule of $M[U^{-1}]$.

Proposition 4.1.5. *Let* M *be an* R-*module* N, P *be submodules of* M *and* $U \subset R$ *a multiplicative subset. Then*

1) $(N + P)[U^{-1}] = N[U^{-1}] + P[U^{-1}]$.
2) $(N \cap P)[U^{-1}] = N[U^{-1}] \cap P[U^{-1}]$.
3) $(M/N)[U^{-1}] \cong M[U^{-1}]/N[U^{-1}]$.

Proof. 1) follows from $n/s + p/t = (tn + sp)/st$.
2): If $n/s = p/t$, then there exists a $u \in U$ with $utn = usp \in N \cap P$.
3) follows from Proposition 4.1.4 applied to the exact sequence

$$0 \to N \to M \to M/N \to 0.$$

□

Notation. Let $\mathfrak{p} \subset R$ be a prime ideal and M an R-module. Then

$$M_\mathfrak{p} = M[U^{-1}],$$

where $U = R \setminus \mathfrak{p}$ is called the **localization of M in $\mathfrak{p}$**. For $f \in R$ the **localization of M in f** is

$$M_f = M[U^{-1}]$$

for $U = \{f^k \mid k \in \mathbb{N}\}$.
Example.

$$\mathbb{Z}_{(2)} = \{\frac{a}{b} \in \mathbb{Q} \mid 2 \text{ does not divide } b\}$$

and

$$\mathbb{Z}_2 = \{\frac{a}{b} \in \mathbb{Q} \mid b \text{ is a power of } 2\}$$

are quite different.

Exercise 4.1.6. 1) Let $R = K[A]$ be the coordinate ring of a variety and $f \in R$ be an element which is not a unit. Prove that R_f is isomorphic to the coordinate ring of a variety.
  2) Prove that $K[x]_{(x)}$, the localization of the polynomial ring in one variable at the maximal ideal (x), is not isomorphic to the coordinate ring of a variety.

Exercise 4.1.7. Let $A = B \cup C$ be a decomposition of an algebraic set into proper algebraic subsets. Let $p \in A \setminus C$ be a point and $\mathfrak{m} = I(p)$ be the corresponding maximal ideal. Prove

$$K[A]_\mathfrak{m} \cong K[B]_\mathfrak{m}.$$

4.2 Some local properties

Some properties of modules are local properties. For example:

Theorem 4.2.1. *Let M be an R-module. The following are equivalent:*

1) $M = 0$.

4.2 Some local properties

2) $M_\mathfrak{p} = 0$ *for all prime ideals* $\mathfrak{p} \subset R$.
3) $M_\mathfrak{m} = 0$ *for all maximal ideals* $\mathfrak{m} \subset R$.

Proof. Only the implication 3) $\implies$ 1) is non-trivial. Let $M \neq 0$ be a non-zero module and $m \in M$ a non-zero element. Then $I = \text{ann}(m) \subsetneq R$ is a proper ideal since $1 \notin I$. The set of ideals $\mathcal{M} = \{J \text{ ideal in } R \mid I \subset J \subsetneq R\}$ contains a maximal element $\mathfrak{m}$ with respect to inclusion. (This is clear for Noetherian rings. For more general rings one applies Zorn's lemma, see Lemma 6.3.2 below.) The ideal $\mathfrak{m}$ is a maximal ideal of R, and $M_\mathfrak{m} \neq 0$ because

$$\frac{m}{1} \neq 0.$$

No element of $R \setminus \mathfrak{m}$ annihilates m because $\mathfrak{m} \supset I = \text{ann}(m)$. □

Further local properties:

Theorem 4.2.2. *Let* $\varphi \colon M \to N$ *be an R-module homomorphism. The following are equivalent:*

1) *φ is injective.*
2) *$\varphi_\mathfrak{p}$ is injective for all prime ideals $\mathfrak{p}$ of R.*
3) *$\varphi_\mathfrak{m}$ is injective for all maximal ideals $\mathfrak{m}$ of R.*

A similar result holds for 'injective' replaced by 'surjective'.

Proof. Consider the sequence

$$0 \to \ker(\varphi) \to M \to N,$$

which is exact at M and $\ker(\varphi)$. By the exactness of localization

$$\ker(\varphi_\mathfrak{p}) = (\ker(\varphi))_\mathfrak{p}.$$

Thus the result follows because being the zero-module is a local property. For the second version we consider the exact sequence

$$M \to N \to \text{coker}(\varphi) \to 0.$$

□

Exercise 4.2.3. A non-zero element a in a ring R is called **nilpotent** if a power $a^n = 0$. Suppose that for each prime ideal $\mathfrak{p} \subset R$, the local ring $R_\mathfrak{p}$ contains no nilpotent elements. Show that R contains no nilpotent elements.

If each $R_\mathfrak{p}$ is an integral domain, is R necessarily an integral domain?

4.3 Extended and contracted ideals

Let $\varphi: A \to B$ a ring homomorphism, $\mathfrak{a}$ an ideal in A and $\mathfrak{b}$ an ideal in B. Then

$$\mathfrak{a}^e = \mathfrak{a}B = \{\sum_i b_i \varphi(a_i) \mid \text{a finite sum with } b_i \in B \text{ and } a_i \in \mathfrak{a}\}$$

is called the **extended** ideal of $\mathfrak{a}$, and

$$\mathfrak{b}^c = \varphi^{-1}(\mathfrak{b})$$

is called the **contracted** ideal of $\mathfrak{b}$.

Primary decompositions behave well under contractions:

1. If $\mathfrak{b}$ is a prime ideal or a primary ideal, then $\mathfrak{b}^c$ is respectively prime or primary as well.
2. $(\mathfrak{b}_1 \cap \mathfrak{b}_2)^c = \mathfrak{b}_1^c \cap \mathfrak{b}_2^c$.
3. $(\mathrm{rad}(\mathfrak{b}))^c = \mathrm{rad}(\mathfrak{b}^c)$.

The behavior under extensions can be complicated:

Example 4.3.1. Consider $\mathbb{Z} \hookrightarrow \mathbb{Z}[\sqrt{-1}]$. Then the prime ideals $(p) \subset \mathbb{Z}$ extend as follows:

1) $(2)^e = (1 + \sqrt{-1})^2$ is a square of a prime ideal.
2) If $p \equiv 1 \bmod 4$, then $(p)^e$ is the product of two distinct prime ideals, for example $(5)^e = (2 + \sqrt{-1})(2 - \sqrt{-1})$.
3) If $p \equiv 3 \bmod 4$, then $(p)^e$ is a prime ideal.

Only 2) is a non-trivial statement. It follows from a theorem of Fermat which says that a prime p is a sum of two squares if and only if $p \equiv 1 \bmod 4$. ($5 = 2^2 + 1^2$, $13 = 3^2 + 2^2, \ldots, 89 = 8^2 + 5^2$, etc.)

Proposition 4.3.2. *Let $A \to B$ be a ring homomorphism and let $\mathfrak{a} \subset A$ and $\mathfrak{b} \subset B$ be ideals. Then*

1) $\mathfrak{a}^{ec} \supset \mathfrak{a}$ *and* $\mathfrak{b}^{ce} \subset \mathfrak{b}$.
2) $\mathfrak{a}^e = \mathfrak{a}^{ece}$ *and* $\mathfrak{b}^{cec} = \mathfrak{b}^c$.
3) *The set of contracted ideals is* $C = \{\mathfrak{a} \mid \mathfrak{a} = \mathfrak{a}^{ec}\}$, *and the set of extended ideals is* $E = \{\mathfrak{b} \mid \mathfrak{b} = \mathfrak{b}^{ce}\}$. *These sets are in bijection via* $\mathfrak{a} \mapsto \mathfrak{a}^e$ *and* $\mathfrak{b} \mapsto \mathfrak{b}^c$.

Proof. 1) is clear. 2) follows from 1): The inclusion $\mathfrak{a}^{ec} \supset \mathfrak{a}$ implies $\mathfrak{a}^{ece} \supset \mathfrak{a}^e$, and $\mathfrak{b}^{ce} \subset \mathfrak{b}$ applied to $\mathfrak{b} = \mathfrak{a}^e$ gives the other inclusion. 3) follows from 2). □

4.4 Ideal theory of localizations

The situation is better for localization from R to $R[U^{-1}]$. Passing from a ring to a localization makes things easier at least from a theoretical point of view. For example, the ideal theory of $R[U^{-1}]$ is a simplified version of the ideal theory of R.

4.4 Ideal theory of localizations

Theorem 4.4.1 (Ideal theory of localizations). *Let $U \subset R$ be a multiplicative subset of a ring and let $\iota: R \to R[U^{-1}]$, $r \mapsto r/1$ denote the natural homomorphism.*

1) If I is an ideal in R, then

$$I^{ec} = \iota^{-1}(IR[U^{-1}]) = \{a \in R \mid \exists u \in U \text{ such that } ua \in I\}.$$

2) If J is an ideal in $R[U^{-1}]$, then

$$J^{ce} = \iota^{-1}(J)R[U^{-1}] = J.$$

Thus the map $J \mapsto \iota^{-1}(J)$ gives an injection of the set of ideals of $R[U^{-1}]$ into the set of ideals of R.

3) If R is Noetherian, then $R[U^{-1}]$ is Noetherian.

4) ι^{-1} induces a bijections between the set of prime (respectively primary) ideals of $R[U^{-1}]$ and the set of prime (respectively primary) ideals $\mathfrak{q}$ of R with $U \cap \mathfrak{q} = \emptyset$.

Proof. 1) If $a \in R$, then $a \in \iota^{-1}(IR[U^{-1}]) \Leftrightarrow a/1 \in IR[U^{-1}] \Leftrightarrow ua \in I$ for some $u \in U$.

2) Let $b/u \in R[U^{-1}]$. Then $b/u \in J \Leftrightarrow b/1 \in J \Leftrightarrow b \in \iota^{-1}(J) \Leftrightarrow b/u \in \iota^{-1}(J)R[U^{-1}]$. 3) follows from 2).

4) Let $\mathfrak{q}$ be a primary ideal of $R[U^{-1}]$. Then $\mathfrak{q}^c = \iota^{-1}(\mathfrak{q})$ is a primary ideal of R which does not intersect U because $\mathfrak{q}$ contains no units. Conversely, let $\mathfrak{q}$ be a primary ideal in R with $\mathfrak{q} \cap U = \emptyset$. Then $\mathfrak{q}^e = \mathfrak{q}R[U^{-1}]$ is a proper ideal because $\mathfrak{q}^{ec} = \iota^{-1}(\mathfrak{q}^e) = \mathfrak{q}$ follows from 1): $ua \in \mathfrak{q}$ and $u^n \notin \mathfrak{q}$ implies $a \in \mathfrak{q}$ since $\mathfrak{q}$ is primary. It remains to prove that $\mathfrak{q}^e$ is a primary ideal. Suppose $a/u \cdot b/v \in \mathfrak{q}^e$, then $wab \in \mathfrak{q}$ for some $w \in U$ by 1). Hence $wa \in \mathfrak{q}$ or $b^n \in \mathfrak{q}$ for some n since $\mathfrak{q}$ is primary. It follows that $a/u \in \mathfrak{q}^e$ or $(b/v)^n \in \mathfrak{q}^e$ because wu and v are units in $R[U^{-1}]$. In the case of prime ideals we have $n = 1$ in the argument above. □

Corollary 4.4.2. *Let U be a multiplicative subset of a ring R and let*

$$I = \mathfrak{q}_1 \cap \ldots \cap \mathfrak{q}_r$$

be a primary decomposition of an ideal $I \subset R$. Then

$$I^e = \bigcap_{\mathfrak{q}_i : \mathfrak{q}_i \cap U = \emptyset} \mathfrak{q}_i^e$$

is a primary decomposition of the extended ideal $I^e \subset R[U^{-1}]$ and

$$I^{ec} = \bigcap_{\mathfrak{q}_i : \mathfrak{q}_i \cap U = \emptyset} \mathfrak{q}_i.$$

In particular, the last intersection does not depend on the choice of the primary decomposition.

Proof. Primary ideals $\mathfrak{q}_j$ with $\mathfrak{q}_j \cap U \neq \emptyset$ extend to $\mathfrak{q}_j^e = (1)$, since elements of U become units in $R[U^{-1}]$. Thus these can be dropped in the intersection, and

$$I^e = \bigcap_{\mathfrak{q}_i : \mathfrak{q}_i \cap U = \emptyset} \mathfrak{q}_i^e$$

follows from Proposition 4.1.5. The rest of the theorem is clear, because contraction commutes with intersections and $\mathfrak{q}_i^{ec} = \mathfrak{q}_i$ for primary ideals with $\mathfrak{q}_i \cap U \neq \emptyset$. □

Corollary 4.4.3. *Let $\mathfrak{p}_i$ be a minimal associated prime of a minimal primary decomposition*

$$I = \mathfrak{q}_1 \cap \ldots \cap \mathfrak{q}_r.$$

Then $\mathfrak{q}_i$ is uniquely determined by I.

Proof. Consider the localization in $\mathfrak{p}_i$, i.e., with respect to $U = R \setminus \mathfrak{p}_i$. Since $\mathfrak{p}_i$ is minimal, all other associated primes $\mathfrak{p}_j = \mathrm{rad}(\mathfrak{q}_j)$ intersect U:

$$(R \setminus \mathfrak{p}_i) \cap \mathfrak{p}_j = \emptyset \iff \mathfrak{p}_j \subset \mathfrak{p}_i$$

and $\mathfrak{p}_j$ equals $\mathfrak{p}_i$. Since U is multiplicative $\mathfrak{p}_j \cap U \neq \emptyset \iff \mathfrak{q}_j \cap U \neq \emptyset$ holds. Thus

$$I^{ec} = \mathfrak{q}_i$$

holds by the Corollary 4.4.2. □

Examples 4.4.4. 1) Consider $R = \mathbb{Z}$. The ideals of $\mathbb{Z}$ are principal and

$$(n) = (p_1^{e_1}) \cap \ldots \cap (p_r^{e_r})$$

is the primary decomposition if

$$n = p_1^{e_1} \cdot \ldots \cdot p_r^{e_r}$$

is the prime factorization.

2) The polynomial ring $k[x_1, \ldots, x_n]$ over any field k is factorial. As above the primary decomposition of a principal ideal (f) corresponds to factorizations: If

$$f = u f_1^{e_1} \cdot \ldots \cdot f_r^{e_r}$$

with $u \in k^*$ a unit and f_j irreducible, then $(f) = (f_1^{e_1}) \cap \ldots \cap (f_r^{e_r})$ is the primary decomposition.

Remark 4.4.5. It is possible to factor polynomials in $\mathbb{Q}[x_1, \ldots, x_n]$ or $\mathbb{F}_q[x_1, \ldots, x_n]$ algorithmically, see [86]. Based on this there are algorithms for primary decompositions of ideals in $\mathbb{Q}[x_1, \ldots, x_n]$ or $\mathbb{F}_q[x_1, \ldots, x_n]$. They are implemented in Macaulay2 and OSCAR.

Primary decomposition however does not behave well under field extensions: The ideal

4.4 Ideal theory of localizations

$$(x_1^2 + x_2^2) \subset \mathbb{Q}[x_1, x_2]$$

is a prime ideal, but $x^2 + y^2 = (x_1 + ix_2)(x_1 - ix_2) \in \mathbb{C}[x_1, x_2]$ factors.

A prime (primary) ideal $\mathfrak{p} \subset k[x_1, \ldots, x_n]$ is called **absolutely prime (primary)** if its extension $\mathfrak{p}^e \subset K[x_1, \ldots, x_n]$ is prime (primary) where $K \supset k$ is an algebraically closed extension field.

An absolute primary decomposition of an ideal $I \subset k[x_1, \ldots, x_n]$ consist of the construction of an algebraic field extension $L \supset k$ and a primary decomposition

$$I^e = \mathfrak{q}_1 \cap \ldots \cap \mathfrak{q}_r \subset L[x_1, \ldots, x_n]$$

such that the $\mathfrak{q}_i$ are absolutely primary.

$$(x_1^2 + x_2^2) = (x_1 + ix_2) \cap (x_1 - ix_2) \subset \mathbb{Q}[i][x_1, x_2]$$

is an absolute primary decomposition of the ideal above.

Exercise 4.4.6. Consider $I = (y - x^2, x^3 - 4x^2y + 4y^3 - xy + y) \subset \mathbb{Q}[x, y]$.

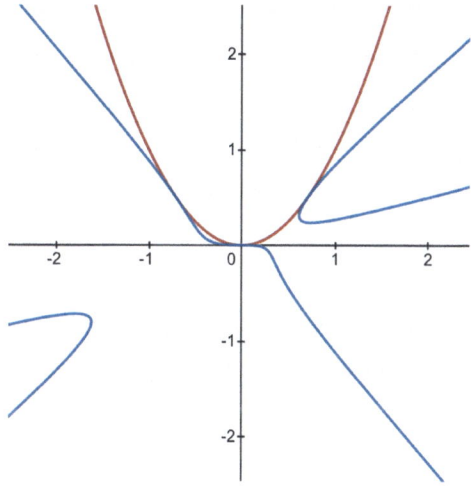

Compute a primary decomposition of I in $\mathbb{Q}[x, y]$ and an absolute primary decomposition.

Exercise 4.4.7. Make the following experiment in Macaulay2: Choose a finite prime field $\mathbb{F}_p$ and a degree d. Choose 1000 random polynomials $f \in \mathbb{F}_p[x]$ of degree d and count how many of them have an $\mathbb{F}_p$-rational root.

Do you have a hypothesis for the probability that a polynomial $f \in \mathbb{F}_p[x]$ of degree d has an $\mathbb{F}_p$-rational root? Can you prove your conjecture?

Exercise 4.4.8. What is the probability that a random polynomial $f \in \mathbb{F}_p[x, y]$ of degree two has a linear factor?

Chapter 5
Rational functions and dimension

In this chapter we consider the rational function field $K(A)$ of an affine variety A. We interpret an element $f \in K(A)$ as a partially defined map $A \dashrightarrow \mathbb{A}^1$ and define the domain of definition of f as the largest open subset where f can be defined. The ring $\mathcal{O}_{A,p} \subset K(A)$ of rational function which are defined at p coincides with the localization $K[A]_{\mathfrak{m}_p}$ of the coordinate ring $K[A]$ at the maximal ideal $\mathfrak{m}_p$ corresponding to p. We prove that a rational function which is defined everywhere is actually a polynomial function.

Rational maps $\varphi \colon A \dashrightarrow B \subset \mathbb{A}^m$ between varieties are defined by an m-tuple of rational functions. However, in general they cannot be composed, because the image of the rational map φ might be completely contained in the complement of the domain of definition of a rational map $\psi \colon B \dashrightarrow C$. This does not happen if $\varphi \colon A \dashrightarrow B$ is dominant, i.e., if φ has a dense image. A dominant rational map corresponds to an inclusion of the function fields $K(B) \hookrightarrow K(A)$ over K. A birational map $\varphi \colon A \dashrightarrow B$ is a dominant rational map which corresponds to a K-isomorphism $K(B) \cong K(A)$, This gives the category of algebraic varieties over K with dominant rational maps as morphisms a field-theoretic interpretation.

Finally, we define the dimension of an algebraic variety A rigorously as $\dim A = \mathrm{trdeg}_K K(A)$, i.e., as the transcendence degree of its function field over K. For an arbitrary affine algebraic set A we define the dimension as the maximum of the dimensions of its irreducible components.

5.1 The rational function field of a variety

Let $A \subset \mathbb{A}^n$ be an algebraic set. The **Zariski topology** on A is the topology induced on A from the Zariski topology of $\mathbb{A}^n$.

Thus the closed subsets of A are the algebraic subsets $B \subset A$. These are in one-to-one correspondence with radical ideals $J \supset I(A)$ which in turn are in bijection with radical ideals $\overline{J} = J/I(A)$ in $K[A]$:

$$\{\text{algebraic subsets of } A\} \xleftrightarrow{1-1} \{\text{radical ideals of } K[A]\}$$

with
$$B \mapsto I_A(B) = \{\overline{f} \in K[A] \mid \overline{f}(p) = 0 \ \forall p \in B\}$$

and
$$V_A(\overline{J}) = \{p \in A \mid \overline{f}(p) = 0 \ \forall \overline{f} \in \overline{J}\} \hookleftarrow \overline{J}.$$

In particular, we have
$$V_A(\overline{J}) = \emptyset \iff \overline{J} = (1).$$

From now on in this section A denotes a **variety**, i.e., an **irreducible algebraic set** unless explicitly stated otherwise. Thus $K[A]$ is an integral domain. We will also drop the overline from $\overline{f}$ in the notation of elements and ideals of $K[A]$.

Definition 5.1.1. The **field of rational functions** on A is the quotient field

$$K(A) = Q(K[A]) = \{f = \frac{g}{h} \mid g, h \in K[A], h \neq 0 \in K[A]\}.$$

We interpret $f \in K(A)$ as a partially defined function

$$f : A \dashrightarrow K.$$

Clearly, if $f = g/h$ and $p \in A$ is a point with $h(p) \neq 0$, then $f(p) = g(p)/h(p)$ makes sense. However, f has many representatives as fraction. Thus from $h(p) = 0$ we cannot conclude that f is not defined at p.

Example 5.1.2 (A non-factorial coordinate ring)**.** Consider $A = V(wx - yz)$. The coordinate ring $K[A] = K[w, x, y, z]/(wx - yz)$ is not factorial. The rational function

$$f = \frac{w}{z} = \frac{y}{x} \in K(A)$$

is defined for all points $p \notin V_A(x, z)$.

Localization of $K[A]$ for an irreducible algebraic set A is simpler than in general.

Proposition 5.1.3. *Let $U \subset K[A]$ be a multiplicative set with $0 \notin U$. Then two fractions*

$$\frac{g_1}{h_1}, \frac{g_2}{h_2} \in K[A][U^{-1}]$$

are equal if and only if $g_1/h_1 = g_2/h_2 \in K(A)$.

Proof. $u(h_2 g_1 - h_1 g_2) = 0 \in K[A] \iff h_2 g_1 - h_1 g_2 = 0 \in K[A]$ because $K[A]$ is an integral domain. □

Corollary 5.1.4. *Let $\mathfrak{p}$ be a prime ideal in $K[A]$. Then $K[A]_\mathfrak{p} \subset K(A)$.* □

Definition 5.1.5. Let $p \in A$ be a point in an algebraic set and $\mathfrak{m}_p \subset K[A]$ be the corresponding maximal ideal. Then

5.1 The rational function field of a variety

$$\mathcal{O}_{A,p} = K[A]_{\mathfrak{m}_p}$$

denotes **the local ring of A at p**. A rational function $f \in K(A)$ is defined at p if and only if $f \in \mathcal{O}_{A,p}$.

Theorem 5.1.6 (Everywhere defined rational functions)**.** *Let A be an irreducible algebraic set. Then*

$$K[A] = \bigcap_{p \in A} \mathcal{O}_{A,p} \subset K(A).$$

Proof. Let $f \in K(A)$. Consider the ideal of denominators of f:

$$I_f = \{h \in K[A] \mid hf \in K[A]\}$$
$$= \{h \in K[A] \mid f = \frac{g}{h}\} \cup \{0\}.$$

Remark. It might be a little surprising that the set in the second line is an ideal. It says: if h_1 and h_2 are denominators of f and $h_1 + h_2 \neq 0$, then $h_1 + h_2$ also occurs as a denominator of f. Indeed

$$f = \frac{g_1}{h_1} = \frac{g_2}{h_2} \Rightarrow f = \frac{g_1 + g_2}{h_1 + h_2}.$$

Now f is defined at p if and only if $f \in \mathcal{O}_{A,p} \Leftrightarrow p \in A \setminus V(I_f)$, since the elements of $\mathcal{O}_{A,p} = K[A]_{\mathfrak{m}_p}$ are fractions with denominator h with $h \notin \mathfrak{m}_p \Leftrightarrow h(p) \neq 0$. If f is everywhere defined, then $V_A(I_f) = \emptyset$ and the Nullstellensatz implies $1 \in I_f$. Hence $f \in K[A]$. $\square$

Definition 5.1.7. Let $f \in K(A)$ be a rational function. Then its **domain of definition** is the Zariski open set

$$\mathrm{dom}(f) = A \setminus V_A(I_f),$$

where $I_f = \{h \in K[A] \mid hf \in K[A]\}$ is the **ideal of denominators of f**. This is a Zariski-dense open subset of A on which f defines a K-valued function

$$A \supset \mathrm{dom}(f) \xrightarrow{f} K, \ a \mapsto f(a).$$

Remark. The fact that $\mathrm{dom}(f)$ is Zariski dense is less spectacular than it might seem at first glance. Actually, every non-empty Zariski open subset of A is Zariski-dense:

Proposition 5.1.8. *Let D_1, D_2 be Zariski open subsets of an irreducible algebraic set A. Then*

$$D_1 \cap D_2 = \emptyset \iff D_1 = \emptyset \text{ or } D_2 = \emptyset.$$

Proof. Let $A_j = A \setminus D_j$ for $j = 1, 2$ be the corresponding closed sets. Then

$$D_1 \cap D_2 = \emptyset \iff A_1 \cup A_2 = A \implies A = A_1 \text{ or } A = A_2$$

because A is irreducible. Thus $D_1 = \emptyset$ or $D_2 = \emptyset$. □

5.2 Dominant rational maps

Definition 5.2.1. Let $A \subset \mathbb{A}^n$ and $B \subset \mathbb{A}^m$ be irreducible algebraic sets. A **rational map** $\varphi \colon A \dashrightarrow B$ is given by an m-tuple of rational functions $f_1, \ldots, f_m \in K(A)$ such that
$$\varphi(p) = (f_1(p), \ldots, f_m(p)) \in B \text{ for all } p \in \bigcap_{j=1}^{m} \mathrm{dom}(f_j).$$

Note that the **domain of definition of** φ defined by $\mathrm{dom}(\varphi) = \bigcap_{j=1}^{m} \mathrm{dom}(f_j)$ is not empty by Proposition 5.1.8.

Example 5.2.2. The map
$$\varphi \colon \mathbb{A}^1 \dashrightarrow V(x^2 + y^2 - 1), \, t \mapsto \left(\frac{2t}{t^2+1}, \frac{t^2-1}{t^2+1}\right)$$
is a rational map. Indeed
$$\left(\frac{2t}{t^2+1}\right)^2 + \left(\frac{t^2-1}{t^2+1}\right)^2 - 1 = \frac{(2t)^2 + (t^2-1)^2 - (t^2+1)^2}{(t^2+1)^2} = 0.$$

Two rational maps $\varphi \colon A \dashrightarrow B$ and $\psi \colon B \dashrightarrow C$ might not be composable because it is possible that the image of φ, i.e., $\varphi(\mathrm{dom}(\varphi))$, may lie entirely in the complement of $\mathrm{dom}(\psi)$. This does not happen if $\varphi(\mathrm{dom}(\varphi))$ is dense in B.

Definition 5.2.3. A **dominant rational map** is a rational map $\varphi \colon A \dashrightarrow B$ such that $\varphi(\mathrm{dom}(\varphi))$ is dense in B.

Thus two dominant rational maps $\varphi \colon A \dashrightarrow B$ and $\psi \colon B \dashrightarrow C$ can be composed, and the composition $\psi \circ \varphi \colon A \dashrightarrow C$ is dominant as well.

The category of affine varieties over an algebraically closed field with dominant rational maps as morphisms has the following field-theoretic description.

Let
$$\varphi \colon A \dashrightarrow B \subset \mathbb{A}^m, \, p \mapsto (f_1(p), \ldots, f_m(p))$$
be a dominant rational map. Then
$$\varphi^* \colon K(B) \to K(A), F = \frac{G}{H} \mapsto F(f_1, \ldots, f_m) = \frac{G(f_1, \ldots, f_m)}{H(f_1, \ldots, f_m)}$$
is an injective K-algebra homomorphism between fields. Note that $H(f_1, \ldots, f_m) \in K(A)$ is not the zero element of $K(A)$ because otherwise $\varphi(\mathrm{dom}(\varphi))$ would be contained in $V_B(H)$, contradicting the assumption that the map is dominant. By the same argument φ^* is injective.

5.2 Dominant rational maps

Conversely, if $\phi: K(B) \to K(A)$ is a non-zero K-algebra homomorphism between fields and if $\overline{y}_1, \ldots, \overline{y}_m$ denote the coordinate functions on B, then $f_1 = \phi(\overline{y}_1), \ldots, f_m = \phi(\overline{y}_m)$ gives a tuple of rational functions which defines a rational map $\varphi: A \dashrightarrow B$. It is dominant because the map $\phi: K(B) \to K(A)$ is injective, and the composition $K[B] \hookrightarrow K(B) \to K(A)$ is injective as well. Since $\phi(F) = F(f_1, \ldots, f_m)$ we have $\varphi^* = \phi$.

Theorem 5.2.4. *Let K be an algebraically closed field. The category of affine varieties over K with dominant rational maps as morphisms and the category of finitely generated field extensions of K with K-algebra injections as morphisms are equivalent via*

$$A \mapsto K(A)$$

and

$$\varphi: A \dashrightarrow B \mapsto \varphi^*: K(B) \hookrightarrow K(A).$$

Proof. Most of the theorem has already been established. It remains to prove that every finitely generated extension field

$$K \subset L$$

arises as $L = K(A)$ for some variety A. Indeed, if

$$L = K(g_1, \ldots, g_n)$$

is generated by elements $g_1, \ldots, g_n$, then the substitution homomorphism

$$K[x_1, \ldots x_n] \to L, \ x_i \mapsto g_i$$

has a prime ideal J as a kernel because the image as a subring of a field is an integral domain. Then

$$A = V(J) \subset \mathbb{A}^n$$

is an affine variety with $K(A) \cong L$. $\square$

Remark. The variety A with $L \cong K(A)$ is not uniquely determined. Choosing different generators gives different varieties.

Example 5.2.5. For $A = V(xy - 1) \subset \mathbb{A}^2$ we have $L = K(A) = K(\overline{x}, \overline{y})$, and these generators give A back again. Since $\overline{y} = 1/\overline{x}$ we have $K(\overline{x}, \overline{y}) = K(\overline{x})$ and the second choice leads to $B = \mathbb{A}^1$.

Definition 5.2.6. A dominant rational map $\varphi: A \dashrightarrow B$ is called **birational** if there exists a dominant rational map $\psi: B \dashrightarrow A$ such that $\psi \circ \varphi = \text{id}_A$ holds, by which we mean that $(\psi \circ \varphi)|_D = \text{id}_D$ holds on the (non-empty) open subset $D \subset A$ on which $\psi \circ \varphi$ is defined as an honest map.

By the theorem φ is birational if and only if $\varphi^*: K(B) \to K(A)$ is an isomorphism. The rational map $\psi: B \dashrightarrow A$ induces the inverse isomorphism $\psi^* = (\varphi^*)^{-1}$, and $\varphi \circ \psi = \text{id}_B$ holds automatically as well.

In the example above $\varphi: V(xy - 1) \to \mathbb{A}^1$ is the projection onto the y-axis, while $\psi: \mathbb{A}^1 \dashrightarrow V(xy - 1), x \mapsto (x, 1/x)$.

Definition 5.2.7. Let A be an irreducible algebraic set. Then the **dimension of** A is

$$\dim A = \operatorname{trdeg}_K K(A).$$

If A is an algebraic set, then we define

$$\dim A = \max\{\dim A_i \mid i = 1, \ldots, r\},$$

where $A = A_1 \cup \ldots \cup A_r$ is the decomposition into irreducible algebraic subsets.

Exercise 5.2.8. Prove: There does not exist a dominant rational map

$$\mathbb{A}^1 \dashrightarrow V(y^2 - x^3 + x) \subset \mathbb{A}^2.$$

Hint: Use that polynomials in $K[t]$ factor into linear forms, and consider a hypothetical rational parametrization given by fractions of coprime polynomials

$$x(t) = \frac{f_1(t)}{g_1(t)}, \quad y(t) = \frac{f_2(t)}{g_2(t)}$$

such that $\max\{\deg f_1, \deg g_1, \deg f_2, \deg g_2\}$ is minimal among all parametrizations.

5.3 Appendix: The transcendence degree

Let $k \subset L$ be a field extension. For $g_1, \ldots, g_n \in L$ we denote by $k(g_1, \ldots, g_n) \subset L$ the smallest subfield of L containing $k \cup \{g_1, \ldots, g_n\}$. In contrast,

$$k[g_1, \ldots, g_n] \subset L$$

denotes the smallest subring of L containing $k \cup \{g_1, \ldots, g_n\}$. This is the image under the substitution homomorphism

$$k[x_1, \ldots, x_n] \to L, \, x_i \mapsto g_i.$$

An element $g \in L$ is called **algebraic over** k if $k[x] \to L, x \mapsto g$ has a nontrivial kernel. In this case the normed generator f of the kernel is called the **minimal polynomial** of g over k, and

$$k[g] \cong k[x]/(f)$$

is a finite-dimensional k-vector space and a field, i.e., $k(g) = k[g]$. If g is not algebraic over k, then g is called **transcendental over** k. In this case $k[g] \cong k[x]$ is an infinite-dimensional k-vector space and not a field: $k[g] \subsetneq k(g)$.

Elements $g_1, \ldots, g_d \in L$ are called **algebraically independent** over k if

5.3 Appendix: The transcendence degree

$$k[x_1, \ldots, x_d] \to L, x_i \mapsto g_i$$

has a trivial kernel.

A maximal subset of L of algebraic independent elements over k is called a **transcendence basis** for L over k. If $k \subset L$ is finitely generated, then by dropping elements from a generating set one can arrive at a transcendence basis.

Suppose $L = k(g_1, \ldots, g_n)$ and the elements $g_1, \ldots, g_d$ form a maximal subset of algebraically independent elements. Then every element $g \in L$ is algebraic over $k(g_1, \ldots, g_d)$, i.e., $\{g_1, \ldots, g_d, g\}$ are algebraically dependent. In particular, $L = k(g_1, \ldots, g_n)$ is a finite-dimensional $k(g_1, \ldots, g_d)$-vector space.

Theorem 5.3.1. *Let $k \subset L$ be a field extension. Any two transcendence bases of L over k have the same cardinality.*

Definition 5.3.2. The common cardinality of all transcendence bases is called the **transcendence degree of L over k**. It is denoted by $\mathrm{trdeg}_k(L)$.

We will prove this only in the case when $L \supset k$ is finitely generated over k. The proof is similar to the proof that the dimension of a vector space is well-defined.

Lemma 5.3.3 (Exchange Lemma). *Let $\{g_1, \ldots, g_d\}$ be a transcendence basis of L over k and let $h \in L$ be transcendental over k. Then there exists an index i such that*

$$\{g_1, \ldots, g_{i-1}, h, g_{i+1}, \ldots, g_d\}$$

is a transcendence basis as well.

Proof. Consider an irreducible polynomial $F \in k[x_1, \ldots, x_d, y]$ in the kernel of the map

$$k[x_1, \ldots, x_d, y] \to L, x_j \mapsto g_j, y \mapsto h$$

of smallest total degree. Such an F exists because the kernel is a prime ideal. F involves y because $g_1, \ldots, g_d$ are algebraically independent, and it involves some variable x_i because h is not algebraic over k. After renumbering the x_i's we may assume that $i = d$.

The polynomial $F(g_1, \ldots, g_{d-1}, x_d, h)$ does not vanish because the leading coefficient of F as a polynomial in x_d has smaller degree than F, hence does not vanish under the substitution by our choice of F. Thus g_d is algebraic over $k(g_1, \ldots, g_{d-1}, h)$. Every element of L is algebraic over $k(g_1, \ldots, g_{d-1}, h)$ because

$$k(g_1, \ldots, g_{d-1}, h) \subset k(g_1, \ldots, g_d, h) \subset L$$

is a tower of algebraic field extensions. Finally, the elements $g_1, \ldots, g_{d-1}, h$ are algebraically independent because otherwise h would be algebraic over $k(g_1, \ldots, g_{d-1})$. This would imply that also

$$k(g_1, \ldots, g_{d-1}) \subset k(g_1, \ldots, g_d, h)$$

is an algebraic field extension and g_d would be algebraic over $k(g_1, \ldots, g_{d-1})$, contradicting our assumption. □

We prove the following proposition by induction on c. It immediately implies Theorem 5.3.1 in the case of finitely generated field extensions.

Proposition 5.3.4. *Let $\{g_1, \ldots, g_d\}$ be a transcendence basis of L over k, and let $h_1, \ldots, h_c$ in L be elements which are algebraically independent over k. Then after a suitable reordering of $g_1, \ldots, g_d$ the set $\{h_1, \ldots, h_c, g_{c+1}, \ldots, g_d\}$ is a transcendence basis as well. In particular, $c \leq d$.*

Proof. The case $c = 1$ is the exchange lemma above after renumbering. By the induction hypothesis we may assume that $\{h_1, \ldots, h_{c-1}, g_c, \ldots, g_d\}$ is a transcendence basis. By the Exchange lemma we can replace one of these elements by h_c and from the proof we see that this element can be chosen to be different from $h_1, \ldots, h_{c-1}$ because $h_1, \ldots, h_c$ are algebraically independent. After reordering we may assume that this element is g_c. □

Exercise 5.3.5. Let $L = \mathbb{Q}(a_1, \ldots, a_n)$ be a finitely generated field extension of $\mathbb{Q}$. Prove that there exists an embedding $L \hookrightarrow \mathbb{C}$.

Chapter 6
Integral ring extensions and Krull dimension

After the Commutative Algebra preparations of Chapters 3, 4 and 5, we are now ready to prove a Gröbner basis criterion for the dimension, establishing that our naive definition from Remark 1.4.9 leads to the correct concept. Moreover we give a Gröbner basis condition which implies that every associated prime of an ideal I generated by polynomials in $k[x_1, \ldots, x_n]$ has the same dimension.

The proof of this theorem relies on the concept of integral ring extensions. We prove the lying-over and the going-up theorem. Finally, we introduce the Krull dimension of a (commutative) ring. To prove that $\dim k[x_1, \ldots, x_n] = n$ holds, we use the dimension theory of algebraic subsets of $\mathbb{A}^n$ instead of the more frequently used concept of a refined Noether normalization.

6.1 A Gröbner basis criterion for the dimension

As always K denotes an algebraically closed extension field of k. For an irreducible algebraic set $A \subset \mathbb{A}^n$ the dimension is defined as

$$\dim A = \operatorname{trdeg}_K K(A),$$

i.e., as the transcendence degree of its function field. For general algebraic subsets A we have defined

$$\dim A = \max\{\dim A_i\},$$

where $A = A_1 \cup \ldots \cup A_r$ is its component decomposition over K.

Theorem 6.1.1 (Gröbner basis criterion for the dimension). *Let I be an ideal in the polynomial ring $k[x_1, \ldots, x_c, y_1, \ldots, y_d]$ and let $A = V(I) \subset \mathbb{A}^{c+d}$ denote the corresponding algebraic set. Let $>$ be a global monomial order. Suppose*

$$\operatorname{rad}(\operatorname{Lt}(I)) = (x_1, \ldots, x_c).$$

Then $\dim A = d$ *and the projection*

$$\pi\colon A \to \mathbb{A}^d,\ (a_1,\ldots,a_c,b_1\ldots,b_d) \mapsto (b_1,\ldots,b_d)$$

onto the last d components is surjective.
Moreover, if $\mathrm{Lt}(I)$ *is generated by monomials in the subring* $k[x_1,\ldots,x_c]$, *then every associated prime of* $(I) \subset K[x_1,\ldots,x_n]$ *defines a variety of dimension d. In particular, every irreducible component of A has dimension d.*

Definition 6.1.2. We call an ideal *I* with all associated primes of the same dimension **unmixed**. A **curve**, **surface** or **3-fold** refers to an unmixed algebraic set of dimension 1, 2 or 3 respectively.

In the situation of the tower of projections theorem we get the desired dimension statement.

Theorem 1.4.8. *Let* $I \subsetneq k[x_1,\ldots,x_n]$ *be a proper ideal. Let* $I_j = I \cap k[x_{j+1},\ldots,x_n]$ *be the j-th elimination ideal. Set*

$$c = \min\{j \mid I_j = (0)\}$$

and suppose that for each j with $0 \le j \le c-1$ *the ideal* I_j *contains an* x_{j+1}-*monic polynomial of degree* d_j. *Then the projection* $\pi_c\colon V(I) \to \mathbb{A}^{n-c}$ *onto the last* $n-c$ *components is surjective, and each fiber*

$$\pi_c^{-1}(a_{c+1},\ldots,a_n)$$

is finite of cardinality $\le \prod_{j=0}^{c-1} d_j$.

Corollary 6.1.3. *With the assumption and notation of Theorem 1.4.8*

$$\dim V(I) = n - c$$

holds.

Proof of the corollary. The assumption of the Gröbner basis criterion is satisfied for $>_{\mathrm{lex}}$ with $d = n - c$ and $y_1 = x_{c+1},\ldots,y_d = x_n$. Indeed, $I_c = 0$ implies

$$\mathrm{rad}(\mathrm{Lt}(I)) \subset (x_1,\ldots x_c)$$

by the key property of $>_{\mathrm{lex}}$ 1.3.24. The existence of the x_j-monic polynomials in I_{j-1} for $j = 1,\ldots,c$ implies that equality holds. □

The key concept for the proof of the dimension criterion is the notion of integral ring extensions. This also played a crucial role in our proof of the Nullstellensatz.

6.2 Integral ring extensions

Definition 6.2.1. Let $R \subset S$ be an inclusion of rings and let $I \subset R$ be an ideal. An element $s \in S$ is **integral over** *I* if it satisfies a monic equation

6.2 Integral ring extensions

$$s^n + r_1 s^{n-1} + \cdots + r_n = 0$$

with $r_i \in I$. An element s is **integral over** R if it is integral over the ideal $(1) = R$. We call $R \subset S$ an **integral ring extension** if every element $s \in S$ is integral over R. The extension $R \subset S$ is called a **finite ring extension** if S is finitely generated as an R-module.

Example. Let $s \in S$ be integral over R. Then $R \subset R[s]$ is a finite ring extension. Indeed, from the monic equation above we see that $R[s]$ is generated by $1, s, \ldots, s^{n-1}$ as an R-module.

Proposition 6.2.2. *Let $R \subset S$ be a ring extension, $s \in S$ an element and $I \subset R$ an ideal. The following are equivalent:*

1) s is integral over R (over I).
2) $R[s]$ is finite over R (and $s \in \mathrm{rad}(IR[s])$).
3) $R[s]$ is contained in a subring $S' \subset S$ which is finite over R (and $s \in \mathrm{rad}(IS')$).

Proof. 1) $\Rightarrow$ 2) was established above. If s is integral over I, then the equation says $s^n \in IR[s]$. 2) $\Rightarrow$ 3) is trivially true. 3) $\Rightarrow$ 1) is the essential direction. Suppose S' is generated by $m_1, \ldots, m_n$ as an R-module. Since $s \in \mathrm{rad}(IS')$, we may write

$$s^N m_i = \sum_{j=1}^{n} r_{ij} m_j$$

with $r_{ij} \in I$ for a suitable power N. In matrix notation we obtain

$$(s^N E_n - B) \begin{pmatrix} m_1 \\ \vdots \\ m_n \end{pmatrix} = 0$$

with $B = (r_{ij})$. Multiplying by the cofactor matrix we obtain

$$\det(s^N E_n - B) m_i = 0$$

for all $i \in \{1, \ldots, n\}$. Since $1 \in R \subset S'$ is a linear combination of $m_1, \ldots, m_n$, we obtain

$$\det(s^N E_n - B) = s^{nN} + r_1 s^{(n-1)N} + \cdots + r_n = 0,$$

i.e., s is integral over I. $\square$

Proposition 6.2.3. *Let $R \subset S \subset T$ be a tower of finite or integral ring extensions. Then $R \subset T$ is a finite respectively integral ring extension as well.*

Proof. Suppose $s_1, \ldots, s_n$ generate S as an R-module and $t_1, \ldots, t_m$ generate T as an S-module. Every $t \in T$ has an expression $t = \sum a_j t_j$ with $a_j \in S$. Every a_j has an expression $a_j = \sum r_{ij} s_i$ with $r_{ij} \in R$. Hence

$$t = \sum_{i=1}^{n}\sum_{j=1}^{m} r_{ij}s_i t_j.$$

So the $n \cdot m$ elements $s_i t_j$ generate T as an R-module.

For the second version consider an element $t \in T$. By assumption t is integral over S, i.e., t satisfies an equation

$$t^n + s_1 t^{n-1} + \cdots + s_n = 0 \text{ with } s_i \in S.$$

Since each s_i is integral over R, the extension

$$R \subset R[s_1,\ldots,s_n]$$

is finite. Hence $R \subset R[s_1,\ldots,s_n,t]$ is finite as well and t is integral over R by the conclusion 3) $\Rightarrow$ 1) of Proposition 6.2.2. □

Proof of the dimension criterion of Theorem 6.1.1. Since $\mathrm{rad}(\mathrm{Lt}(I)) = (x_1,\ldots x_c)$, we have $I \cap k[y_1,\ldots,y_d] = 0$. Thus the induced map

$$k[y_1,\ldots,y_d] \to S = k[x_1,\ldots x_c, y_1,\ldots,y_d]/I$$

is injective. The ring extension $k[y_1,\ldots,y_d] \subset S$ is finite because for each x_i there exists an $x_i^{n_i} \in \mathrm{Lt}(I)$. Hence the $\overline{x}^\alpha$ with $\alpha_i < n_i$ generate S as a $k[y_1,\ldots,y_d]$-module by the division theorem. Consider now a minimal primary decomposition of the extended ideal $I^e \subset K[x_1,\ldots x_c, y_1,\ldots,y_d]$:

$$I^e = \mathfrak{q}_1 \cap \ldots \cap \mathfrak{q}_r.$$

For at least one associated prime $\mathfrak{p}_j = \mathrm{rad}(\mathfrak{q}_j)$ we must have

$$\mathfrak{p}_j \cap K[y_1,\ldots,y_d] = 0.$$

Indeed, if there are non-zero elements $f_i \in \mathfrak{p}_i \cap K[y_1,\ldots,y_d]$ for every i, then their product $\prod f_i \in \mathrm{rad}(I^e) \cap K[y_1,\ldots,y_d]$ and a suitable power $(\prod f_i)^N$ lies in $I^e \cap K[y_1,\ldots,y_d]$, which contradicts $\mathrm{Lt}(I) \cap k[y_1,\ldots,y_d] = 0$ since $I^e \cap K[y_1,\ldots,y_d]$ is generated by elements of $I \cap k[y_1,\ldots,y_d]$.

For $A_j = V(\mathfrak{p}_j)$ with $\mathfrak{p}_j \cap K[y_1,\ldots,y_d] = 0$ we have that $K[y_1,\ldots,y_d] \subset K[A_j] = K[x_1,\ldots x_c, y_1,\ldots,y_d]/\mathfrak{p}_j$ is a finite extension. Hence

$$K(y_1,\ldots,y_d) \subset K(A_j)$$

is an algebraic field extension and

$$\dim A_j = \mathrm{trdeg}_K K(A_j) = \mathrm{trdeg}_K K(y_1,\ldots,y_d) = d.$$

For $\mathfrak{p}_i$ with $\mathfrak{p}_i^c = \mathfrak{p}_i \cap K[y_1,\ldots,y_d] \neq 0$ and $B_i = V(\mathfrak{p}_i^c) \subsetneq \mathbb{A}^d$ we have that

$$K[B_i] \subset K[A_i]$$

is a finite ring extension. Hence

$$\dim A_i = \dim B_i = \operatorname{trdeg}_K K(B_i) < d$$

since $\bar{y}_1, \ldots, \bar{y}_d$ give algebraically dependent generators of $K(B_i)$ over K. Thus

$$\dim A = \max\{\dim A_j\} = d.$$

□

Proof of the unmixedness. If I satisfies the additional assumption, i.e., if $\operatorname{Lt}(I)$ is an $(x_1, \ldots, x_c)$-primary ideal, then $k[x_1, \ldots x_c, y_1, \ldots, y_d]/I$ is actually a free $k[y_1, \ldots, y_d]$-module: The leading term ideal $\operatorname{Lt}(I)$ is generated by monomials in $k[x_1, \ldots, x_c]$ and the monomials $x^\alpha \in k[x_1, \ldots, x_c] \setminus \operatorname{Lt}(I)$ form a basis by the division theorem. If $\mathfrak{p}_i = \operatorname{ann}(m)$ for some $m \in K[x_1, \ldots x_c, y_1, \ldots, y_d]/I^e$ is an associated prime, then

$$\mathfrak{p}_i \cap K[y_1, \ldots, y_d] = \operatorname{ann}_{K[y_1, \ldots, y_d]}(m)$$

is an associated prime of $K[x_1, \ldots x_c, y_1, \ldots, y_d]/I^e$ as a $K[y_1, \ldots, y_d]$-module. A free $K[y_1, \ldots, y_d]$-module has (0) as the only associated prime. Thus $\mathfrak{p}_i \cap K[y_1, \ldots, y_d] = 0$ for all i and every associated prime defines a variety $V(\mathfrak{p}_i)$ of dimension d. □

It remains to prove that the map $\pi \colon A \to \mathbb{A}^d$ is surjective. We prove a more general result.

6.3 The lying-over theorem

Let $R \subset S$ be a ring extension. If $\mathfrak{P}$ is a prime ideal in S, then $\mathfrak{p} = \mathfrak{P} \cap R$ is a prime ideal in R. One says $\mathfrak{P}$ **lies over** $\mathfrak{p}$.

Theorem 6.3.1 (Lying-over). *Let $R \subset S$ be an integral ring extension and let $\mathfrak{p}$ be a prime ideal of R. Then the following holds:*

1) *There exists a prime ideal $\mathfrak{P}$ of S with $\mathfrak{p} = \mathfrak{P} \cap R$.*
2) *There are no strict inclusions between prime ideals lying over $\mathfrak{p}$.*
3) *If $\mathfrak{P}$ is a prime ideal lying over $\mathfrak{p}$, then $\mathfrak{P}$ is a maximal ideal if and only if $\mathfrak{p}$ is a maximal ideal.*
4) *If S is Noetherian, then the prime ideals lying over $\mathfrak{p}$ are precisely the minimal primes of $\mathfrak{p}S$.*

The surjectivity of $\pi \colon A \to \mathbb{A}^d$ follows from 1) and 3) since maximal ideals in $S = K[x_1, \ldots, x_c, y_1, \ldots y_d]/I^e$ correspond to points $(a_1, \ldots, a_c, b_1, \ldots, b_d) \in A \subset \mathbb{A}^{c+d}$, and maximal ideals of $R = K[y_1, \ldots, y_d]$ correspond to points $(b_1, \ldots, b_d) \in \mathbb{A}^d$.

Lemma 6.3.2 (Krull's Prime Existence Lemma). *Let I be an ideal of the ring R and let $U \subset R$ be a multiplicative subset with $I \cap U = \emptyset$. Then there exists a prime ideal $\mathfrak{p}$ of R with $I \subset \mathfrak{p}$ and $U \cap \mathfrak{p} = \emptyset$.*

Proof. Consider the set

$$\mathcal{M} = \{J \subset R \mid J \text{ is an ideal with } I \subset J \text{ and } J \cap U = \emptyset\}.$$

It is non-empty because $I \in \mathcal{M}$ and it consists of proper ideals because $1 \in U$. Let $\mathfrak{p}$ be a maximal element of $\mathcal{M}$ with respect to inclusion. We claim that $\mathfrak{p}$ is a prime ideal.

Indeed, suppose $r_1, r_2 \notin \mathfrak{p}$. Then $(\mathfrak{p} + (r_j)) \cap U \neq \emptyset$ because $\mathfrak{p}$ is maximal in $\mathcal{M}$. Thus there are $p_j \in \mathfrak{p}$ and $a_j \in R$ such that $p_j + a_j r_j \in U$. Since U is multiplicative we have

$$(p_1 + a_1 r_1)(p_2 + a_2 r_2) \in U \subset R \setminus \mathfrak{p}.$$

Hence $a_1 a_2 r_1 r_2 \notin \mathfrak{p}$. In particular $r_1 r_2 \notin \mathfrak{p}$, as desired.

The existence of a maximal element $\mathfrak{p}$ in $\mathcal{M}$ is clear if R is Noetherian. For more general rings we apply Zorn's lemma: The set $\mathcal{M}$ is partially ordered by inclusion. If $\{J_\lambda\}$ is a totally ordered subset set of $\mathcal{M}$, then $\bigcup_\lambda J_\lambda$ is an upper bound. Thus the assumptions of Zorn's lemma are satisfied, and $\mathcal{M}$ contains maximal elements. □

Corollary 6.3.3. *Every proper ideal I in a ring R is contained in a maximal ideal.*

Proof. We apply Krull's lemma to $I \subset R$ and $U = \{1\}$. The prime ideal is a maximal ideal in this case. □

Proof of the lying over theorem. 1): Consider the ideal $\mathfrak{p}S$ of S and the multiplicative subset $U = R \setminus \mathfrak{p}$ of S. Using that $R \subset S$ is an integral extension we verify that $\mathfrak{p}S \cap U = \emptyset$: Every $s \in \mathfrak{p}S$ has an expression $s = \sum_{i=1}^n a_i s_i$ with $a_i \in \mathfrak{p}$ and $s_i \in S$. Thus s is integral over $\mathfrak{p}$. Consider an integral equation

$$s^d + r_1 s^{d-1} + \cdots + r_d = 0 \text{ with } r_i \in \mathfrak{p}.$$

We have to show that $s \notin U = R \setminus \mathfrak{p}$. Assume the contrary, then $s^d \in \mathfrak{p}$, hence $s \in \mathfrak{p}$ since $\mathfrak{p}$ is a prime ideal. This contradicts $s \in U = R \setminus \mathfrak{p}$.

We can now apply Krull's lemma to the ideal $I = \mathfrak{p}S$ of S and the multiplicative subset U. There exists a prime ideal $\mathfrak{P}$ of S with $\mathfrak{p} \subset \mathfrak{p}S \subset \mathfrak{P}$ and $\mathfrak{P} \cap U = \emptyset$. Hence $\mathfrak{P} \cap R \subset \mathfrak{p}$ and equality holds.

2): Consider prime ideals $\mathfrak{P}_1 \subset \mathfrak{P}_2$ of S, both lying over $\mathfrak{p}$. Then $\overline{R} = R/\mathfrak{p} \subset \overline{S} = S/\mathfrak{P}_1$ is an integral ring extension of domains and $\mathfrak{P}_2/\mathfrak{P}_1 \subset \overline{S}$ is a prime ideal which lies over $(0) \subset \overline{R}$. We have to prove that $\mathfrak{P}_2/\mathfrak{P}_1 = (0)$. Suppose $\overline{s} \in \mathfrak{P}_2/\mathfrak{P}_1$ is non-zero. Let

$$\overline{s}^d + \overline{r}_1 \overline{s}^{d-1} + \cdots + \overline{r}_d = 0$$

be an integral equation of minimal degree. Then $\overline{r}_d \in \mathfrak{P}_2/\mathfrak{P}_1 \cap \overline{R} = (0)$. Thus $\overline{r}_d = 0$. If $d = 1$, then this says $s = 0$. If $d > 1$, then we can divide the integral equation by

$\bar{s}$ since $\bar{S}$ is a domain, and we obtain an equation of smaller degree. Thus we get a contradiction in any case.

3): If $\mathfrak{p}$ is a maximal ideal in R, then $\mathfrak{P}$ is a maximal ideal as well by 2). Any prime ideal $\mathfrak{P}' \supset \mathfrak{P}$ lies over $\mathfrak{p}$ as well because $\mathfrak{p}$ is maximal. Hence $\mathfrak{P}' = \mathfrak{P}$ by assertion 2). Conversely, suppose $\mathfrak{P} \subset S$ is maximal. Then $R/\mathfrak{p} \subset S/\mathfrak{P}$ is an integral extension. Since $S/\mathfrak{P}$ is a field, a non-zero prime ideal of $R/\mathfrak{p}$ has no prime in $S/\mathfrak{P}$ lying over it. This contradicts assertion 1). Using Corollary 6.3.3 we see that (0) is the only proper ideal in $R/\mathfrak{p}$, i.e., $R/\mathfrak{p}$ is a field and $\mathfrak{p}$ a maximal ideal.

4): If $\mathfrak{P}$ lies over $\mathfrak{p}$, then $\mathfrak{p}S \subset \mathfrak{P}$ and $\mathfrak{P}$ is a minimal prime containing $\mathfrak{p}S$ by 2). Since S is Noetherian, $\mathfrak{p}S$ has a primary decomposition

$$\mathfrak{p}S = \mathfrak{Q}_1 \cap \ldots \cap \mathfrak{Q}_r$$

and $\mathrm{rad}(\mathfrak{p}S) = \mathfrak{P}_1 \cap \ldots \cap \mathfrak{P}_r$ with $\mathfrak{P}_i = \mathrm{rad}(\mathfrak{Q}_i)$. Since $\mathfrak{p}S \subset \mathfrak{P}$ implies $\mathrm{rad}(\mathfrak{p}S) \subset \mathfrak{P}$ we conclude that $\mathfrak{P} \supset \mathfrak{P}_i$ for some i because otherwise the product of elements $f_j \in \mathfrak{P}_j \setminus \mathfrak{P}$ would be an element of $\mathrm{rad}(\mathfrak{p}S)$ whose factors do not lie in $\mathfrak{P}$. This is impossible since $\mathfrak{P}$ is prime. Since $\mathfrak{P}$ is a minimal prime over $\mathfrak{p}S$, we have $\mathfrak{P} = \mathfrak{P}_i$. Thus $\mathfrak{P}$ coincides with an associated prime of $\mathfrak{p}S$ which is minimal among the associated primes of $\mathfrak{p}S$. □

6.4 Krull dimension

Definition 6.4.1. Let R be a ring. A **chain of prime ideals** in R is a sequence

$$\mathfrak{p}_0 \subsetneq \mathfrak{p}_1 \subsetneq \ldots \subsetneq \mathfrak{p}_\ell$$

of prime ideals. We call ℓ the **length** of the chain. The **Krull dimension** of R

$$\dim R = \sup\{\ell \mid \exists \text{ a chain of prime ideals of length } \ell \text{ in } R\}$$

is the maximal length of a chain of prime ideals in R.

Theorem 6.4.2. $\dim k[x_1, \ldots, x_n] = n$.

Proof. Indeed,

$$(0) \subsetneq (x_1) \subsetneq (x_1, x_2) \subsetneq \ldots \subsetneq (x_1, \ldots, x_n)$$

is a chain of prime ideals of length n. Thus

$$\dim k[x_1, \ldots, x_n] \geq n.$$

To show equality, we note the following.

Claim. *If $\mathfrak{p} \subsetneq \mathfrak{p}' \subset k[x_1, \ldots, x_n]$ are prime ideals, then $d = \dim V(\mathfrak{p}) > \dim V(\mathfrak{p}')$.*

Indeed, if we change coordinates such that the assumption of the tower of projections theorem 1.4.8 is satisfied for $I = \mathfrak{p}$, then

$$k[x_{n-d+1},\ldots x_n] \hookrightarrow k[x_1,\ldots,x_n]/\mathfrak{p}$$

is an integral ring extension. If $k[x_{n-d+1},\ldots x_n] \cap \mathfrak{p}' = (0)$, then both $(0) \subsetneq \mathfrak{p}'/\mathfrak{p}$ would lie over $(0) \subset k[x_{n-d+1},\ldots x_n]$, contradicting assertion 2) of the lying-over theorem. Thus

$$\mathfrak{q} = k[x_{n-d+1},\ldots x_n] \cap \mathfrak{p}' \neq (0),$$

and $\dim V(\mathfrak{p}') = \dim V(\mathfrak{q}) < \dim \mathbb{A}^d = \dim V(\mathfrak{p})$.

This proves that any chain of prime ideals $\mathfrak{p}_0 \subsetneq \mathfrak{p}_1 \subsetneq \ldots \subsetneq \mathfrak{p}_\ell$ in $k[x_1,\ldots,x_n]$ gives a chain of algebraic sets of strictly descending dimensions

$$\dim V(\mathfrak{p}_0) > \dim V(\mathfrak{p}_1) > \ldots > \dim V(\mathfrak{p}_\ell)$$

and the length cannot be larger than n. □

Remark 6.4.3. A **maximal chain of prime ideals** is a chain of prime ideals which cannot be extended by inserting a further prime. A maximal chain of prime ideals in $k[x_1,\ldots,x_n]$ starts with $\mathfrak{p}_0 = (0)$ and ends with a maximal ideal whose zero locus in $\mathbb{A}^n$ consist of a finite collection of points which are conjugated under the Galois action of $\overline{k}$ over k, where $\overline{k} \subset K$ denotes the algebraic closure of k in K. Actually every maximal chain of prime ideals in $k[x_1,\ldots,x_n]$ has length n. This is usually proved with the so-called refined version of the Noether normalization.

More generally, one has the following result.

Theorem 6.4.4. *Let $A \subset \mathbb{A}^n$ be an irreducible algebraic set. Then every maximal chain of prime ideals in $K[A]$ has length $\dim A$.*

This statement fails for algebraic sets which have components of different dimensions.

Definition 6.4.5. Let $\mathfrak{p}$ be a prime ideal in a ring R. The **height** of $\mathfrak{p}$ is

$$\mathrm{height}(\mathfrak{p}) = \sup\{\ell \mid \exists \text{ a chain of prime ideals}$$
$$\mathfrak{p}_0 \subsetneq \mathfrak{p}_1 \subsetneq \ldots \subsetneq \mathfrak{p}_\ell \text{ with } \mathfrak{p}_\ell = \mathfrak{p}\}.$$

Thus $\mathrm{height}(\mathfrak{p}) = \dim R_\mathfrak{p}$.

Theorem 6.4.4 implies

$$\mathrm{height}(\mathfrak{p}) + \dim R/\mathfrak{p} = \dim R$$

for affine domains $R = K[A]$.

Theorem 6.4.6 (Going-up)**.** *Let $R \subset S$ be an integral ring extension. For every chain*

$$\mathfrak{p}_0 \subsetneq \mathfrak{p}_1 \subsetneq \ldots \subsetneq \mathfrak{p}_\ell$$

of prime ideals in R there exists a chain of prime ideals

6.4 Krull dimension

$$\mathfrak{P}_0 \subsetneq \mathfrak{P}_1 \subsetneq \ldots \subsetneq \mathfrak{P}_\ell$$

in S with $\mathfrak{P}_i \cap R = \mathfrak{p}_i$. In particular, $\dim R = \dim S$.

Proof. By Theorem 6.3.1 1) there exists a prime ideal $\mathfrak{P}_0$ over $\mathfrak{p}_0$. Next we notice that $R/\mathfrak{p}_0 \subset S/\mathfrak{P}_0$ is again an integral extension. Thus there exist a prime ideal $\overline{\mathfrak{P}}_1 \subset S/\mathfrak{P}_0$ over $\overline{\mathfrak{p}}_1 = \mathfrak{p}_1/\mathfrak{p}_0 \subset R/\mathfrak{p}_0$. The pre-image $\mathfrak{P}_1$ of $\overline{\mathfrak{P}}_1$ in S is a prime ideal $\mathfrak{P}_1 \supsetneq \mathfrak{P}_0$ over $\mathfrak{p}_1$. Thus going-up the chain step by step from $\mathfrak{p}_0$ to $\mathfrak{p}_\ell$ gives a chain $\mathfrak{P}_0 \subsetneq \mathfrak{P}_1 \subsetneq \ldots \subsetneq \mathfrak{P}_\ell$ over $\mathfrak{p}_0 \subsetneq \mathfrak{p}_1 \subsetneq \ldots \subsetneq \mathfrak{p}_\ell$. This proves $\dim R \leq \dim S$.

For the converse inequality we notice that for any chain $\mathfrak{P}_0 \subsetneq \mathfrak{P}_1 \subsetneq \ldots \subsetneq \mathfrak{P}_\ell$ the contracted ideals $\mathfrak{p}_i = \mathfrak{P}_i \cap R$ form a chain of prime ideals in R which remains strict by Theorem 6.3.1 2). □

Corollary 6.4.7. *Let $A \subset \mathbb{A}^n$ be an algebraic set and let $K[A]$ be its coordinate ring. Then*

$$\dim A = \dim K[A].$$

Proof. Consider a projection $\pi: A \to \mathbb{A}^d$ as in Theorem 1.4.8. Then $K[x_{n-d+1}, \ldots, x_n] \subset K[A]$ is an integral ring extension. Hence

$$\dim A = \dim \mathbb{A}^d = d = \dim K[x_{n-d+1}, \ldots, x_n] = \dim K[A].$$

□

Exercise 6.4.8. Describe all prime ideals for

- $R = \mathbb{R}[x]$,
- $R = K[x, y]$ with K algebraically closed,
- $R = \mathbb{Z}[i] \subset \mathbb{C}$.

Exercise 6.4.9. Consider the ring extension

$$R = \mathbb{R}[e_2, e_3] \hookrightarrow T = \mathbb{R}[t_1, t_2]$$

defined by $e_2 \mapsto t_1 t_2 - (t_1 + t_2)^2$, $e_3 \mapsto t_1 t_2 (t_1 + t_2)$.

1. Prove that $S = R[t_1] \cong R[x]/(x^3 + e_2 x + e_3)$ and conclude that

$$R \subset S \subset T$$

   is a tower of finite extensions.
2. Compute the degrees of the field extensions

$$Q(R) \subset Q(S) \subset Q(T).$$

3. Prove: $(t_1 - t_2) \cap R = (4e_2^3 + 27e_3^2)$.
4. Let $(b_2, b_3) \in \mathbb{A}^2(\mathbb{R})$ be a point. How many maximal ideals $\mathfrak{P}$ in S can lie over the maximal ideal of $\mathfrak{p} = (e_2 - b_2, e_3 - b_3) \subset R$? How many maximal ideals $\mathfrak{P}'$ in T can lie over $\mathfrak{p}$?

5. What residue fields $S/\mathfrak{P}$ and $T/\mathfrak{P}'$ occur?

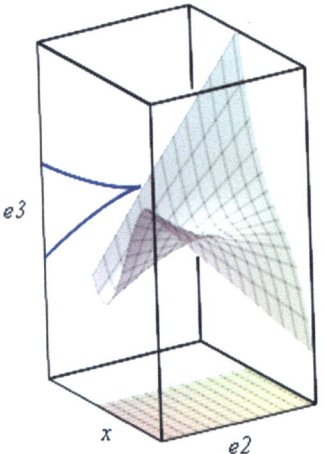 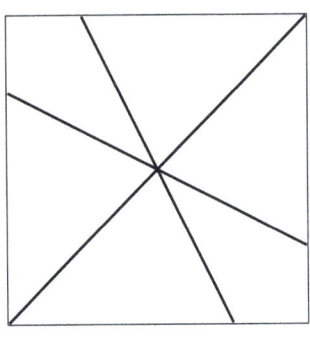

$V(4e_2^3 + 27e_3^2), V(x^3 + e_2 x + e_3)$ and $V((t_1 - t_2)(t_1 + 2t_2)(2t_1 + t_2))$

Chapter 7
Constructive ideal and module theory

In this chapter we give various applications of Gröbner basis computations. We describe algorithms for ideals and modules over polynomial rings which compute
  - intersections of ideals,
  - syzygies among a tuple $f_1, \ldots, f_r$ of polynomial vectors,
  - colon ideals $I : J$,
  - elimination ideals,
  - kernels of ring homomorphism between affine algebras,
  - the image, cokernel and kernel of a homomorphism $\varphi \colon M \to N$ between finitely presented modules,

and finally in the exercises
  - the module $\mathrm{Hom}(M, N)$ for two finitely presented modules M and N.

7.1 Syzygies and applications

Let $I, J \subset S = k[x_1, \ldots, x_n]$ be ideals. We want to compute their intersection.

Algorithm 7.1.1 (Intersection of ideals).
Input. $f_1, \ldots, f_r$ generators of the ideal I,
  $g_1, \ldots, g_s$ generators of the ideal J.
Output. Generators of the ideal $I \cap J$.

1. Form the matrix
$$\varphi = \begin{pmatrix} 1 & f_1 & \ldots & f_r & 0 & \ldots & 0 \\ 1 & 0 & \ldots & 0 & g_1 & \ldots & g_s \end{pmatrix}.$$

2. Compute the syzygy matrix $\psi = (h_{ij})$ whose columns generate the kernel
$$\ker(\varphi \colon S^{r+s+1} \to S^2).$$

3. Return the entries of the first row

$$h_{11}, h_{12}, \ldots, h_{1t}$$

of the $(r+s+1) \times t$-matrix ψ.

Proof of correctness. The equation

$$\begin{pmatrix} 1 & f_1 & \cdots & f_r & 0 & \cdots & 0 \\ 1 & 0 & \cdots & 0 & g_1 & \cdots & g_s \end{pmatrix} \begin{pmatrix} h_{1j} \\ h_{2j} \\ \vdots \\ h_{(r+s+1),j} \end{pmatrix} = 0$$

shows that h_{1j} is both a linear combination of the f_i's and the g_i's. Hence $h_{1j} \in I \cap J$. Conversely, if $h \in I \cap J$, then

$$h = h_1 f_1 + \ldots + h_r f_r = h'_1 g_1 + \ldots + h'_s g_s$$

for suitable h_i and h'_j. Hence the vector

$$(h, -h_1, \ldots, -h_r, -h'_1, \ldots, -h'_s)^t \in \ker(\varphi).$$

Since the kernel is generated by the columns of ψ, we obtain that h is a linear combination of $h_{11}, h_{12}, \ldots, h_{1t}$. □

Let $S = k[x_1, \ldots, x_n]$ be the polynomial ring and $F = S^s$ be a free S-module.

Algorithm 7.1.2 (Computation of syzygies).
Input. Vectors $f_1, \ldots, f_r \in F$
Output. A matrix $\psi \in S^{r \times t}$ whose columns generate the kernel of the S-module homomorphism

$$\varphi : S^r \to F, e_i \mapsto f_i.$$

1. Choose a global monomial order on F and compute a Gröbner basis

$$f_1, \ldots, f_r, f_{r+1}, \ldots, f_{r'}$$

of $(f_1, \ldots, f_r)$, while keeping track of the Buchberger test syzygies $G^{(i,\alpha)}$.
2. Sort the $G^{(i,\alpha)}$ such that the test syzygies which produced new Gröbner basis elements come first.
3. The matrix with columns $G^{(i,\alpha)}$ now has shape

$$\psi' = \begin{pmatrix} A & B \\ C & D \end{pmatrix} \text{ with } C = \begin{pmatrix} 1 & & * \\ & \ddots & \\ 0 & & 1 \end{pmatrix}$$

an $(r' - r) \times (r' - r)$ upper triangular square matrix with 1's on the diagonal. Return

$$\psi = B - AC^{-1}D.$$

7.1 Syzygies and applications

Note that one can compute C^{-1} by applying row operations to the matrix $(E|C)$ to obtain $(C'|E)$. The inverse matrix $C' = C^{-1}$ has entries in S.

Proof of correctness. The matrix ψ' is an $r' \times (r' - r + t)$-matrix whose columns generate the kernel of the map

$$\varphi' : S^{r'} \to F, e_i \mapsto f_i$$

since the $G^{(i,\alpha)}$ form a Gröbner basis of $\ker(\varphi')$. Multiplying

$$\begin{pmatrix} A & B \\ C & D \end{pmatrix} \text{ by } \begin{pmatrix} E_{r'-r} & -C^{-1}D \\ 0 & E_t \end{pmatrix}$$

yields

$$\widetilde{\psi}' = \begin{pmatrix} A & B - AC^{-1}D \\ C & 0 \end{pmatrix},$$

whose columns still generate $\ker(\varphi')$. Elements of $\ker(\varphi)$ correspond to elements of $\ker(\varphi')$ of shape

$$(h_1, \ldots, h_r, 0 \ldots, 0)^t.$$

Such an element is a linear combination of the last t columns of $\widetilde{\psi}'$ because of the upper triangular shape of C. Thus the columns of $\psi = B - AC^{-1}D$ generate $\ker(\varphi)$. □

Algorithm 7.1.3 $(I : J)$.
Input. $f_1, \ldots, f_r$ generators of the ideal $I \subset S$,
$g_1, \ldots, g_s$ generators of the ideal $J \subset S$.
Output. Generators of the ideal $I : J$.

1. Form the $s \times (rs + 1)$-matrix

$$\varphi = \begin{pmatrix} g_1 & f_1 & \cdots & f_r & & & & 0 \\ g_2 & & & & f_1 & \cdots & f_r & \\ \vdots & & & & & & & \ddots \\ g_s & 0 & & & & & & f_1 & \cdots & f_r \end{pmatrix}.$$

2. Compute the syzygy matrix $\psi = (h_{ij})$ whose columns generate the kernel

$$\ker(\varphi : S^{rs+1} \to S^s).$$

3. Return the entries of the first row $h_{11}, h_{12}, \ldots, h_{1t}$ of the $(rs + 1) \times t$-matrix ψ.

□

Exercise 7.1.4. A 3SAT formula with m clauses and n logical variables is an expression of the form

$$(a_1 \vee b_1 \vee c_1) \wedge (a_2 \vee b_2 \vee c_2) \wedge \ldots \wedge (a_m \vee b_m \vee c_m)$$

with
$$a_i, b_i, c_i \in \{z_1, \neg z_1, \ldots, z_n, \neg z_n\}.$$

It is satisfiable if there exist values $z_i \in \{true, false\}$ which make the formula true.

We translate this formula into the square-free monomial ideal $I \subset K[x_1, \ldots, y_n]$ in $2n$ variables as follows. There are n degree two monomials $x_1 y_1, \ldots, x_n y_n$ and m monomials of degree three. Each clause gives a degree three monomial, where x_i and y_i correspond to z_i and $\neg z_i$, respectively. For example
$$(z_1 \vee \neg z_3 \vee z_4) \longleftrightarrow x_1 y_3 x_4.$$

Thus altogether I has $n + m$ generators. Prove:

1) $\dim V(I) = n$ if and only if the formula is satisfiable. Here a solution with $x_i = 0$ corresponds to a SAT-solution with $z_i = true$, and $y_i = 0$ corresponds to a solution with $z_i = false$.
2) The number of points in $V(I + J)$, where $J = (x_1 + y_1 - 1, \ldots, x_n + y_n - 1)$, coincides with the number of solutions of the 3SAT formula.
3) Define $I_0 = I$ and recursively
$$I_\nu = I_{\nu-1} : (x_\nu + y_\nu).$$

Then the formula is not satisfiable if and only if $I_n = (1)$.

Thus computing the dimension of monomial ideals is NP-hard. The algorithm in 3) is a variant of the well-known resolution algorithm of J.A. Robinson (1963) for logical formulas [73].

In general, the complexity of Gröbner basis computation is very bad. A doubly exponential bound was given Grete Hermann [43]. That these bounds are essentially optimal was proved by Mayr and Meyer [59]. The examples of Mayr and Meyer are binomial ideals. All the complexity in these examples comes from the exponents of the binomials.

Notice, however, that by Kollár [53] the geometric question of 'radical ideal membership' is much better behaved than ideal membership.

7.2 Elimination and the kernel of a ring homomorphism

Given an ideal $I \subset k[x_1, \ldots, x_n, y_1, \ldots, y_m]$ we want to compute $I \cap k[y_1, \ldots, y_m]$. This can be done by computing a Gröbner basis with respect to $>_{lex}$. However this computes the whole flag of elimination ideals. Using a product order is often cheaper.

Definition 7.2.1. Let $>_1$ be a global monomial order on $k[x_1, \ldots, x_n]$ and let $>_2$ be a global monomial order on $k[y_1, \ldots, y_m]$. Then we define on $k[x_1, \ldots, x_n, y_1, \ldots, y_m]$ the **product order** ($>_{12}$) by

7.2 Elimination and the kernel of a ring homomorphism

$$x^\alpha y^\beta >_{12} x^{\alpha'} y^{\beta'} \text{ if and only if } x^\alpha >_1 x^{\alpha'} \text{ or}$$
$$x^\alpha = x^{\alpha'} \text{ and } y^\beta >_2 y^{\beta'}.$$

This order has the key property that

$$\mathrm{Lt}(f) \in k[y_1, \ldots, y_m] \implies f \in k[y_1, \ldots, y_m]$$

holds.

Algorithm 7.2.2 (Elimination).
Input. $f_1, \ldots, f_r$, generators of an ideal $I \subset k[x_1, \ldots, x_n, y_1, \ldots, y_m]$.
Output. A Gröbner basis of $I \cap k[y_1, \ldots, y_m]$.

1. Compute a Gröbner basis $f_1 \ldots, f_{r'}$ of $(f_1, \ldots, f_r)$ with respect to a product order.
2. Return all Gröbner basis elements f_j with

$$\mathrm{Lt}(f_j) \in k[y_1, \ldots, y_m].$$

Proof. An element $f \in k[y_1, \ldots, y_m]$ lies in I if and only if the remainder under division by $f_1 \ldots, f_{r'}$ is zero. This division involves only the Gröbner basis elements which we return. □

Let $\varphi \colon k[y_1, \ldots, y_m] \to K[x_1, \ldots, x_n]/I, y_i \mapsto \overline{g}_i$ be a substitution homomorphism. We want to compute $\ker(\varphi)$.

Algorithm 7.2.3 (Kernel of a ring homomorphism).
Input. $f_1, \ldots, f_r$ generators of the ideal I,
$g_1, \ldots, g_m$ representatives of the $\overline{g}_i$.
Output. A Gröbner basis of the ideal $\ker(\varphi)$.

1. Consider the ideal $J \subset k[x_1, \ldots, x_n, y_1, \ldots, y_m]$ generated by $f_1, \ldots, f_r$ and $y_1 - g_1, \ldots, y_m - g_m$.
2. Compute a Gröbner basis of J with respect to a product order and return the Gröbner basis elements with leading terms in $k[y_1, \ldots, y_m]$.

Proof of correctness. Let $F \in k[y_1, \ldots, y_m]$ be an element of the kernel, i.e.,

$$F(g_1, \ldots, g_m) \in I \iff F \in J \subset k[x_1, \ldots, x_n, y_1, \ldots, y_m].$$

Thus $\ker(\varphi) = J \cap k[y_1, \ldots, y_m]$ and a Gröbner basis is obtained by computing a Gröbner basis of J with respect to $>_{12}$. □

Geometric interpretation. Suppose $K[x_1, \ldots, x_n]/I = K[A]$ is the coordinate ring of an algebraic set $A \subset \mathbb{A}^n$ and $(\overline{g}_1, \ldots, \overline{g}_r)$ are the components of a morphism

$$\phi \colon A \to \mathbb{A}^m.$$

Then the kernel J of

$$\varphi \colon K[y_1, \ldots, y_m] \to K[x_1, \ldots, x_n]/I, y_i \mapsto \overline{g}_i$$

is a radical ideal. Indeed,

$$\begin{aligned}
F \in \mathrm{rad}(J) &\implies F^N \in J \text{ for some N}\\
&\implies \varphi(F^N) = 0\\
&\implies (F(g_1,\ldots,g_m))^N \in I\\
&\implies F(g_1,\ldots,g_m) \in I \text{ because } I \text{ is a radical ideal}\\
&\implies F \in \ker(\varphi) = J.
\end{aligned}$$

The algebraic set $B = V(J) \subset \mathbb{A}^m$ is the Zariski closure $B = \overline{\phi(A)}$ of the image $\phi(A)$.

Exercise 7.2.4. Let $A = V(I) \subset \mathbb{A}^n$ be a variety, and let

$$\varphi: A \dashrightarrow \mathbb{A}^m, p \mapsto (f_1(p),\ldots,f_m(p))$$

be the rational map defined by the rational functions

$$f_1 = \frac{g_1}{h_1}, \ldots, f_m = \frac{g_m}{h_m}.$$

Design an algorithm which computes a defining ideal J of the algebraic closure

$$B = \overline{\varphi(\mathrm{dom}\,\varphi)} \subset \mathbb{A}^m$$

of the image.

7.3 Homomorphisms between modules

Let $\varphi: M \to N$ be a homomorphism between two finitely presented $R = k[x_1,\ldots,x_n]$-modules. Then φ can be lifted to a commutative diagram between the presentations

$$\begin{array}{ccccccc}
R^{r_1} & \xrightarrow{\phi} & R^{r_0} & \longrightarrow & M & \longrightarrow & 0\\
\downarrow{\varphi_1} & & \downarrow{\varphi_0} & & \downarrow{\varphi} & &\\
R^{s_1} & \xrightarrow{\psi} & R^{s_0} & \longrightarrow & N & \longrightarrow & 0.
\end{array}$$

Here M is a module with r_0 generators $m_1,\ldots,m_{r_0}$ which are the images of the basis elements $e_1,\ldots,e_{r_0}$ and the columns of the matrix ϕ generate the kernel $\ker(R^{r_0} \to M)$. Thus $M = \mathrm{coker}(\phi)$. Similarly, $N = \mathrm{coker}(\psi)$.

To obtain φ_0 we choose preimages $f_i \in R^{s_0}$ of $\varphi(m_i)$ for $i \in \{1,\ldots,t_0\}$ and define

$$\varphi_0 = (f_1 | \ldots | f_{r_0})$$

to be the $s_0 \times r_0$-matrix with column vectors f_i.

7.3 Homomorphisms between modules

Proposition 7.3.1. *An $s_0 \times r_0$-matrix φ_0 induces a well-defined R-module homomorphism $\varphi : M \to N$ if and only if φ_0 can be completed to a commutative diagram*

$$\begin{array}{ccc} R^{r_1} & \xrightarrow{\phi} & R^{r_0} \\ {\scriptstyle \exists \varphi_1} \downarrow & & \downarrow {\scriptstyle \varphi_0} \\ R^{s_1} & \xrightarrow{\psi} & R^{s_0} \end{array}$$

Proof. The matrix φ_0 induces a well-defined map $\varphi : M \to N$ if and only if the composition

$$\begin{array}{ccc} R^{r_1} & \xrightarrow{\phi} & R^{r_0} \\ & & \downarrow {\scriptstyle \varphi_0} \\ & & R^{s_0} \longrightarrow N \end{array}$$

is zero. Since

$$R^{s_1} \xrightarrow{\psi} R^{s_0} \longrightarrow N \longrightarrow 0$$

is exact at R^{s_0}, this is the case if and only if $\operatorname{im}(\varphi_0 \circ \phi) \subset \operatorname{im}(\psi)$. Since R^{r_1} is free, this is the case if and only if there exists a matrix φ_1 with $\psi \circ \varphi_1 = \varphi_0 \circ \phi$. □

Given two matrices A and B we want to decide whether A can be factored over B, i.e., whether there exists a matrix C with $A = BC$

$$\begin{array}{ccc} & & R^r \\ {\scriptstyle \exists C\,?} \nearrow & & \downarrow {\scriptstyle A} \\ R^t & \xrightarrow{B} & R^s \end{array}$$

If C exists, then C is called a **lifting of A along B**.

Algorithm 7.3.2 (Lifting). Can A be factored over B?
Input. Matrices $A \in R^{s \times r}$ and $B \in R^{s \times t}$ over $R = k[x_1, \ldots, x_n]$.
Output. A boolean value, and in case of *true* a matrix $C \in R^{t \times r}$ such that $A = BC$.

1. Compute a Gröbner basis of the column vectors $a_1, \ldots, a_r$ of A.
2. Divide each column vector b_j of B by the Gröbner basis. If one of the remainders is non-zero return *false*.
3. If all remainders are zero, express the b_i as a linear combination of the original generators $a_1, \ldots, a_r$ of the image $\operatorname{im}(A)$:

$$b_i = \sum_{j=1}^{r} c_{ij} a_j.$$

4. Return *true* and $C = (c_{ij})$.

Using this algorithm we can decide whether a matrix φ_0 induces a well-defined homomorphism $\varphi\colon M \to N$

$$\begin{array}{ccccccc} R^{r_1} & \xrightarrow{\phi} & R^{r_0} & \longrightarrow & M & \longrightarrow & 0 \\ & \searrow & \downarrow{\varphi_0} & & & & \\ R^{s_1} & \xrightarrow{\psi} & R^{s_0} & \longrightarrow & N & \longrightarrow & 0 \end{array}$$

by computing a lifting φ_1 of $\varphi_0 \circ \phi$ along ψ.

7.4 Cokernel, image and kernel of an R-module homomorphism

Given a homomorphism $\varphi\colon M \to N$ represented by a matrix φ_0

$$\begin{array}{ccccccc} R^{r_1} & \xrightarrow{\phi} & R^{r_0} & \longrightarrow & M & \longrightarrow & 0 \\ \downarrow{\varphi_1} & & \downarrow{\varphi_0} & & \downarrow{\varphi} & & \\ R^{s_1} & \xrightarrow{\psi} & R^{s_0} & \longrightarrow & N & \longrightarrow & 0 \end{array}$$

we will describe presentations of $\operatorname{coker}(\varphi)$, $\operatorname{im}(\varphi)$ and $\operatorname{ker}(\varphi)$. We have presentations

$$R^{r_0} \oplus R^{s_1} \xrightarrow{(\varphi_0|\psi)} R^{s_0} \longrightarrow \operatorname{coker}(\varphi) \longrightarrow 0$$

and

$$R^{t_0} \oplus R^{r_1} \xrightarrow{(A|\phi)} R^{r_0} \longrightarrow \operatorname{im}(\varphi) \longrightarrow 0,$$

where A is part of the syzygy matrix $\begin{pmatrix} A \\ B \end{pmatrix}$ of $(\varphi_0|\psi)$:

$$R^{t_0} \xrightarrow{\begin{pmatrix} A \\ B \end{pmatrix}} R^{r_0} \oplus R^{s_1} \xrightarrow{(\varphi_0|\psi)} R^{s_0}.$$

The computation of the presentation of $\operatorname{ker}(\varphi)$ takes more steps.

Algorithm 7.4.1 (Computation of the kernel).
Input. An R-module homomorphism $\varphi\colon M \to N$ given by a commutative diagram

$$\begin{array}{ccccccc} R^{r_1} & \xrightarrow{\phi} & R^{r_0} & \longrightarrow & M & \longrightarrow & 0 \\ \downarrow{\varphi_1} & & \downarrow{\varphi_0} & & \downarrow{\varphi} & & \\ R^{s_1} & \xrightarrow{\psi} & R^{s_0} & \longrightarrow & N & \longrightarrow & 0 \end{array}$$

7.4 Cokernel, image and kernel of an R-module homomorphism

Output. A presentation matrix C of $\ker(\varphi)$

1. Compute the syzygy matrix $\begin{pmatrix} A \\ B \end{pmatrix}$ of $(\varphi_0|\psi)$:

$$R^{t_0} \xrightarrow{\begin{pmatrix} A \\ B \end{pmatrix}} R^{r_0} \oplus R^{s_1} \xrightarrow{(\varphi_0|\psi)} R^{s_0}.$$

2. Compute the syzygy matrix $\begin{pmatrix} C \\ D \end{pmatrix}$ of $(A|\phi)$:

$$R^{t_1} \xrightarrow{\begin{pmatrix} C \\ D \end{pmatrix}} R^{t_0} \oplus R^{r_1} \xrightarrow{(A|\phi)} R^{r_0}.$$

3. Then C is the presentation matrix of $\ker(\varphi)$:

$$R^{t_1} \xrightarrow{C} R^{t_0} \longrightarrow \ker(\varphi) \longrightarrow 0.$$

Proof of correctness. We have a commutative diagram

$$\begin{array}{ccccccc}
R^{t_1} & \xrightarrow{C} & R^{t_0} & \longrightarrow & \operatorname{coker}(C) & \longrightarrow & 0 \\
{\scriptstyle -D}\downarrow & & {\scriptstyle A}\downarrow & & {\scriptstyle \iota}\downarrow & & \\
R^{r_1} & \xrightarrow{\phi} & R^{r_0} & \longrightarrow & M & \longrightarrow & 0 \\
{\scriptstyle \varphi_1}\downarrow & & {\scriptstyle \varphi_0}\downarrow & & {\scriptstyle \varphi}\downarrow & & \\
R^{s_1} & \xrightarrow{\psi} & R^{s_0} & \longrightarrow & N & \longrightarrow & 0
\end{array}$$

The map ι induced by A maps into the $\ker(\varphi)$ because $\varphi_0 A$ induces the zero map as $\varphi_0 A = -\psi B$.

The map $\iota: \operatorname{coker}(C) \to \ker(\varphi)$ is surjective, because an element of $f \in R^{r_0}$ maps to $0 \in N$ if and only if $\varphi_0(f)$ lies in the image of ψ. Such an element is of the form $f = Ag$ for some $g \in R^{t_0}$ since $\begin{pmatrix} A \\ B \end{pmatrix}$ is the syzygy matrix of $(\varphi_0|\psi)$. This also shows that the presentation of $\operatorname{im}(\varphi)$ as $\operatorname{coker}(A|\phi)$ above is correct.

Finally, we prove that the map $\iota: \operatorname{coker}(C) \to \ker(\varphi)$ is injective and conclude that this map is an isomorphism.

An element $g \in R^{t_0}$ maps to $0 \in M$ if and only if $Ag \in \operatorname{im}(\phi)$. These elements are of the form Ch for some $h \in R^{t_1}$ since $\begin{pmatrix} C \\ D \end{pmatrix}$ is the syzygy matrix of $(A|\phi)$. Hence $g \mapsto 0 \in \operatorname{coker}(C)$. □

Exercise 7.4.2. If $\psi: L \to M$ and $\varphi: M \to N$ are two R-module homomorphisms with $\varphi \circ \psi = 0$, then

$$H = \frac{\ker(\varphi)}{\operatorname{im} \psi}$$

is called the **homology** of the complex

$$L \xrightarrow{\psi} M \xrightarrow{\varphi} N$$

at M. Let

$$\begin{array}{ccccccc}
E_1 & \xrightarrow{a} & E_0 & \longrightarrow & L & \longrightarrow & 0 \\
\downarrow{\psi_1} & & \downarrow{\psi_0} & & \downarrow{\psi} & & \\
F_1 & \xrightarrow{b} & F_0 & \longrightarrow & M & \longrightarrow & 0 \\
\downarrow{\varphi_1} & & \downarrow{\varphi_0} & & \downarrow{\varphi} & & \\
G_1 & \xrightarrow{c} & G_0 & \longrightarrow & N & \longrightarrow & 0
\end{array}$$

be free presentations of ψ and φ. Prove the correctness of the following algorithm.

1. Compute the syzygy matrix of $(c|\varphi_0)$

$$H_0 \xrightarrow{\binom{g_0}{h_0}} G_1 \oplus F_0 \xrightarrow{(c|\varphi_0)} G_0 \ .$$

2. Compute the syzygy matrix of $(h_0|b|\psi_0)$

$$H_1 \xrightarrow{\binom{h_1}{g_1}{f_1}} H_0 \oplus F_1 \oplus E_0 \xrightarrow{(h_0|b|\psi_0)} F_0 \ .$$

3. Then

$$H_1 \xrightarrow{h_1} H_0 \longrightarrow H \longrightarrow 0$$

is a presentation of H.

Exercise 7.4.3. Prove: Two commutative diagrams

$$\begin{array}{ccc}
R^{r_1} & \xrightarrow{\phi_1} & R^{r_0} \\
\downarrow{\varphi_1} & & \downarrow{\varphi_0} \\
R^{s_1} & \xrightarrow{\psi_1} & R^{s_0}
\end{array}
\quad \text{and} \quad
\begin{array}{ccc}
R^{r_1} & \xrightarrow{\phi_1} & R^{r_0} \\
\downarrow{\varphi'_1} & & \downarrow{\varphi'_0} \\
R^{s_1} & \xrightarrow{\psi_1} & R^{s_0}
\end{array}$$

induce the same homomorphism $\varphi: M \to N$ for $M = \operatorname{coker} \phi$ and $N = \operatorname{coker} \psi$ if and only if there exist homomorphisms h_0 and h_1 as in the diagram

7.4 Cokernel, image and kernel of an R-module homomorphism

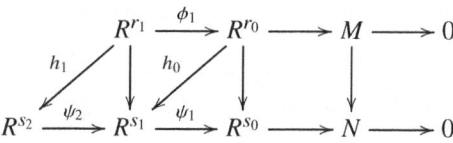

such that
$$\varphi_0 - \varphi_0' = \psi_1 h_0 \text{ and } \varphi_1 - \varphi_1' = \psi_2 h_1 + h_0 \phi_1.$$

Here ψ_2 is a syzygy matrix of ψ_1.

Exercise 7.4.4. Design an algorithm which computes a presentation of $\mathrm{Hom}(M,N)$ from presentations of M and N by building on Exercises 7.4.3 and 7.4.2.

Chapter 8
Projective algebraic geometry

Higher-dimensional affine varieties $A \subset \mathbb{A}^n(\mathbb{C})$ are never compact in the euclidean topology. The projective n-space $\mathbb{P}^n(\mathbb{C})$ is a compactification of $\mathbb{A}^n(\mathbb{C})$, and the closure $\overline{A} \subset \mathbb{P}^n(\mathbb{C})$ is a compactification of A.

For an affine zero-dimensional algebraic set the number of solutions is a numerical invariant. Introducing projective algebraic sets will allow us to generalize the number of points and the degree of a hypersurface, defined as the degree of the defining equation, into a concept of degree for an arbitrary projective algebraic set. These concepts are geometrically related by Corollary 14.1.3 of Bertini's theorem.

We start by defining the projective space, its standard charts and the projective closure $\overline{A}$ of an affine algebraic set. They are defined by the homogeneous ideals in the standard graded polynomial ring $S = K[x_0, \ldots, x_n]$. The homogeneous coordinate ring of projective algebraic set $A \subset \mathbb{P}^n$ is the graded ring $S_A = S/I(A)$, where $I(A)$ denotes the homogeneous vanishing ideal of A.

Section 8.3 contains an elementary proof of Hilbert's syzygy theorem. Hilbert's original motivation for the syzygy theorem was to prove the polynomial nature of the Hilbert function. We prove that the dimension of a projective algebraic set $V(I)$ coincides with the degree of its Hilbert polynomial $p_{S/I}(t) \in \mathbb{Q}[t]$ and define the degree of a homogeneous ideal I via the leading coefficient of the Hilbert polynomial.

8.1 The projective space

The projective space was discovered by artists in the Renaissance. Two parallel lines in $\mathbb{A}^2$ do not intersect. However, in perspective drawing they do intersect at a point on the horizon.

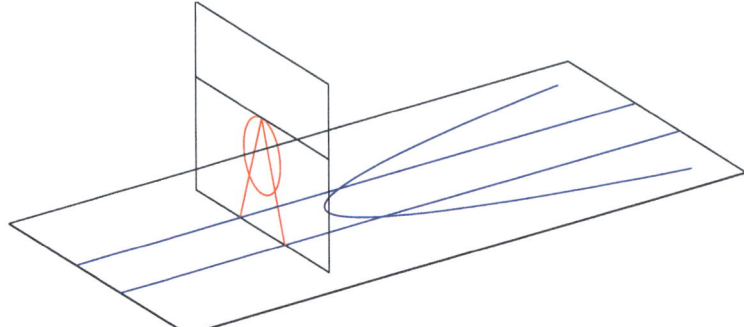

To put this into the right framework, we define $\mathbb{P}^2(\mathbb{R})$ as the lines through the origin of $\mathbb{R}^3$. Then each point in the plane $\{z = 1\}$ gives a point of $\mathbb{P}^2(\mathbb{R})$, and in addition we have the horizon corresponding to one-dimensional subvector spaces of $\mathbb{R}^3$ contained in the subvector space $\{(x, y, z) \in \mathbb{R}^3 \mid z = 0\}$.

Definition 8.1.1. Let k be any field and W be a finite-dimensional k vector space. The **projective space of** W is

$$\mathbb{P}(W) = \{\text{1-dimensional subvector spaces of } W\}.$$

In particular, we define

$$\mathbb{P}^n(k) = \mathbb{P}(k^{n+1}).$$

The projective space $\mathbb{P}^n$ refers to $\mathbb{P}^n(K)$ over an algebraic closed extension field K of k, and we call $\mathbb{P}^n(k)$ the set of k-rational points of $\mathbb{P}^n$.

A different way to define $\mathbb{P}^n$ is via an equivalence relation: Two points $a = (a_0, \ldots, a_n)$ and $b = (b_0, \ldots, b_n) \in K^{n+1} \setminus \{0\}$ are equivalent, i.e., $a \sim b$, if there exists a $\lambda \in K^*$ with $\lambda a = b$. Then

$$\mathbb{P}^n = (K^{n+1} \setminus \{0\})/\sim$$

identifies the equivalence class $[a]$ with the one-dimensional subspace spanned by a. We write $[a] = (a_0 : a_1 : \ldots : a_n)$ and refer to the a_i as the **homogeneous coordinates of the point** $p = [a] \in \mathbb{P}^n$. Note that the ratios $a_i : a_j$ for $a_j \neq 0$ are well-defined.

Given a polynomial $f \in K[x_0, \ldots, x_n]$ the value $f(p)$ does not make sense. However, for a **homogeneous polynomial** of degree d we have

$$f(\lambda a) = \lambda^d f(a).$$

Here f is called homogeneous if each term of f has the same total degree d. Thus

$$V(f) = \{p \in \mathbb{P}^n \mid f(p) = 0\},$$

where f is homogeneous, is a well-defined subset of $\mathbb{P}^n$.

8.1 The projective space

Definition 8.1.2. A **projective algebraic set** is a subset of the form

$$V(f_1,\ldots,f_r) = \bigcap V(f_i),$$

where the f_i are homogeneous of degree d_i. These sets form the closed sets of the Zariski topology of $\mathbb{P}^n$.

The standard atlas of $\mathbb{P}^n$. The (Zariski) open subsets

$$U_i = \{(a_0 : \ldots : a_n) \in \mathbb{P}^n \mid a_i \neq 0\} = \mathbb{P}^n \setminus V(x_i)$$

cover $\mathbb{P}^n$ because each point in $\mathbb{P}^n$ has homogeneous coordinates $(a_0 : \ldots : a_n)$ with at least one $a_i \neq 0$. The maps

$$\varphi_i : U_i \to \mathbb{A}^n, (a_0 : \ldots : a_i : \ldots a_n) \mapsto (\frac{a_0}{a_i},\ldots,\frac{a_{i-1}}{a_i},\frac{a_{i+1}}{a_i},\ldots,\frac{a_n}{a_i})$$

are well-defined bijections. For example, the inverse of φ_0 is

$$\varphi_0^{-1} : \mathbb{A}^n \to U_0 \subset \mathbb{P}^n, (b_1,\ldots,b_n) \mapsto (1 : b_1 : \ldots : b_n).$$

More generally, φ_i^{-1} inserts 1 into the i-th position. The pair (U_i,φ_i) is called a **chart** of the standard atlas of $\mathbb{P}^n$. The change of charts maps

$$\varphi_{ij} = \varphi_i \circ \varphi_j^{-1} : \varphi_j(U_i \cap U_j) \to \varphi_i(U_i \cap U_j)$$

are given by rational maps. For example

$$\varphi_{i0} : \mathbb{A}^n \dashrightarrow \mathbb{A}^n, (a_1,\ldots,a_n) \mapsto (\frac{1}{a_i},\ldots,\frac{a_{i-1}}{a_i},\frac{a_{i+1}}{a_i},\ldots,\frac{a_n}{a_i}).$$

The projective space $\mathbb{P}^n$ as a manifold. The atlas

$$\mathcal{A} = \{(U_i,\varphi_i) \mid i = 0,\ldots,n\}$$

gives $\mathbb{P}^n(\mathbb{R})$ and $\mathbb{P}^n(\mathbb{C})$ the structure of a compact differentiable or compact complex manifold, respectively, because rational functions are differentiable and holomorphic on their domain of definition.

The real projective spaces are obtained as

$$\mathbb{P}^n(\mathbb{R}) = S^n/\sim,$$

where $\sim$ identifies antipodal points of the unit sphere $S^n \subset \mathbb{R}^{n+1}$. So $\mathbb{P}^2(\mathbb{R})$ is a non-orientable surface. It is the union of a Möbius strip M and a disc D glued along their common boundary $\partial M \cong \partial D = S^1$.

As a real manifold $\mathbb{P}^1(\mathbb{C})$ is homeomorphic to S^2 since both spaces are one-point compactifications of $U_0 \cong \mathbb{C} \cong \mathbb{R}^2$.

The complex projective space $\mathbb{P}^n(\mathbb{C})$ with the euclidean topology is compact since the map from the unit sphere $S^{2n+1} \subset \mathbb{C}^{n+1} \setminus \{0\}$ to $\mathbb{P}^n(\mathbb{C})$ is continuous. The map

$$h \colon S^{2n+1} \to \mathbb{P}^n(\mathbb{C})$$

is called the **Hopf fibration**. The fibers of h are isomorphic to circles

$$S^1 = \{\lambda \in \mathbb{C} \mid |\lambda| = 1\}.$$

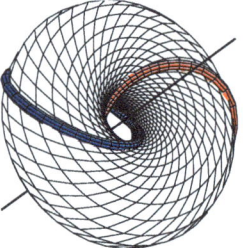

Identifying S^3 with the one-point compactification of $\mathbb{R}^3$ we see that $\mathbb{R}^3$ is a disjoint union of linked circles and one line.

We can regard $\mathbb{P}^n$ as a compactification of $\mathbb{A}^n$:

$$\mathbb{P}^n = \mathbb{A}^n \cup \mathbb{P}^{n-1} = \mathbb{A}^n \cup \mathbb{A}^{n-1} \cup \ldots \cup \mathbb{A}^0,$$

where we identify $\mathbb{A}^n \cong U_0$ with a Zariski open subset via φ_0. For this reason we call $\mathbb{P}^{n-1} = V(x_0) \subset \mathbb{P}^n$ the **hyperplane at infinity**.

Let $A = V(f) \subset \mathbb{A}^n$ be a hypersurface defined by $f \in K[x_1, \ldots, x_n]$. Then the Zariski closure $\overline{A} \subset \mathbb{P}^n$ is defined by $\overline{A} = V(f^h)$ where

$$f^h = x_0^{\deg f} f\left(\frac{x_1}{x_0}, \ldots, \frac{x_n}{x_0}\right) \in K[x_0, \ldots, x_n]$$

denotes the **homogenization of** f. Conversely, for a homogeneous polynomial $f \in K[x_0, \ldots, x_n]$ we denote by

$$f^a = f(1, x_1, \ldots, x_n)$$

the corresponding affine polynomial. Clearly $(f^h)^a = f$. Conversely we have

$$x_0^{\deg f - \deg f^a} (f^a)^h = f.$$

Thus $(f^a)^h$ coincides with f if and only if x_0 is not a factor of f.

Example 8.1.3 (A plane cubic curve in all three charts). Consider the curve

$$C = V(y^2 z + x^3 - x^2 z) \subset \mathbb{P}^2$$

with homogeneous coordinates $(x : y : z)$.

8.1 The projective space

We obtain

$$y^2 = x^2 - x^3$$

in the (x,y)-plane $U_2 = \{z = 1\}$. The picture shows C together with all tangent lines to C at points on the line $V(y)$.

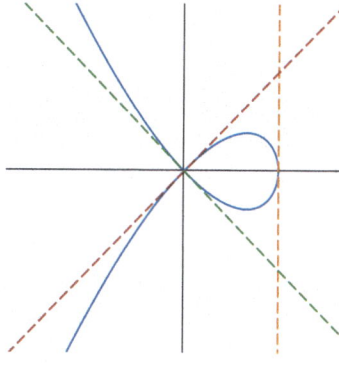

The equation

$$z = \frac{x^3}{x^2 - 1}$$

defines C in the (x,z)-plane $U_1 = \{y = 1\}$. The intersection points $C \cap V(y)$ are not visible in this chart. Their tangents give asymptotes.

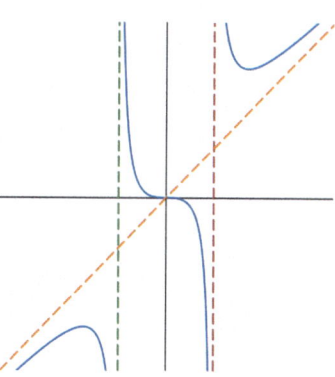

Finally we get

$$z = \frac{1}{1 - y^2}$$

in the (y,z)-plane $U_0 = \{x = 1\}$. Notice that points (z, y) on C in this chart with large absolute $|y|$-value lie on the same side of $V(z)$ in contrast to the situation at the origin in U_1. This comes from the fact that $\mathbb{P}^2(\mathbb{R})$ is not orientable.

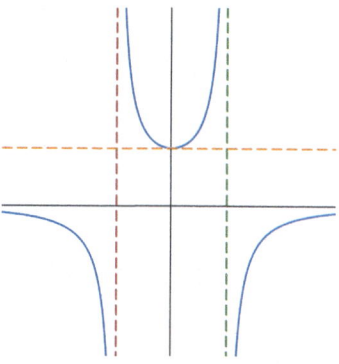

The linear map $A \colon K^{n+1} \to K^{n+1}$ defined by a matrix $A \in \mathrm{GL}(n+1, K)$ induces an automorphism

$$\phi_A \colon \mathbb{P}^n \to \mathbb{P}^n.$$

Since $\lambda \operatorname{id}_{K^{n+1}}$ for $\lambda \in K^*$ induces the identical map on $\mathbb{P}^n$, we see that the quotient group

$$\mathrm{PGL}(n+1, K) = \mathrm{GL}(n+1, K)/K^* \subset \mathbb{P}^{n^2+2n}$$

acts on $\mathbb{P}^n$. Notice that $\mathrm{PGL}(n+1, K)$ is also a quasi-projective variety in the sense of Definition 11.2.3. It is the complement of the hypersurface $V(\det A) \subset \mathbb{P}^{n^2+2n}$.

Exercise 8.1.4. Let $p_0 = (1 : 0 : \ldots : 0), \ldots, p_n = (0 : \ldots : 0 : 1)$ denote the **coordinate points** and let $p_{n+1} = (1 : 1 : \ldots : 1)$ be the **scaling point**. Suppose $q_0, \ldots, q_{n+1} \in \mathbb{P}^n$ are $n+2$ points such that no subset of $n+1$ of them are contained in a hyperplane. Prove there exists a unique automorphism ϕ_A of $\mathbb{P}^n$ with $\phi_A(p_j) = q_j$ for $j = 0, \ldots, n+1$. For this reason, p_{n+1} is called the scaling point.

Exercise 8.1.5. With the notation of Exercise 8.1.4 and $K = \mathbb{C}$, let $\phi_A \colon \mathbb{P}^2 \to \mathbb{P}^2$ be the automorphism which maps $p_0, \ldots, p_3$ to points $q_0, \ldots, q_3 \in U_0 \cap \mathbb{P}^2(\mathbb{R}) = \mathbb{A}^2(\mathbb{R}) = \mathbb{R}^2$ where $q_3 = \frac{1}{3}(q_0 + q_1 + q_2)$ is the center of gravity of the triangle $\Delta = \overline{q_0 q_1 q_2} \subset \mathbb{R}^2$. Let $(\lambda_0 : \lambda_1 : \lambda_2) \in \mathbb{P}^2(\mathbb{R})$ be a point with non-negative homogeneous coordinates. After rescaling we may assume that $\sum_{i=0}^2 \lambda_i = 1$. Prove:

$$\phi_A((\lambda_0 : \lambda_1 : \lambda_2)) = \sum_{i=0}^2 \lambda_i q_i$$

is the convex combination of the points $q_0, q_1, q_2 \in \mathbb{R}^2$ and any point inside the triangle Δ arises this way.

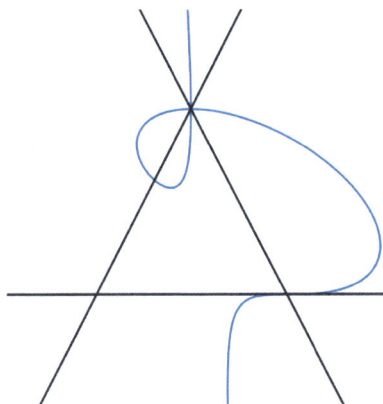

The blue curve C shows Example 8.1.3 in convex coordinates. Compute the equations of C in this affine plane.

Exercise 8.1.6. Draw the ellipse

$$V((x-1)^2 + \frac{1}{4}(y-1)^2 - 1) \subset \mathbb{A}^2 \cong U_0 \subset \mathbb{P}^2$$

in all three charts of $\mathbb{P}^2(\mathbb{R})$.

Exercise 8.1.7. Consider the map

$$S^2 \to \mathbb{R}^3 = \mathbb{A}^3(\mathbb{R}), (x, y, z) \mapsto (yz, xz, xy).$$

Prove that the map factors over $\mathbb{P}^2(\mathbb{R})$ and compute the equation of the algebraic closure in $\mathbb{A}^3$. The image

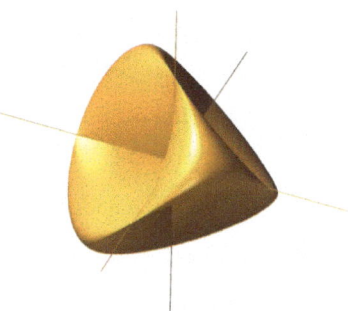

was created with the program surfer https://imaginary.org/de/program/surfer. This surface is known under the name Steiner surface or Roman surface.

8.2 The algebra-geometry dictionary in the projective case

Definition 8.2.1. A **graded ring** R is a ring together with a decomposition

$$R = \bigoplus_{d \in \mathbb{Z}} R_d$$

as abelian groups satisfying

$$R_d \cdot R_e \subset R_{d+e}$$

for the multiplication. An ideal J in a graded ring is called homogeneous if

$$J = \bigoplus_{d \in \mathbb{Z}} J_d \text{ with } J_d = J \cap R_d,$$

equivalently, if J is generated by homogeneous elements. In that case

$$R/J = \bigoplus R_d/J_d$$

is again a graded ring. For example, $S = K[x_0, \ldots, x_n]$ with

$$S_d = \{f \in S \mid f \text{ is homogeneous of degree } d\}$$

is a graded ring. It is called the **standard graded polynomial ring** in $n+1$ variables over K.

Definition 8.2.2. Let $A \subset \mathbb{P}^n$. Then
$$I(A) = (\{f \in S_d \mid f(p) = 0 \; \forall p \in A\})$$
is called the **homogeneous ideal of** A and
$$S/I(A) = \bigoplus_{d \geq 0} (S/I(A))_d = \bigoplus_{d \geq 0} S_d/I(A)_d$$
is called the **homogeneous coordinate ring** of A.

Conversely, for a homogeneous ideal $J \subset S$ we define
$$V(J) = \{p \in \mathbb{P}^n \mid f(p) = 0 \; \forall \text{ homogeneous } f \in J\}.$$

The correspondences
$$\{\text{subsets of } \mathbb{P}^n\} \leftrightarrow \{\text{homogeneous ideals of } S = K[x_0, \ldots, x_n]\}$$
$$A \mapsto I(A), V(J) \leftarrow J$$
induce bijections
$$\{\text{algebraic subsets of } \mathbb{P}^n\} \xleftrightarrow{1:1} \{\text{homogeneous radical ideals of } S\}$$
and
$$\{\text{projective subvarieties of } \mathbb{P}^n\} \xleftrightarrow{1:1} \{\text{homogeneous prime ideals of } S\}.$$

The homogeneous maximal ideal $\mathfrak{m} = (x_0, \ldots, x_n)$ corresponds to the empty set $\emptyset$. For this reason $\mathfrak{m}$ is sometimes called the **irrelevant ideal**.

Proposition 8.2.3 (The projective Nullstellensatz). *Let $J \subsetneq S$ be a homogeneous ideal of the standard graded polynomial ring $S = K[x_0, \ldots, x_n]$ over an algebraically closed field K. Then*
$$V(J) = \emptyset \subset \mathbb{P}^n \iff \mathrm{rad}(J) = (x_0, \ldots, x_n).$$

Proof. We denote by
$$C(J) = \{a \in \mathbb{A}^{n+1} \mid f(a) = 0 \; \forall f \in J\}$$
the zero locus of J in $\mathbb{A}^{n+1}$. This is a cone whose vertex is the origin $o = (0, \ldots, 0) \in \mathbb{A}^{n+1}$. The zero locus $C(J)$ is non-empty, because J is a proper homogeneous ideal. If $C(J)$ contains a point $a = (a_0, \ldots, a_n)$ different from the origin, then $[a] = (a_0 : \ldots : a_n)$ lies in $V(J) \subset \mathbb{P}^n$. Thus
$$V(J) = \emptyset \iff C(J) = \{o\}$$
$$\iff \mathrm{rad}(J) = I(\{o\}) = (x_0, \ldots, x_n)$$

8.2 The algebra-geometry dictionary in the projective case

by the Nullstellensatz for $\mathbb{A}^{n+1}$. □

Example 8.2.4. Consider the curve $C = V(y - x^2, z - xy) \subset \mathbb{A}^3$, the image of

$$\varphi : \mathbb{A}^1 \to \mathbb{A}^3, t \mapsto (t, t^2, t^3).$$

Using homogeneous coordinates $(w : x : y : z)$ on $\mathbb{P}^3$ and homogenizing both equations yields

$$(wy - x^2, wz - xy) = (wy - x^2, wz - xy, y^2 - xz) \cap (w, x).$$

The line $V(w, x) \cong \mathbb{P}^1$ is completely contained in the hyperplane at infinity $\mathbb{P}^2 = V(w)$. It does not belong to the projective closure

$$\overline{C} = V(wy - x^2, wz - xy, y^2 - xz)$$

of C in $\mathbb{P}^3$. The closure $\overline{C}$ intersects the hyperplane at infinity in a single point: We have

$$V(wy - x^2, wz - xy, y^2 - xz, w) = V(w, x^2, xy, y^2 - xz)$$
$$= V(w, x, y) = \{(0 : 0 : 0 : 1)\},$$

which is the limit of the points

$$(1 : t : t^2 : t^3) = (\frac{1}{t^3} : \frac{1}{t^2} : \frac{1}{t} : 1) \text{ for } t \to \infty.$$

Let $J \subset k[x_1, \ldots, x_n]$. Then

$$J^h = (\{f^h \mid f \in J\}) \subset k[x_0, \ldots, x_n]$$

is called the **homogenization** of J.

Algorithm 8.2.5 (Computation of the homogenization).
Input. Generators $f_1, \ldots, f_r$ of an ideal $J \subset k[x_1, \ldots, x_n]$.
Output. Generators of $J^h \subset k[x_0, \ldots, x_n]$.

1. Choose a global monomial order $>$ on $k[x_1, \ldots, x_n]$ which refines the total degree, for example, $>_{\text{rdlex}}$.
2. Compute a Gröbner basis $f_1, \ldots f_{r'}$ of $(f_1, \ldots, f_r)$ with respect to $>$.
3. Return $f_1^h, \ldots, f_{r'}^h$.

Example 8.2.4 continued. The computation

$x^2 - y$	$-y$	z
$xy - z$	x	$-y$
$y^2 - xz$	-1	x

shows that $x^2 - y, xy - z, y^2 - xz$ is a Gröbner basis. Thus

$$(y - x^2, z - xy)^h = (x^2 - wy, xy - wz, y^2 - xz).$$

Proof of correctness. Let $f_1, \ldots, f_{r'}$ be a Gröbner basis with respect to $>$ and let $f \in J$ be an arbitrary element. Consider the division expression

$$f = g_1 f_1 + \ldots + g_{r'} f_{r'}$$

for f. Since the leading terms $\mathrm{Lt}(g_i f_i)$ are disjoint and $>$ refines the total degree, we have $d = \deg f \geq \deg(g_i f_i) = d_i$ and equality holds for at least one $j \in \{1, \ldots, r\}$. Thus

$$f^h = x_0^{d-d_1} g_1^h f_1^h + \ldots + x_0^{d-d_{r'}} g_{r'}^h f_{r'}^h$$

lies in $(f_1^h, \ldots, f_{r'}^h)$. $\square$

Exercise 8.2.6. Compute the homogeneous ideal of the projective closure of the affine curve parametrized by

$$\mathbb{A}^1 \to \mathbb{A}^3, t \mapsto (t, t^3, t^4).$$

8.3 Hilbert's syzygy theorem

Theorem 8.3.1 (Hilbert's syzygy theorem). *Let M be a finitely generated $S = k[x_1, \ldots, x_n]$-module. Then M has a finite free resolution*

$$0 \longleftarrow M \longleftarrow F_0 \xleftarrow{\varphi_1} F_1 \xleftarrow{\varphi_2} \ldots \xleftarrow{\varphi_{c-1}} F_{c-1} \xleftarrow{\varphi_c} F_c \longleftarrow 0$$

of length $c \leq n$.

Here the $F_i = S^{b_i}$ are free S-modules and the maps $\varphi_i : F_i \to F_{i-1}$ satisfy

$$\ker(\varphi_i) = \mathrm{im}(\varphi_{i+1})$$

and the map φ_1 gives a free presentation of $M \cong \mathrm{coker}(\varphi_1)$:

$$0 \longleftarrow M \longleftarrow F_0 \xleftarrow{\varphi_1} F_1.$$

Proof of the syzygy theorem following [75]. Every finitely generated S-module M has a finite presentation

$$0 \longleftarrow M \longleftarrow F_0 \xleftarrow{\varphi_1'} F_1.$$

We give an algorithm which computes a finite free resolution of M starting from a finite presentation. Choose a global monomial order on F_0 and compute a Gröbner basis $f_1, \ldots, f_{b_1}$ of $\mathrm{im}(\varphi_1')$. In the first step we replace φ_1' by $\varphi_1 = (f_1|f_2|\ldots|f_{b_1})$.

8.3 Hilbert's syzygy theorem

The Buchberger test syzygies $G^{(i,\alpha)}$ form a Gröbner basis of $\ker(\varphi_1)$ with respect to the induced order, and we take φ_2 as the matrix which has these test syzygies as columns. Computing the Buchberger test syzygies of the $G^{(i,\alpha)}$ yields the φ_3 and continuing in this way produces a free resolution.

We still have a lot of choice in this process. We will show that under a suitable ordering of the Gröbner basis elements the process will stop after $c \leq n$ steps with a matrix φ_c which has a trivial kernel.

Choose ℓ minimal such that

$$\mathrm{Lt}(f_1), \ldots, \mathrm{Lt}(f_{b_1}) \in k[x_1, \ldots, x_\ell]^{b_0} \subset k[x_1, \ldots, x_n]^{b_0}.$$

In the worst case $\ell = n$. Now sort $f_1, \ldots, f_{b_1}$ such that for every p

$$x_\ell^p \mid \mathrm{Lt}(f_i) \implies x_\ell^p \mid \mathrm{Lt}(f_j) \text{ for } i < j$$

holds. Then

$$\mathrm{Lt}(G^{(j,\alpha)}) \in k[x_1, \ldots, x_{\ell-1}]^{b_1} \subset k[x_1, \ldots, x_n]^{b_1}$$

because the power of x_ℓ in $\mathrm{Lt}(f_j)$ is at least as large as the power of x_ℓ in any $\mathrm{Lt}(f_i)$ with $i < j$. Sorting the $G^{(i,\alpha)}$ and the higher test syzygies similarly we obtain for the columns $H_\nu = H^{(j,\alpha)}$ of φ_{c+1} that

$$\mathrm{Lt}(H^{(j,\alpha)}) \subset k^{b_c} \subset k[x_1, \ldots, x_n]^{b_c}$$

holds after $c \leq \ell \leq n$. But if such $H^{(j,\alpha)}$ exists, this means that two columns of φ_{c-1} have the same leading term, which is not the case since we choose minimal generators of the M_i's in each step. Thus there are no elements of this kind and $b_{c+1} = 0$. □

Example 8.3.2. We consider the ideal $J \subset S = k[w, x, y, z]$ generated by the entries of the first column in the following table.

$w^2 - xz$	$-x$	y	0	$-z$	0	$-y^2 + wz$	
$wx - yz$	w	$-x$	$-y$	0	z	z^2	
$x^2 - wy$	$-z$	w	0	$-y$	0	0	
$xy - z^2$	0	0	w	x	$-y$	$-yz$	
$y^2 - wz$	0	0	$-z$	$-w$	x	w^2	
	0	y	$-x$	w	$-z$	1	
	$-y^2 + wz$	z^2	$-wy$	yz	$-w^2$	x	

The original generators turn out to be a Gröbner basis, and the algorithm produces a free resolution of shape

$$0 \longleftarrow S/J \longleftarrow S \xleftarrow{\varphi_1} S^5 \xleftarrow{\varphi_2} S^6 \xleftarrow{\varphi_3} S^2 \longleftarrow 0$$

with matrices $\begin{array}{|c|c|} \hline \varphi_1^t & \varphi_2 \\ \hline & \varphi_3^t \\ \hline \end{array}$ as above.

Remark 8.3.3 (Free resolution over Noetherian rings). Let R be a Noetherian ring and let M be a finitely generated R-module. Then M has a free resolution

$$0 \leftarrow M \leftarrow R^{b_0} \leftarrow R^{b_1} \leftarrow \ldots \leftarrow R^{b_j} \leftarrow \ldots$$

where b_0 is the number of generators, b_1 the number of generators of the kernel of $R^{b_0} \to M$ and so on. What is so remarkable about $k[x_1,\ldots,x_n]$ is that the free resolution ends after finitely many steps. In general, this is not true.

Example. Consider $R = k[x,y]/(xy)$ and the R-module $M = R/(\bar{x})$. The kernel of the presentation matrix

$$0 \leftarrow M \leftarrow R \xleftarrow{\bar{x}} R$$

is generated by $\bar{y}$. The kernel of the matrix $(\bar{y})$ is generated by $\bar{x}$, and the free resolution becomes periodic

$$0 \leftarrow M \leftarrow R \xleftarrow{\bar{x}} R \xleftarrow{\bar{y}} R \xleftarrow{\bar{x}} R \xleftarrow{\bar{y}} \ldots$$

Definition 8.3.4. Let $R = \bigoplus_{d \in \mathbb{Z}} R_d$ be a graded ring. A **graded R-module** is an R-module with a decomposition

$$M = \bigoplus_{d \in \mathbb{Z}} M_d$$

as an abelian group satisfying

$$R_e \cdot M_d \subset M_{e+d}$$

for the multiplication. A **homomorphism** $\varphi : M \to N$ **of graded R-modules** is an R-module homomorphism which respects the degree, i.e.,

$$\varphi(M_d) \subset N_d.$$

With this notation the R-module homomorphism

$$R \xrightarrow{f} R$$

given by multiplication by a homogeneous element $f \in R_d$ of degree $d \neq 0$ is not a homomorphism of graded R-modules. To remedy this situation, we define the **degree shifted module** $M(d)$ as the graded R-module with $M(d)_e = M_{d+e}$. The multiplication by a homogeneous element $f \in R_d$ induces graded R-module homomorphisms

$$M \xrightarrow{f} M(d), \quad M(-d) \xrightarrow{f} M \quad \text{and, more general,} \quad M(e-d) \xrightarrow{f} M(e).$$

8.3 Hilbert's syzygy theorem

Example. Let $S = k[x_0, \ldots, x_n]$ be the standard graded polynomial ring in $n + 1$ variables. Then $S(-j)$ is the free graded S-module with generator in degree j:

$$1 \in S(-j)_j = S_{-j+j} = S_0.$$

Theorem 8.3.5 (Hilbert's syzygy theorem in the graded case). *Let $S = k[x_0, \ldots, x_n]$ be the standard graded polynomial ring in $n + 1$ variables and let M be a finitely generated graded S-module. Then M has a finite free resolution*

$$0 \longleftarrow M \longleftarrow F_0 \xleftarrow{\varphi_1} F_1 \xleftarrow{\varphi_2} \cdots \xleftarrow{\varphi_{c-1}} F_{c-1} \xleftarrow{\varphi_c} F_c \longleftarrow 0$$

of length $c \leq n + 1$, where

$$F_i = \bigoplus_j S(-j)^{\beta_{ij}}$$

is a free graded S-module with β_{ij} generators in degree j.

The β_{ij} are called **graded Betti numbers** of the resolution $F_\bullet$.

Proof. The same procedure as before, we just keep track of the degrees in addition. □

Example 8.3.2 with grading. The ideal $J \subset S = k[w, x, y, z]$ from above is generated by homogeneous forms of degree 2

$w^2 - xz$	$-x$	y	0	$-z$	0	$-y^2 + wz$
$wx - yz$	w	$-x$	$-y$	0	z	z^2
$x^2 - wy$	$-z$	w	0	$-y$	0	0
$xy - z^2$	0	0	w	x	$-y$	$-yz$
$y^2 - wz$	0	0	$-z$	$-w$	x	w^2
	0	y	$-x$	w	$-z$	1
	$-y^2 + wz$	z^2	$-wy$	yz	$-w^2$	x

and the resolution is graded:

$$0 \leftarrow S/J \leftarrow S \leftarrow S(-2)^5 \leftarrow S(-3)^5 \oplus S(-4) \leftarrow S(-4) \oplus S(-5) \leftarrow 0.$$

Exercise 8.3.6. Compute the ranks of the free modules F_i and the maps between them in the free resolution F of the following $R = K[x_1, \ldots, x_n]$-modules:

1) $K \cong K[x_1, \ldots, x_n]/(x_1, \ldots, x_n)$ and
2) $K[x_1, \ldots, x_n]/(x_1, \ldots, x_n)^2$.

Hint: Treat the special cases $n = 2, 3, 4$ first to guess the ranks in the general case, and/or make experiments with Macaulay2.

Exercise 8.3.7. Write pseudo code for the following algorithm

Algorithm 8.3.8 (Finite free resolution).
Input. A presentation matrix φ_0 of a finitely presented $k[x_1, \ldots, x_n]$-module M

Output. A finite free resolution $F_\bullet$ of M

Specify the monom order, the choice of the Gröbner basis elements and their ordering.

Introduce a concept of lower order terms. Can they be used to speed up the computation? See [23] for a solution.

Exercise 8.3.9. Let R be a Noetherian ring, and let M and N be finitely generated R-modules, and let $F_\bullet$ and $G_\bullet$ free resolutions of M and N, respectively. Prove:

1) Every R-module homomorphism $\varphi \colon M \to N$ extends to a **map of complexes**

$$\begin{array}{ccccccccccc}
\cdots & \xrightarrow{\partial_3} & F_2 & \xrightarrow{\partial_2} & F_1 & \xrightarrow{\partial_1} & F_0 & \longrightarrow & M & \longrightarrow & 0 \\
& & \downarrow{\varphi_2} & & \downarrow{\varphi_1} & & \downarrow{\varphi_0} & & \downarrow{\varphi} & & \\
\cdots & \xrightarrow{\partial'_3} & G_2 & \xrightarrow{\partial'_2} & G_1 & \xrightarrow{\partial'_1} & G_0 & \longrightarrow & N & \longrightarrow & 0,
\end{array}$$

i.e., all squares in this diagram commute. In particular, we have $\varphi_{i-1}\partial_i = \partial'_i\varphi_i$ for all $i \geq 1$.

2) Two extensions $(\varphi_i)_{i\in\mathbb{N}}$ and $(\varphi'_i)_{i\in\mathbb{N}}$ of φ differ by a **homotopy**, i.e., there exists $(h_i)_{i\in\mathbb{N}}$

$$\begin{array}{ccccccccccc}
\cdots & \xrightarrow{\partial_3} & F_2 & \xrightarrow{\partial_2} & F_1 & \xrightarrow{\partial_1} & F_0 & \longrightarrow & M & \longrightarrow & 0 \\
& {}^{h_2}\swarrow & \downarrow & {}^{h_1}\swarrow & \downarrow & {}^{h_0}\swarrow & \downarrow & & \downarrow{\varphi} & & \\
\cdots & \xrightarrow{\partial'_3} & G_2 & \xrightarrow{\partial'_2} & G_1 & \xrightarrow{\partial'_1} & G_0 & \longrightarrow & N & \longrightarrow & 0
\end{array}$$

such that

$$\varphi_0 - \varphi'_0 = \partial'_1 h_0 \quad \text{and} \quad \varphi_i - \varphi'_i = h_{i-1}\partial_i + \partial'_{i+1}h_i \quad \text{for } i \geq 1.$$

Definition 8.3.10. A **(chain) complex** of R-modules is a sequence

$$C_\bullet : \quad \cdots \xrightarrow{\partial_{i+2}} C_{i+1} \xrightarrow{\partial_{i+1}} C_i \xrightarrow{\partial_i} C_{i-1} \xrightarrow{\partial_{i-1}} \cdots$$

of R-module homomorphism such that $\partial_i \circ \partial_{i+1} = 0$ for all i. The submodules $Z_i = \ker \partial_i$ and $B_i = \operatorname{im} \partial_{i+1}$ of C_i are called the submodule of **cycles** and **boundaries**, respectively, and

$$H_i(C_\bullet) = Z_i/B_i$$

is called the **homology of the complex** $C_\bullet$ at C_i. The complex $C_\bullet$ is **exact at** C_i if $H_i(C_\bullet) = 0$. $C_\bullet$ is an **exact complex** if $H_i(C_\bullet) = 0$ for all i. The complex is **bounded** if only finitely many C_i are non-zero.

Remark 8.3.11. A non-exact complexes arise for example in the following way: Let

$$F_\bullet : \quad \cdots \xrightarrow{\partial_3} F_2 \xrightarrow{\partial_2} F_1 \xrightarrow{\partial_1} F_0 \longrightarrow M \longrightarrow 0$$

8.4 The Hilbert function

be a free resolution of an R-module M. Let N be a further R-module. Then $F_\bullet \otimes_R N$ is a complex which might have homology. One defines the **Tor groups** as

$$\operatorname{Tor}_i^R(M,N) = H_i(F_\bullet \otimes N).$$

Then $\operatorname{Tor}_0^R(M,N) = M \otimes_R N$.

Exercise 8.3.12. Use Exercise 8.3.9 to prove that the R-modules $\operatorname{Tor}_i^R(M,N)$ are well-defined, i.e., independent from the choice of a free resolution $F_\bullet$ of M.

8.4 The Hilbert function

Let $S = k[x_0, \ldots, x_n]$ be the standard graded polynomial ring in $n+1$ variables and let M be a finitely generated graded S-module. Then each M_d is a finite-dimensional k-vector space.

Definition 8.4.1. The function

$$h_M : \mathbb{Z} \to \mathbb{Z},\ d \mapsto h_M(d) = \dim_k M_d$$

is called the **Hilbert function** of M.

Example. The standard graded polynomial ring in $n+1$ variables has the Hilbert function

$$h_S(d) = \binom{d+n}{n}.$$

Proof. We have a bijection between subsets of cardinality n of a set with $n+d$ elements and monomials of degree d in $n+1$ variables.

$$\underbrace{\circ \ldots \circ}_{\alpha_0} \bullet \underbrace{\circ \ldots \circ}_{\alpha_1} \bullet \cdots \bullet \underbrace{\circ \ldots \circ}_{\alpha_n} \longleftrightarrow x^\alpha = x_0^{\alpha_0} \cdot \ldots \cdot x_n^{\alpha_n}.$$

$\square$

Theorem 8.4.2 (Polynomial nature of the Hilbert function). *Let $S = k[x_0, \ldots, x_n]$ be the standard graded polynomial ring in $n+1$ variables and let M be a finitely generated graded S-module. Then there exists a polynomial $p_M(t) \in \mathbb{Q}[t]$ and an integer d_0 such that*

$$h_M(d) = p_M(d)\ \text{for all}\ d \geq d_0.$$

The polynomial $p_M(t)$ is called the **Hilbert polynomial** of M.

Example.

$$p_S(t) = \frac{(t+n)(t+n-1) \cdot \ldots \cdot (t+1)}{n!}$$

and $p_S(t) = h_S(t) = \binom{t+n}{n}$ for all $t \geq -n$.

Proof. Let

$$0 \longleftarrow M \longleftarrow F_0 \xleftarrow{\varphi_1} F_1 \xleftarrow{\varphi_2} \cdots \xleftarrow{\varphi_{c-1}} F_{c-1} \xleftarrow{\varphi_c} F_c \longleftarrow 0$$

be a finite free resolution of M with $F_i = \oplus_j S(-j)^{\beta_{ij}}$. Then for each $d \in \mathbb{Z}$ the sequence

$$0 \leftarrow M_d \leftarrow (F_0)_d \leftarrow (F_1)_d \leftarrow \cdots \leftarrow (F_c)_d \leftarrow 0$$

is an exact complex of finite-dimensional k-vector spaces. Thus

$$\dim M_d = \sum_{i=0}^{c} (-1)^i \dim(F_i)_d$$

$$= \sum_{i=0}^{c} (-1)^i \sum_j \beta_{ij} \binom{d-j+n}{n}.$$

Interpreting the binomial coefficients as polynomials

$$\binom{t-j+n}{n} = \frac{(t-j+n) \cdot \ldots \cdot (t-j+1)}{n!} \in \mathbb{Q}[t],$$

the formula

$$p_M(t) = \sum_{i=0}^{c} (-1)^i \sum_j \beta_{ij} \binom{t-j+n}{n} \in \mathbb{Q}[t]$$

defines the Hilbert polynomial, and $h_M(d) = p_M(d)$ holds for all $d \geq d_0$ with

$$d_0 = \min\{j \mid \exists i \text{ with } \beta_{ij} \neq 0\}.$$

$\square$

Corollary 8.4.3. *The graded S-modules S/J and $S/\mathrm{Lt}(J)$ have the same Hilbert function and Hilbert polynomial.*

Proof. The graded Betti numbers of our resolution of S/J depend only on $\mathrm{Lt}(J)$. $\square$

Example 8.4.4 (Hypersurfaces). Let $X = V(f) \subset \mathbb{P}^n$ be a hypersurface defined by a (square free) homogeneous polynomial of degree d. Then

$$0 \longleftarrow S/(f) \longleftarrow S \xleftarrow{f} S(-d) \longleftarrow 0$$

is a free resolution and

8.4 The Hilbert function

$$p_{S/(f)}(t) = \binom{t+n}{n} - \binom{t-d+n}{n}$$

$$= \frac{t^n + \frac{n^2+n}{2}t^{n-1}}{n!} - \frac{t^n + (\frac{n^2+n}{2} - dn)t^{n-1}}{n!} + O(t^{n-2})$$

$$= d\frac{t^{n-1}}{(n-1)!} + \text{lower terms.}$$

In particular, we have

$$\deg p_{S/(f)} = n - 1 = \dim X$$

and the leading coefficient has the form $\frac{d}{(n-1)!}$.

Theorem 8.4.5. *Let $J \subset S = k[x_0, \ldots, x_n]$ be a homogeneous ideal, and denote by $X = V(J) \subset \mathbb{P}^n$ the algebraic set defined by J. The Hilbert polynomial of S/J has degree $r = \dim X$ and leading term*

$$d\frac{t^r}{r!}$$

for some positive integer d.

We call d the **degree** of J.

Definition 8.4.6. For a projective algebraic set $X \subset \mathbb{P}^n$ the **degree** is defined by

$$\deg X = \deg \mathrm{I}(X),$$

where $\mathrm{I}(X) \subset K[x_0, \ldots, x_n]$ denotes its homogeneous ideal.

Proof. Let $C(J) \subset \mathbb{A}^{n+1}$ be the cone defined by J. Since the Hilbert function of S/J depends only on $\mathrm{Lt}(J)$ we may assume that $k = K$ is algebraically closed. In particular, we may assume that k is an infinite field. Then there exists a triangular linear change of coordinates such that in these new coordinates J satisfies the assumption of the tower of projections theorem: There exists an r such that the projection $\mathbb{A}^{n+1} \to \mathbb{A}^{r+1}$ onto the last $r+1$ coordinates induces a finite surjection

$$C(J) \to \mathbb{A}^{r+1},$$

and each elimination ideal $J_\nu = K[x_\nu, \ldots, x_n] \cap J$ contains an x_ν-monic polynomial for $\nu = 0, \ldots, n-r-1$. Thus S/J is a finite $T = k[x_{n-r}, \ldots, x_n]$-module. As a graded T-module S/J has a finite free resolution

$$0 \longleftarrow S/J \longleftarrow G_0 \overset{\varphi_1}{\longleftarrow} G_1 \overset{\varphi_2}{\longleftarrow} \cdots \overset{\varphi_{c-1}}{\longleftarrow} G_{c'-1} \overset{\varphi_c}{\longleftarrow} G_{c'} \longleftarrow 0$$

of length $c' \le r+1$, where

$$G_i = \bigoplus_j T(-j)^{\beta'_{ij}}$$

is a free graded T-module with β'_{ij} generators in degree j. Hence

$$p_{S/J}(t) = \sum_{i=0}^{c'} (-1)^i \sum_j \beta'_{ij} \binom{t-j+r}{r}$$

is an alternating sum of polynomials of degree r and

$$p_{S/J}(t) = d\frac{t^r}{r!} + \text{lower terms}$$

with $d \in \mathbb{Z}$. To see that $d > 0$ holds, we note that $T \cdot 1 \subset S/J$ is a T-submodule. Thus

$$h_{S/J}(t) \geq h_T(t) = \binom{t+r}{r}$$

grows at least as fast as a polynomial of degree r for $t \to \infty$. It remains to identify r with the dimension of X. Consider the charts $U_i = \{x_i \neq 0\} \cong \mathbb{A}^n$ for $i = n-r, \ldots, n$ and the corresponding substitution homomorphism

$$\varphi_i : S \to K[x_0, \ldots, x_{i-1}, x_{i+1}, \ldots, x_n], \quad x_i \mapsto 1.$$

The ideal $\varphi_i(J)$ satisfies the assumption of the tower of projections theorem. Thus $X \cap U_i \to \mathbb{A}^r$ is a finite surjection and all the affine algebraic sets $X \cap U_i$ have dimension r. Since $\mathrm{rad}(J + (x_{n-r}, \ldots, x_n)) = (x_0, \ldots, x_n)$ due to the monic polynomials in the elimination ideals, we see that

$$V(J) \cap V(x_{n-r}, \ldots, x_n) = \emptyset, \text{ equivalently, } X \subset U_{n-r} \cup \ldots \cup U_n.$$

Thus $\dim X = r$ if we define $\dim X = \max\{\dim X \cap U_j \mid j = 0, \ldots, n\}$. □

Corollary 8.4.7. *Let $J \subsetneq K[x_0, \ldots, x_n]$ be a proper homogeneous ideal. The dimensions of the projective algebraic set $V(J) \subset \mathbb{P}^n$ and the affine cone $C(J) \subset \mathbb{A}^{n+1}$ differ by one, i.e.,*

$$\dim C(J) = \dim V(J) + 1.$$

Here we use the convention that $\dim \emptyset = -1$. □

Example 8.2.4 continued. Our computation (homogenized) of the projective closure $C \subset \mathbb{P}^3$ of the twisted cubic

$$\begin{vmatrix} x^2 - wy \\ xy - wz \\ y^2 - xz \end{vmatrix} \begin{vmatrix} -y & z \\ x & -y \\ -w & x \end{vmatrix} = (\varphi_1^t || \varphi_2)$$

shows that the homogeneous coordinate ring S_C of C has a free resolution

$$0 \longleftarrow S_C \longleftarrow S \xleftarrow{\varphi_1} S^3(-2) \xleftarrow{\varphi_2} S^2(-3) \longleftarrow 0.$$

Thus

8.4 The Hilbert function

$$p_C(t) = \binom{t+3}{3} - 3\binom{t+1}{3} + 2\binom{t}{3}$$

$$= \frac{1}{3!}[(1 - 3 + 2)t^3 + (1 \cdot (3 + 2 + 1) - 3 \cdot (1 - 1) + 2 \cdot (-1 - 2))t^2$$

$$+ ((3 \cdot 2 + 3 \cdot 1 + 2 \cdot 1) - 3(1 \cdot (-1)) + 2((-1) \cdot (-2)))t + 3 \cdot 2 \cdot 1]$$

$$= \frac{1}{3!}[(6 + 3 + 2 + 3 + 4)t + 6]$$

$$= 3t + 1$$

and deg $C = 3$.

An easier way to compute the Hilbert polynomial of C is by counting the monomials of degree d not contained in $\mathrm{Lt}(I_C) = (x^2, xy, y^2)$. By Macaulay's theorem 1.2.16 the monomials in $k[w, z] + xk[w, z] + yk[w, z]$ form a k-vector space basis of S/I_C. There are $(d + 1) + 2d = 3d + 1$ monomials of degree d in this set.

Definition 8.4.8. The coefficients of the Hilbert polynomial $p_X(t) \in \mathbb{Q}[t]$ are numerical invariants of a projective algebraic set X. Apart from the leading coefficient, the constant term $p_X(0)$ plays an important role. The values $p_X(d)$ for $d \in \mathbb{Z}$ are integers. This is clear for $d \gg 0$ by Theorem 8.4.2, from which it follows for all $d \in \mathbb{Z}$ by Exercise 8.4.9. In particular, $p_X(0) \in \mathbb{Z}$. For a variety $X \subset \mathbb{P}^n$ of dimension r we define the **arithmetic genus** $p_a(X)$ of X as

$$p_a(X) = (-1)^r(p_X(0) - 1), \text{ i.e., } p_X(0) = 1 + (-1)^r p_a(X).$$

The reason for this peculiar notation of the constant term becomes clear when one interprets the values $p_M(d)$ of the Hilbert polynomial of a module M for all $d \in \mathbb{Z}$ in terms Euler characteristics of coherent sheaves, see Proposition A.2.9. This is beyond the scope of this book.

Exercise 8.4.9 (Numerical polynomials). Let $p(t) \in \mathbb{Q}[t]$ be a polynomial which takes integral values $p(d)$ for all $d \in \mathbb{Z}$ with $d \geq d_0$ for some bound d_0. Prove that $p(t)$ can be written in the form

$$p(t) = a_r\binom{t+r}{r} + a_{r-1}\binom{t+r-1}{r-1} + \ldots + a_1\binom{t+1}{1} + a_0$$

with integers $a_r, \ldots, a_0$. Conversely, any polynomial of this form takes integral values for all $d \in \mathbb{Z}$. Hint: Consider the difference polynomial $(\Delta p)(t) = p(t) - p(t-1)$.

Chapter 9
Bézout's theorem

In $\mathbb{P}^2$ any two lines meet, which was one of the motivations for its construction. Much more is true. Bézout's theorem for plane curves says:

Any two curves C and H in $\mathbb{P}^2$ of degree d and e without a common component intersect counted with multiplicities in precisely $d \cdot e$ points.

In this chapter we define the intersection multiplicity $i(C, H; p)$ of two plane curves at a point p and prove two versions Bézout's theorem. The first is the one above, whose proof via resultants is delegated to the exercises. We formulate a lower bound for the intersection multiplicity by the product of the multiplicities of the defining equations of C and H at p.

Applications of this Bézout's theorem includes rational parametrization of plane curves with enough singularities, a topic which will finally be settled in Chapter 15.

A second version of a Bézout's theorem concerns the intersection of a projective variety $X \subset \mathbb{P}^n$ with a hypersurface $H = V(f)$. The intersection multiplicity of X and H along one of the components Z of $X \cap H$ is defined via the frequency in which the homogeneous prime ideal $\mathfrak{p} = I(Z)$ occurs in any filtration of the module $M = S/(I(X) + f)$.

9.1 Rational functions and regular functions on projective varieties

Definition 9.1.1. Let $X \subset \mathbb{P}^n$ be a projective variety, i.e., an irreducible algebraic subset of $\mathbb{P}^n$. Let $I(X) \subset S = K[x_0, \ldots, x_n]$ denote its homogeneous ideal and $S_X = S/I(X)$ its homogeneous coordinate ring. Then

$$K(X) = \{f = \frac{g}{h} \mid g \in S_X, h \in S_X \setminus \{0\} \text{ and } \deg g = \deg h\} \subset Q(S_X)$$

is called the **rational function field** of X. Notice that since $\deg f = \deg g$, the fraction $f = g/h$ defines a well-defined function

$$X \setminus V(h) \to K, p = (a_0 : \ldots : a_n) \mapsto \frac{g(a)}{h(a)}.$$

The rational function $f \in K(X)$ is defined at $p \in X$ if f has a representative g/h with $h(p) \neq 0$. We define the local ring of X at p by

$$\mathcal{O}_{X,p} = \{f \in K(X) \mid f = \frac{g}{h} \text{ with } h(p) \neq 0\}.$$

This is the ring of rational functions which are well-defined functions in some open neighborhood of $p \in X$.

Proposition 9.1.2. *Let $U_i \cong \mathbb{A}^n$ be an affine chart of $\mathbb{P}^n$ which intersects X. Then*

$$K(X \cap U_i) \cong K(X)$$

via dehomogenization and homogenization.

Proof. If $i = 0$, we have

$$f(x_0, \ldots, x_n) = \frac{g(x_0, \ldots, x_n)}{h(x_0, \ldots, x_n)} \mapsto f^a = \frac{g(1, x_1, \ldots, x_n)}{h(1, x_1, \ldots, x_n)}$$

and conversely

$$f = \frac{g(x_1, \ldots, x_n)}{h(x_1, \ldots, x_n)} \mapsto f^h = \frac{x_0^{\deg g + \deg h} g(x_1/x_0, \ldots, x_n/x_0)}{x_0^{\deg g + \deg h} h(x_1/x_0, \ldots, x_n/x_0)}.$$

Hence $(f^h)^a = f$ is clear, and $(f^a)^h = f$ holds because a possible common factor x_0 of the nominator and the denominator cancels. □

Corollary 9.1.3. *Let $p \in X \cap U_i$ and let $q = \varphi_i(p) \in A = \varphi_i(X \cap U_i) \subset \mathbb{A}^n$ be the corresponding point in the corresponding affine variety. Then*

$$\mathcal{O}_{X,p} \cong \mathcal{O}_{A,q}$$

via dehomogenization and homogenization. □

9.2 Intersection multiplicities for plane curves

Let $C = V(f)$ and $H = V(g)$ in $\mathbb{P}^2$ be two plane algebraic curves without a common component. For $p \in C \cap H$ we define the **intersection multiplicity of C and H at p** by

$$i(C, H; p) = i(f, g; p) = \dim_K \mathcal{O}_{\mathbb{P}^2, p} / (\frac{f}{\ell^d}, \frac{g}{\ell^e}) \mathcal{O}_{\mathbb{P}^2, p},$$

9.2 Intersection multiplicities for plane curves

i.e., as the K-vector space dimension of the quotient of the local ring by the ideal generated by $f/\ell^d, g/\ell^e$, where $d = \deg f$, $e = \deg g$ and $\ell \in K[x_0, x_1, x_2]$ is a linear form which does not vanish at p. This is well-defined because $\ell/\ell' \in \mathcal{O}_{\mathbb{P}^2, p}$ is a unit for any two linear forms ℓ, ℓ' which do not vanish at p. If $p \in U_0$ then we may take $\ell = x_0$ and

$$\mathcal{O}_{\mathbb{P}^2, p}/(\frac{f}{\ell^d}, \frac{g}{\ell^e})\mathcal{O}_{\mathbb{P}^2, p} \cong \mathcal{O}_{\mathbb{A}^2, q}/(f^a, g^a)\mathcal{O}_{\mathbb{A}^2, q}$$

for $q = \varphi_0(p)$ the corresponding point in the affine chart.

Example 9.2.1. Consider the plane affine curves defined by $f = y$ and $g = y - x^n$. The intersection number at the origin is

$$i(f, g; o) = \dim_K K[x, y]_{(x,y)}/(f, g) = \dim_K K[x, y]_{(x,y)}/(y, x^n)$$
$$= \dim_K (K[x, y]/(y, x^n))_{(x,y)} = \dim_K K[x, y]/(y, x^n) = n.$$

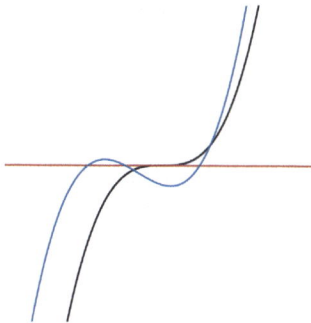

This makes a lot of sense, because g can be deformed to a function

$$g_t = y - (x - t_1) \cdot \ldots \cdot (x - t_n)$$

with n roots. In Section 14.3 we will prove that such a dynamical interpretation of the intersection numbers is always possible in the case of the ground field $K = \mathbb{C}$.

Example 9.2.2. For $f = y^2 - x^3$ and $g = x^2 - y^3$ we obtain

$$\mathcal{O}_{\mathbb{A}^2, o}/(f, g) \cong \mathcal{O}_{\mathbb{A}^2, o}/(y^2 - x^3, x^2 - yx^3)$$
$$\cong \mathcal{O}_{\mathbb{A}^2, o}/(y^2 - x^3, x^2(1 - yx))$$
$$\cong \mathcal{O}_{\mathbb{A}^2, o}/(y^2 - x^3, x^2)$$
$$\cong \mathcal{O}_{\mathbb{A}^2, o}/(y^2, x^2) = K[x, y]/(y^2, x^2).$$

Hence $i(f, g; o) = 4$.

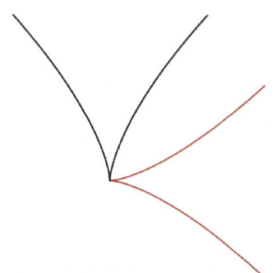
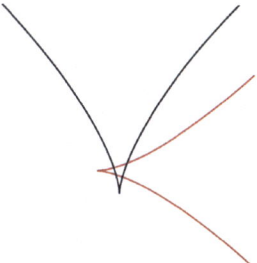

Example 9.2.3. For $f = y^2 - x^3$ and $g = y^2 - 8x^3$ we obtain

$$\mathcal{O}_{\mathbb{A}^2,o}/(f,g) \cong \mathcal{O}_{\mathbb{A}^2,o}/(y^2 - x^3, y^2 - 8x^3)$$
$$\cong \mathcal{O}_{\mathbb{A}^2,o}/(y^2, x^3).$$

Hence $i(f,g;o) = 6$. If we perturb g a little bit $g_t = y^2 - 8(x - t^2)(x - t)^2$, then $V(f, g_t)$ has six intersection points which approach the origin o for $t \to 0$.

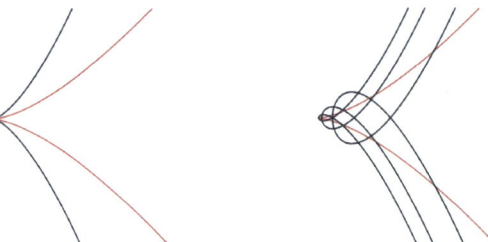

Definition 9.2.4. Let $p \in V(f) \subset \mathbb{P}^2$ be a point. After a change of coordinates we may assume that p corresponds to the origin $o \in \mathbb{A}^2 \cong U_0 \subset \mathbb{P}^2$. Suppose

$$f^a = f_m + \ldots + f_d \text{ with } f_j \in K[x,y]_j,$$

i.e., f_j is homogeneous of degree j, and suppose that f_m is not the zero polynomial. Then we say that f has **multiplicity** m at p,

$$\text{mult}_p(f) = m.$$

The homogeneous binary polynomial f_m factors into linear forms ℓ_k. We have

$$f_m = \prod_{k=1}^{m} \ell_k.$$

We call the lines $L_k = V(\ell_k)$ the **tangent lines** of $V(f)$ at p. If they are pairwise distinct, then we call p an **ordinary m-fold point** of $V(f)$.

Example 9.2.5. The curve $V(y^2 - x^2 - x^3)$ has an ordinary double point at o with tangent lines $L_1 = V(y - x)$ and $L_2 = V(y + x)$. The curve $V(y^2 - x^3)$ has a non-ordinary double point.

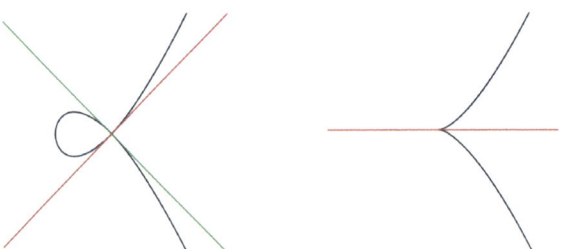

Definition 9.2.6. If $\text{mult}_p(f) = 1$, then $p \in C = V(f)$ is called a **smooth point** of C. Otherwise p is called a **singular point** of C.

9.2 Intersection multiplicities for plane curves

Remark. Suppose $K = \mathbb{C}$. If $\text{mult}_p(f) = 1$, then the zero locus coincides locally in the euclidean topology of $\mathbb{A}^2(\mathbb{C}) = \mathbb{C}^2$ with the graph of a holomorphic function by the implicit function theorem.

Exercise 9.2.7. Solve the following modification of Exercise I.5.1 from Hartshorne [42]. Consider the plane curves defined by

$$y^2 = (1 - x^2)^3, \quad y^2 = x^4 - x^6, \quad y^3 - 3x^2 y = (x^2 + y^2)^2, \quad y^2 = x^2 - x^4$$

Their real points are one of the following:

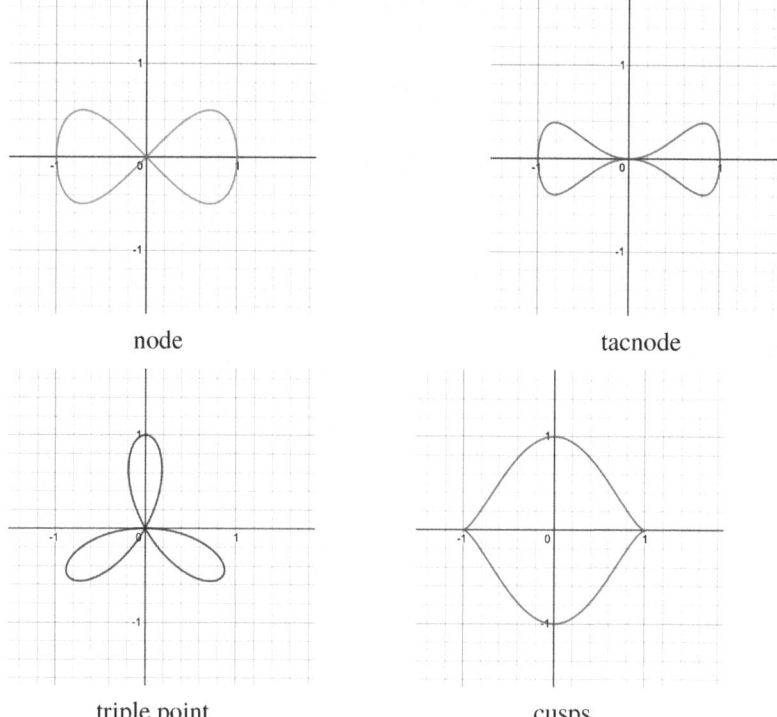

node   tacnode

triple point   cusps

Which is which?

Exercise 9.2.8. Let $f, g, h \in K[x, y]$ be polynomials such that f and gh have no common factor and let $p \in \mathbb{A}^2$ be a point. Prove:

$$i(f, gh; p) = i(f, g; p) + i(f, h; p).$$

Hint: Establish that the sequence

$$0 \to \mathcal{O}_{\mathbb{A}^2, p}/(f, h)\mathcal{O}_{\mathbb{A}^2, p} \to \mathcal{O}_{\mathbb{A}^2, p}/(f, gh)\mathcal{O}_{\mathbb{A}^2, p} \to \mathcal{O}_{\mathbb{A}^2, p}/(f, g)\mathcal{O}_{\mathbb{A}^2, p} \to 0$$

is exact, where the right-hand map is the natural projection and the left-hand map is given by

$$\bar{z} \mapsto \overline{zg}.$$

9.3 Bézout's theorem for plane curves

Theorem 9.3.1 (Bézout's theorem). *Let $C = V(f)$ and $H = V(g)$ in $\mathbb{P}^2$ be two plane curves of degree d and e. Counted with multiplicities C and D intersect in precisely $d \cdot e$ points:*

$$\sum_{p \in C \cap H} i(C, H; p) = d \cdot e.$$

Example 9.3.2. In Exercise 4.4.6 we saw a conic and a cubic which intersect in 3 points each with multiplicity 2. So there are no further intersection points by Bézout.

A proof of this theorem using resultants is outlined in the exercises.

Remark 9.3.3. 1) If $p \notin C \cap H$, then $i(C, H; p) = 0$ because f or g gives a unit in $\mathcal{O}_{\mathbb{P}^2, p}$.
2) If $i(C, H; p) = 1$, then we say C and H **intersect transversally at** p. In that case both C and H are smooth at p and have different tangent lines, because $\dim K[x, y]_1$ is two-dimensional.

Example 9.2.2 continued. For $f = y^2 - x^3$ and $g = x^2 - y^3$ we have intersection multiplicity 4 at $o \in \mathbb{A}^2 \cong U_2 = \{z \neq 0\} \subset \mathbb{P}^2$. One further intersection point is $p = (1 : 1 : 1) \in U_2$.

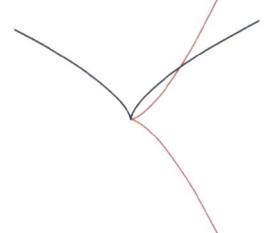

So there should be

$$4 = 3 \cdot 3 - 4 - 1$$

further intersection points. Indeed these are the points with coordinates $(\zeta^2 : \zeta^3 : 1)$, where ζ is any of the four non-trivial fifth roots of unity in $K = \mathbb{C}$.

Example 9.2.3 continued. For $f = y^2 - x^3$ and $g = y^2 - 8x^3$ we obtain intersection multiplicity 6 at $o \in \mathbb{A}^2 \subset \mathbb{P}^2$. So we are missing 3 intersection points. They lie on the line at infinity: In the chart $U_1 = \{y \neq 0\}$ we have the equations $z - x^3, z - 8x^3$, and the intersection multiplicity at $p = (0 : 1 : 0)$ is 3.

Theorem 9.3.4 (A lower bound on the intersection multiplicity). *Let $f, g \in K[x, y]$ be polynomials without a common factor which vanish at the origin $o \in \mathbb{A}^2$. Then*

$$i(f, g; o) \geq \mathrm{mult}_o(f) \, \mathrm{mult}_o(g)$$

and equality holds if and only if $V(f)$ and $V(g)$ have no common tangent line at o.

We will prove this with a Gröber basis computation in local rings in Chapter 10.

Example 9.3.5 (Computation of a rational parametrization of a plane curve). Consider the plane curve C defined by $f = -3x^5 - 2x^4y - 3x^3y^2 + xy^4 + 3y^5 + 6x^4 + 7x^3y + 3x^2y^2 - 2xy^3 - 6y^4 - 3x^3 - 5x^2y + xy^2 + 3y^3$.

9.3 Bézout's theorem for plane curves

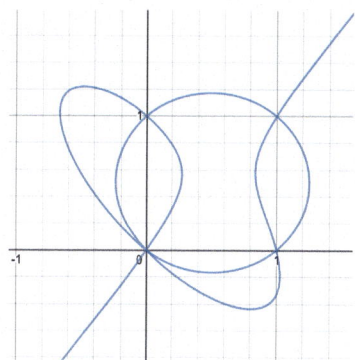

The curve $V(f)$ has a triple point at the origin and double points at the points with coordinates $(0,1), (1,0), (1,1)$.

Consider now the pencil of conics through these four points

$$D_t = V(t(x^2 - x) + y^2 - y).$$

The curve D_t intersects C with intersection multiplicity 3 at the origin and intersection multiplicity 2 at the double points.

Thus by Bézout

$$2 \cdot 5 - 3 - 2 - 2 - 2 = 1,$$

there remains one moving intersection point $p(t)$. Computing the coordinates of this point gives a rational parametrization of C. The final result is $p(t) = (x(t), y(t))$ with

$$x(t) = \frac{9t^5 - 3t^4 - 21t^3 + 11t^2 + 10t - 6}{9t^5 + t^4 - 6t^3 + 3t^2 - 14t + 9}$$

and

$$y(t) = \frac{-3t^5 - 8t^4 + 17t^3 + 9t^2 - 24t + 9}{9t^5 + t^4 - 6t^3 + 3t^2 - 14t + 9}.$$

In Section 15.1.2 we will see that plane curves with sufficiently many singular points can be parametrized rationally.

Exercise 9.3.6. Use Macaulay2 to compute the following.

1) The rational parametrization of the curve from Example 9.3.5.
2) A rational parametrization of the plane quartic curve $V(f) \subset \mathbb{A}^2$ defined by

$$f = -2x^4 - 2x^3y + x^2y^2 + 3xy^3 + 4y^4 + 4x^3 + x^2y - 4xy^2 - 8y^3 - 2x^2 + xy + 4y^2.$$

Hint: The zero locus $V(f)$ contains the points with coordinates $(0,0), (1,0), (0,1)$ and $(1,1)$.

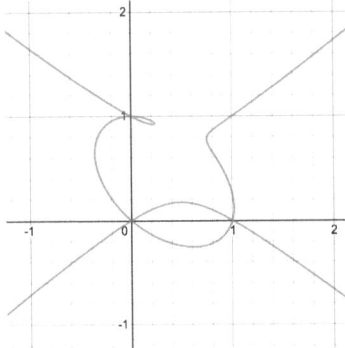

Exercise 9.3.7. Compute a rational parametrization of the curve C from Exercise 1.4.11. Verify the statement about the red points of C_1.
Hint: Start by computing a rational parametrization of C_2.

Exercise 9.3.8. Compute the equation of the image C of
$$\varphi : \mathbb{P}^1 \to \mathbb{P}^2, (t_0 : t_1) \mapsto (t_0^4 : t_0^3 t_1 - t_0 t_1^3 : t_1^4).$$

Where are the singular points of C?

Exercise 9.3.9. Give examples of two smooth plane conics which intersect in points with multiplicities (a) $1, 1, 1, 1$, (b) $2, 1, 1$, (c) $2, 2$, (d) $3, 1$ and (e) 4.

Exercise 9.3.10. Let
$$f = a_0 x^d + a_1 x^{d-1} + \ldots + a_d$$
$$g = b_0 x^e + b_1 x^{e-1} + \ldots + b_e$$

be two polynomials in $K[x]$ of degree d and e. Consider the $(d+e) \times (d+e)$ **Sylvester matrix**

$$\mathrm{Syl}(f,g) = \begin{pmatrix} a_0 & 0 & \cdots & 0 & b_0 & 0 & \cdots & 0 \\ a_1 & a_0 & & \vdots & b_1 & b_0 & & \vdots \\ \vdots & a_1 & \ddots & \vdots & \vdots & b_1 & \ddots & \vdots \\ \vdots & \vdots & \ddots & a_0 & \vdots & \vdots & \ddots & b_0 \\ a_d & & & a_1 & b_e & & & b_1 \\ 0 & a_d & & \vdots & 0 & b_e & & \vdots \\ \vdots & & \ddots & \vdots & \vdots & & \ddots & \vdots \\ 0 & 0 & \cdots & a_d & 0 & 0 & \cdots & b_e \end{pmatrix}.$$

There are e columns with entries a_i's and d columns with entries b_j's. Prove

1) f and g have a common root if and only if the **resultant**

9.3 Bézout's theorem for plane curves

$$\mathrm{Res}(f,g) = \det \mathrm{Syl}(f,g) = 0$$

of f and g vanishes.

2) Suppose that the a_i's and b_j's are independent variables of degree i and j, respectively. Prove that the resultant

$$\mathrm{Res}(f,g) \in \mathbb{Z}[a_i, b_j]$$

is a homogeneous polynomial of degree $d \cdot e$.

Exercise 9.3.11. Let $f, g \in K[x, y, z]$ be homogeneous forms of degree d and e without common factors and let $C = V(f)$ and $D = V(g)$ denote their zero loci. Assume that the coordinate system is chosen general enough such that

1) $(1:0:0) \notin C \cap D$,
2) The intersection points $C \cap D$ map to different points on $\mathbb{P}^1$ under the projection

$$\pi \colon \mathbb{P}^2 \setminus \{(1:0:0)\} \to \mathbb{P}^1, (a:b:c) \mapsto (b:c),$$

3) None of the intersection points lies on the line $V(z)$.

Write $f = a_0 x^d + a_1 x^{d-1} + \ldots + x + a_d$ and $g = b_0 x^e + b_1 x^{e-1} + \ldots + b_e$ with $a_i, b_j \in K[y, z]$ homogeneous of degree i and j, respectively. Then $\mathrm{Res}(f, g) \in K[y, z]$ is a homogeneous polynomial of degree $d \cdot e$ and the intersection points are mapped to zeroes of the resultant by Exercise 9.3.10. Prove that the multiplicities match:

$$i(f, g; p) = \mathrm{mult}_q \mathrm{Res}(f, g) \text{ for } p = (a : b : c) \text{ and } q = (b : c).$$

Hint: By assumption 3) we may restrict our attention to the open set $U_{z=1}$. Prove:

i)
$$K[x, y]/(f^a, g^a) \cong \bigoplus_{p' \in C \cap D} \mathcal{O}_{\mathbb{A}^2, p'}/(f^a, g^a)\mathcal{O}_{\mathbb{A}^2, p'}$$

by using Exercise 3.4.3.

ii) The Sylvester matrix $M = \mathrm{Syl}(f, g)(y, 1)$ is a presentation matrix of $K[x, y]/(f^a, g^a)$ as a $K[y]$-module.

iii) Let $h \in K[y]$ be a polynomial which vanishes at all points $q' = \pi(p') \in \pi(C \cap D)$ except at $q = \pi(p)$. Then

$$\mathrm{coker}\, M[h^{-1}] \cong \mathcal{O}_{\mathbb{A}^2, p}/(f^a, g^a)\mathcal{O}_{\mathbb{A}^2, p}.$$

iv) Consider the Smith normal form of the matrix $M[h^{-1}]$ over the PID $K[y, h^{-1}]$. Every elementary divisor of $M[h^{-1}]$ is up to a unit a power of $t = cy - b$.

Conclude $i(f, g; p) = \dim_K \mathrm{coker}\, M[h^{-1}] = \mathrm{mult}_q \mathrm{Res}(f, g)$.

9.4 Bézout's theorem for the intersection with a hypersurface

Theorem 9.4.1. *Let $X \subset \mathbb{P}^n$ be a projective variety and $H = V(g)$ a hypersurface of degree e which does not contain X. Let $Z_1, \ldots, Z_r$ be the irreducible components of $X \cap H$. Then*

$$\deg X \cdot \deg H = \sum_{i=1}^{r} i(X, H; Z_i) \deg Z_i.$$

We will see how to define the **intersection multiplicity** $i(X, H; Z_i)$ of X and H **along** Z_i in the course of the proof.

The proof is built upon the computation of the Hilbert polynomial of S/J for the ideal $J = I(X) + (g)$ in two ways.

First computation of $p_{S/J}(t)$. Since X is a variety, $I(X)$ is a prime ideal and since $g \notin I(X)$, it gives a non-zero-divisor in $S_X = S/I(X)$. Hence

$$0 \longleftarrow S/J \longleftarrow S_X \xleftarrow{g} S_X(-e) \longleftarrow 0$$

is a short exact sequence. Since

$$p_X(t) = p_{S_X}(t) = \deg X \frac{t^r}{r!} + \text{lower terms},$$

where $r = \dim X$, we obtain

$$p_{S/J}(t) = p_X(t) - p_X(t - e)$$
$$= \deg X \frac{ret^{r-1}}{r!} + \text{lower terms}$$
$$= \deg X \deg H \frac{t^{r-1}}{(r-1)!} + \text{lower terms}.$$

Hence $\dim V(J) = r - 1$ and $\deg J = \deg X \cdot \deg H$.

For the second computation we consider the filtration of the S-module $M = S/J$ by quotients of prime ideals from Theorem 3.4.1. Since M is graded, all associated primes turn out to be graded as well.

We start by proving that a non-zero graded module M has at least one homogeneous associated prime.

Let $m \in M_d$ be a non-zero homogeneous element of degree d. Then the ideal $\text{ann}(m)$ is homogeneous as well, and the map

$$S(-d) \to M, f \mapsto fm$$

induces an inclusion $S/\text{ann}(m)(-d) \hookrightarrow M$. A maximal element in the set

$$\mathcal{M} = \{\text{ann}(m) \mid m \in M \setminus \{0\} \mid m \text{ is homogeneous}\}$$

9.4 Bézout's theorem for the intersection with a hypersurface

is a prime ideal. Since S is Noetherian $\mathcal{M}$ contains a maximal element. Hence M has a homogeneous associated prime.

Proposition 9.4.2 (Filtration in the graded case). *Let M be a finitely generated graded S-module. Then M has a filtration*

$$0 = M_0 \subset M_1 \subset \ldots \subset M_N = M$$

by homogeneous submodules with quotients

$$M_i/M_{i-1} \cong S/\mathfrak{p}_i(-d_i)$$

for homogeneous prime ideals $\mathfrak{p}_i$ and integers d_i.

Proof. Let $m_1 \in M_{d_1}$ be a homogeneous element whose annihilator is a prime $\mathfrak{p}_1$. Then we can take $M_1 = Sm_1$. If $M_{\nu-1} \subset M$ is already constructed and $M_{\nu-1} \neq M$, we consider an associated prime $\mathfrak{p}_\nu = \mathrm{ann}(\overline{m}_\nu)$ of a homogeneous element $\overline{m}_\nu \in M/M_{\nu-1}$ and take M_ν as the preimage of $S/\mathfrak{p}_\nu(-d_\nu) \hookrightarrow M/M_{\nu-1}$ in M. This process stops with an $M_N = M$ since M is Noetherian. □

Corollary 9.4.3. *The associated primes of a finitely generated graded S-module are homogeneous.*

Proof. $\mathrm{Ass}(M) \subset \{\mathfrak{p}_1, \ldots, \mathfrak{p}_N\}$ by Proposition 3.3.5. □

Second computation of $p_{S/J}(t)$. Consider $M = S/J$ and a filtration

$$0 = M_0 \subset M_1 \subset \ldots \subset M_N = M$$

with quotients

$$M_i/M_{i-1} \cong S/\mathfrak{p}_i(-d_i)$$

for homogeneous prime ideals $\mathfrak{p}_i$ and integers d_i. The Hilbert functions and Hilbert polynomials are additive in short exact sequences.

Proposition 9.4.4. *If*

$$0 \to M' \to M \to M'' \to 0$$

is a short exact sequence of graded S-modules, then

$$h_M = h_{M'} + h_{M''}.$$

□

Hence we obtain

$$p_M(t) = \sum_{j=1}^{N} p_{S/\mathfrak{p}_j}(t - d_j).$$

Proof of Theorem 9.4.1. Comparing both formulas, we obtain $\dim V(\mathfrak{p}_j) \leq r - 1$ for all $\mathfrak{p}_j$ since $p_M(t)$ has degree $r - 1$. Only those with equality contribute to the leading coefficient. The minimal associated primes correspond to the irreducible components Z_j of $X \cap H$.

Thus
$$\deg X \cdot \deg H = \sum_{Z_j \text{ with } \dim Z_j = r-1} i(X, H; Z_j) \deg Z_j,$$

if we define
$$i(X, H; Z_j) = \begin{cases} |\{\nu \mid \mathfrak{p}_\nu = I(Z_j)\}| & \text{if } \dim Z_j = r - 1 \\ 0 & \text{if } \dim Z_j < r - 1. \end{cases}$$

Actually, $\dim Z_j = r - 1$ holds for every component of $X \cap H$. This follows from Krull's principal ideal theorem 10.5.5. □

Example 9.4.5 (Touching surfaces). Consider

$$f = \det \begin{pmatrix} x_0 & x_1 \\ x_1 & x_2 \end{pmatrix} \text{ and } g = \det \begin{pmatrix} x_0 & x_1 & x_2 \\ x_1 & x_2 & x_3 \\ x_2 & x_3 & 0 \end{pmatrix} \in S = K[x_0, \ldots, x_3].$$

We have $\mathrm{Lt}(f) = -x_1^2$ and $\mathrm{Lt}(g) = -x_2^3$. Thus f, g form a Gröbner basis of $J = (f, g)$, J is unmixed by Theorem 6.1.1 and $A = V(J) \subset \mathbb{P}^3$ is a curve. The twisted cubic C defined by the 2×2-minors of

$$\begin{pmatrix} x_0 & x_1 & x_2 \\ x_1 & x_2 & x_3 \end{pmatrix}$$

is a component of A. Thus $\mathfrak{p}_1 = (x_1^2 - x_0 x_2, x_1 x_2 - x_0 x_3, x_2^2 - x_1 x_3)$ is one of the associated primes of J. Actually, $\mathrm{rad}(J) = \mathfrak{p}_1 = I(C)$. The graded S-module $\mathrm{Hom}(S/\mathfrak{p}_1, S/J)$ is generated in degree 2 with generators corresponding to multiplication by $x_2^2 - x_1 x_3$ and $x_1 x_2 - x_0 x_3$, respectively. Thus we may take

$$M_1 = S/\mathfrak{p}_1(-2) \hookrightarrow M = S/J$$

as the first step of a filtration in the sense of Proposition 9.4.2 with the inclusion induced by multiplication by $x_2^2 - x_1 x_3$. The annihilator of the image of $x_1 x_2 - x_0 x_3$ in M/M_1 is the prime ideal $\mathfrak{p}_2 = (x_1, x_2, x_3)$. Thus we may take $M_2 \subset M$ as the preimage of the image $S/\mathfrak{p}_2(-2) \hookrightarrow M/M_1$ and $M = M_3$ because $M/M_2 \cong S/\mathfrak{p}_1$. So in the filtration

$$0 \subset M_1 \subset M_2 \subset M_3 = M$$

the prime ideal $\mathfrak{p}_1$ occurs twice and $\mathfrak{p}_2$ occurs once. We conclude: the surfaces $Q = V(f)$ and $H = V(g)$ intersect with multiplicity 2 along C and there are no further intersections since $\deg Q \cdot \deg H = 2 \cdot 3$ and $2 \deg C = 2 \cdot 3$ coincide.

Remark 9.4.6. Although $\mathfrak{p}_2$ is not an associated prime of J it is not possible to avoid points in the filtration entirely. The graded module

$$\mathrm{Hom}(S/\mathfrak{p}_1, M/M_1)$$

9.4 Bézout's theorem for the intersection with a hypersurface

is generated by three elements corresponding to multiplication by generators of $\mathfrak{p}_2$. Since

$$p_{S/\mathfrak{p}_1(-2)}(t) + p_{S/\mathfrak{p}_1(-1)}(t) = 3(t-2) + 1 + 3(t-1) + 1 = 6t - 7$$

does not coincide with $p_{S/J}(t) = 6t - 3$, we need 4 further steps corresponding to points to complete the filtration in this case. If we try to work with $S/\mathfrak{p}_1(-d_1)$ and $S/\mathfrak{p}_1(-d_2)$ with $d_1 \geq 2$ and $d_2 \geq 1$ we need even more points.

Since the filtration S/J is not unique, it is not clear that the intersection multiplicity

$$i(X, H; Z_j) = |\{v \mid \mathfrak{p}_v = I(Z_j)\}|$$

is well-defined. This can be established via localization in $I(Z_j)$, see [18].

Exercise 9.4.7. Verify the computations in Example 9.4.5 with Macaulay2.

Chapter 10
Local rings and power series

Local rings are easier to handle because of Nakayama's lemma which we prove in Section 10.1. In Section 10.2 we introduce formal power series rings and completions. We prove Grauert's division theorem for $k[[x_1,\ldots,x_n]]$, the Weierstrass preparation theorem and $\dim k[[x_1,\ldots,x_n]] = n$. Section 10.3 contains a Gröbner basis proof of the lower bound for the intersection multiplicity of plane curves in terms of their multiplicities. Since Grauert's division theorem does not give an algorithm, but rather only a convergent procedure, we introduce Mora's division algorithm for applications in $k[x_1,\ldots,x_n]_{(x_1,\ldots,x_n)} \subset k[[x_1,\ldots,x_n]]$ in Section 10.4. In Section 10.5 we introduce the tangent space $T_p A$ and prove the Jacobian criterion for smoothness using Krull's principal ideal theorem, for which we give no proof but refer to the literature. We compare the tangent space $T_p A$, the tangent cone $gr_\mathfrak{m} \mathcal{O}_{A,p}$, which can be computed by Mora's algorithm, and the completion $\widehat{\mathcal{O}_{A,p}}$. In the final section we speak about discrete valuation rings.

10.1 Local rings

Definition 10.1.1. A **local ring** is a ring R which has a unique maximal ideal $\mathfrak{m}$. The **residue field** of the local ring R is $k = R/\mathfrak{m}$. We write $(R, \mathfrak{m})$ or even $(R, \mathfrak{m}, k)$ if we want to specify the notation for the maximal ideal and residue field of a local ring.

Examples 10.1.2. 1) Let R be a ring and $\mathfrak{p}$ a prime ideal. Then the localization

$$R_\mathfrak{p} = \{\frac{g}{h} \mid h \notin \mathfrak{p}\}$$

is a local ring with maximal ideal

$$\mathfrak{m} = \mathfrak{p} R_\mathfrak{p} = \{\frac{g}{h} \mid g \in \mathfrak{p}, h \notin \mathfrak{p}\}$$

and residue field
$$R_\mathfrak{p}/\mathfrak{p}R_\mathfrak{p} \cong Q(R/\mathfrak{p})$$
the quotient field of the integral domain $R/\mathfrak{p}$.

2) $\mathcal{O}_{\mathbb{A}^n,o} = K[x_1,\ldots,x_n]_{(x_1,\ldots,x_n)}$ has a residue field isomorphic to K. In general, the residue field $R/\mathfrak{m}$ is not a subring of R.

A local Noetherian ring $(R,\mathfrak{m})$ is easier to handle than general rings since every element $f \notin \mathfrak{m}$ is a unit in R

Lemma 10.1.3 (Nakayama's Lemma). *Let $(R,\mathfrak{m})$ be a local ring and let $N \subset M$ be a submodule of a finitely generated R-module M. Then*
$$N + \mathfrak{m}M = M \text{ if and only if } N = M.$$

Proof. By replacing M by M/N, we reduce to the case $N = 0$. So we have to prove $\mathfrak{m}M = M \implies M = 0$. The other direction is trivial. Let $m_1,\ldots,m_r$ denote generators of M. Since $\mathfrak{m}M = M$ we find expressions
$$m_i = \sum_{j=1}^{r} g_{ij} m_j \text{ with } g_{ij} \in \mathfrak{m}.$$

In matrix notation
$$(E - B) \begin{pmatrix} m_1 \\ \vdots \\ m_r \end{pmatrix} = 0 \text{ with } B = (g_{ij}).$$

Multiplication by the cofactor matrix of $E - B$ yields $\det(E - B)m_i = 0$ for all i. Since $\det(E - B) \equiv 1 \mod \mathfrak{m}$, the determinant is a unit. Hence $m_i = 0$ for all i and $M = 0$. □

Corollary 10.1.4. *Let $(R,\mathfrak{m},k)$ be a local ring and let $m_1,\ldots,m_r \in M$ be elements of a finitely generated R-module M. Then $m_1,\ldots,m_r$ generate M if and only if $\overline{m}_1,\ldots,\overline{m}_r$ span the k-vector space $M/\mathfrak{m}M$.*

Proof. We consider the submodule $N = Rm_1 + \cdots + Rm_r \subset M$.
$$N + \mathfrak{m}M = M$$
holds if and only if $\overline{m}_1,\ldots,\overline{m}_r \in M/\mathfrak{m}M$ generate $M/\mathfrak{m}M$. Since $M/\mathfrak{m}M$ is a $k = R/\mathfrak{m}$-vector space, the result follows. In particular, any **minimal set of generators** has precisely $\dim_k M/\mathfrak{m}M$ elements. □

Theorem 10.1.5 (Krull's intersection theorem). *Let $(R,\mathfrak{m})$ be a Noetherian local ring. Then*
$$\bigcap_{i=1}^{\infty} \mathfrak{m}^i = (0).$$

10.1 Local rings

Proof. Consider the subring

$$S = R[\mathfrak{m}t] = R \oplus \mathfrak{m}t \oplus \mathfrak{m}^2 t^2 \oplus \ldots \subset R[t].$$

Since $\mathfrak{m}$ is a finitely generated ideal in R, S is a finitely generated (graded) R-algebra, hence Noetherian as well by Exercise 3.1.10. Consider now $J = \bigcap_{i=1}^{\infty} \mathfrak{m}^i$. Then the ideal

$$J \oplus Jt \oplus Jt^2 \oplus \ldots \subset S$$

is generated by finitely many homogeneous elements. Let r be the maximal degree of a generator. Hence

$$\mathfrak{m}t Jt^r = Jt^{r+1}.$$

Thus $\mathfrak{m}J = J$ and $J = 0$ follows from Nakayama's lemma 10.1.3. □

Exercise 10.1.6. Let $(R, \mathfrak{m})$ be a local Noetherian ring and let M be a finitely generated R-module. A free resolution

$$\ldots \xrightarrow{\partial_3} F_2 \xrightarrow{\partial_2} F_1 \xrightarrow{\partial_1} F_0 \xrightarrow{\partial_0} M \xrightarrow{\partial_{-1}} 0$$

is **minimal** if at each step we choose a minimal set of generators of $\ker \partial_{i-1}$ and a free module F_i whose basis maps to these generators. Prove:

1) A resolution $F_\bullet$ is minimal if and only if the matrices describing ∂_i for $i \geq 1$ have entries in $\mathfrak{m}$.
2) The minimal free resolution of M is uniquely determined up to an isomorphism of complexes.

Exercise 10.1.7. Let $R = \oplus_{d \geq 0} R_d$ be a finitely generated graded k-algebra such that $\mathfrak{m} = R_+ = \oplus_{d > 0} R_d$ is a maximal ideal with residue field $k = R_0$. Let M be a finitely generated graded R-module.

1) Prove Nakayama's lemma in the graded case: If $N \subset M$ is a graded submodule, then

$$M = N + \mathfrak{m}M \implies N = M.$$

2) Conclude that the minimal graded free resolution of M is unique up to isomorphism.

Deduce that the graded Betti numbers β_{ij} of the minimal free resolution $F_\bullet$ of a finitely generated graded module M over the standard graded polynomial ring $S = k[x_0, \ldots, x_n]$ are invariants of M.

Exercise 10.1.8. A monomial ideal $I \subset S = k[x_0, \ldots, x_n]$ is called **Borel fixed** if

$$x_i x^\alpha \in I \implies x_j x^\alpha \in I \quad \text{holds for all monomials } x^\alpha \text{ and all } j < i.$$

Prove that the algorithm from the proof of Theorem 8.3.5 computes the minimal free resolution. Eliahou and Kervaire (1990) [22] even gave a description of the matrices in the minimal free resolution.

In general, the algorithm from Theorem 8.3.5 does not produce a minimal resolution. Give an example where the resolution is not minimal.

Exercise 10.1.9. 1) Consider the ideal I generated by

$$(x_0x_1x_2, x_1x_2x_3, x_0x_1x_4, x_0x_3x_4, x_2x_3x_4, x_0x_2x_5, x_0x_3x_5, x_1x_3x_5, x_1x_4x_5, x_2x_4x_5).$$

Prove that the minimal free resolution of I as a $K[x_0, \ldots, x_5]$-module depends on the characteristic of the ground field. The ranks of the free modules in char$(K) = 2$ and char$(K) = 0$ are different.

An explanation of this phenomenon is given by Hochster's theory (1977) [46] of Stanley-Reisner rings. Let Δ be a simplicial complex with $n + 1$ vertices. We may regard Δ as a subcomplex of the standard n-simplex Δ_n, which by definition is the convex hull of the coordinate vectors $e_i \in \mathbb{R}^{n+1} = \mathbb{A}^{n+1}(\mathbb{R})$. The square-free monomial ideal I_Δ of Δ is the vanishing ideal of the cone over Δ with vertex $0 \in \mathbb{R}^{n+1}$. Since I_Δ is a monomial ideal its generators generate an ideal $I_\Delta^K \subset K[x_0, \ldots, x_n]$ for any field K. The coordinate ring $K[\Delta] = K[x_0, \ldots, x_n]/I_\Delta^K$ is called the **Stanley-Reisner ring** of Δ over K. Conversely, any square-free monomial ideal $J \subset \mathbb{Q}[x_0, \ldots, x_n]$ defines a simplicial complex by intersecting the cone $C(J) \subset \mathbb{A}^{n+1}(\mathbb{C})$ over $V(J) \subset \mathbb{P}^n(\mathbb{C})$ with the standard n-simplex $\Delta_n \subset \mathbb{R}^{n+1} \subset \mathbb{C}^{n+1} = \mathbb{A}^{n+1}(\mathbb{C})$.

2) Prove

$$\Delta = C(J) \cap \Delta_n$$

is a simplicial complex with $I_\Delta^\mathbb{Q} = J$.

3) Check that the monomial ideal

$$I \subset K[x_0, \ldots, x_5]$$

above corresponds to a triangulation of $\mathbb{P}^2(\mathbb{R})$. Hint: Compute a primary decomposition of I.

The fact that the homology groups $H_i(\mathbb{P}^2(\mathbb{R}), K)$ of $\mathbb{P}^2(\mathbb{R})$ with coefficients in K behave differently for char$(K) = 2$ and char$(K) \neq 2$ explains Example 1) by Hochster's theory.

10.2 Formal power series and completions

We want to compute in $\mathcal{O}_{\mathbb{A}^n,o} = K[x_1, \ldots, x_n]_{(x_1, \ldots, x_n)}$. As a first step we regard $\mathcal{O}_{\mathbb{A}^n,o}$ as a subring of the formal power series ring

$$K[[x_1, \ldots, x_n]] = \{f = \sum_{\alpha \in \mathbb{N}^n} f_\alpha x^\alpha\}.$$

10.2 Formal power series and completions

The product $f = \sum_{\alpha \in \mathbb{N}^n} f_\alpha x^\alpha$ of two elements $g = \sum_{\beta \in \mathbb{N}^n} g_\beta x^\beta$ and $h = \sum_{\gamma \in \mathbb{N}^n} g_\gamma x^\gamma$ in $K[[x_1,\ldots,x_n]]$ is well-defined since the sum

$$f_\alpha = \sum_{\beta+\gamma=\alpha} g_\beta h_\gamma$$

is finite. Every fraction $f \in \mathcal{O}_{\mathbb{A}^n, o}$ may be written in the form $f = \frac{g}{1-h}$ with $h \in (x_1,\ldots,x_n)$. We embed

$$\mathcal{O}_{\mathbb{A}^n, o} \hookrightarrow K[[x_1,\ldots,x_n]], \frac{g}{1-h} \mapsto g\sum_{\nu=0}^{\infty} h^\nu.$$

To make sense of the infinite sum $\sum_{\nu=0}^{\infty} h^\nu \in K[[x_1,\ldots,x_n]]$ we need a bit of topology.

Definition 10.2.1. Let R be a ring and let $\mathfrak{m} \subset R$ be an ideal. We define a system of open neighborhoods of $0 \in R$ as the subsets $\mathfrak{m}^\nu \subset R$.
A sequence of (a_n) of elements of R **converges in the $\mathfrak{m}$-adic topology** to an element $a \in R$ if

$$\forall \nu \in \mathbb{N} \; \exists n_0 \in \mathbb{N} \text{ such that } a_n - a \in \mathfrak{m}^\nu \; \forall n \geq n_0$$

holds. A sequence (a_n) is a **Cauchy sequence** with respect to the $\mathfrak{m}$-adic topology if

$$\forall \nu \in \mathbb{N} \; \exists n_0 \in \mathbb{N} \text{ such that } a_m - a_n \in \mathfrak{m}^\nu \; \forall m,n \geq n_0$$

holds. The ring R is **Hausdorff** with respect to the $\mathfrak{m}$-adic topology if

$$\bigcap_{\nu=1}^{\infty} \mathfrak{m}^\nu = 0.$$

We call R **complete** with respect to the $\mathfrak{m}$-adic topology if R is Hausdorff and every Cauchy sequence converges.

Definition 10.2.2. For a ring R and the $\mathfrak{m}$-adic topology the quotient ring

$$\hat{R} = \{\text{Cauchy sequence}\}/\{\text{zero sequences}\}$$

is called the $\mathfrak{m}$**-adic completion**. This is a ring since the set of zero-sequences is an ideal in the term-wise defined ring of Cauchy sequences. The map

$$R \to \hat{R}, \; a \mapsto [\text{constant sequence } (a)]$$

is a ring homomorphism which is injective if and only if $\bigcap_{\nu=1}^{\infty} \mathfrak{m}^\nu = 0$. The completion $\hat{R}$ is always complete with respect to the $\hat{\mathfrak{m}} = \mathfrak{m}\hat{R}$-adic topology.

Thus we may regard the power series ring $k[[x_1,\ldots,x_n]]$ over any field as the completion of the polynomial ring $k[x_1,\ldots,x_n]$ with respect to the $(x_1,\ldots,x_n)$-adic topology and

$$f = \sum_{\alpha \in \mathbb{N}^n} f_\alpha x^\alpha = \lim_{d \to \infty} \sum_{\alpha: |\alpha| \leq d} f_\alpha x^\alpha.$$

The power series ring $k[[x_1, \ldots, x_n]]$ is a local ring. Its maximal ideal is $\mathfrak{m} = (x_1, \ldots, x_n)$. Indeed, every element $u \notin \mathfrak{m}$ has the form $u = \lambda(1 - h)$ with $h \in \mathfrak{m}$ and $\lambda \in k^*$ and

$$u^{-1} = \lambda^{-1} \sum_{\nu=0}^{\infty} h^\nu$$

since this series converges by the following proposition.

Proposition 10.2.3. *Let (h_ν) be a sequence of power series. Then $\sum_{\nu=0}^{\infty} h_\nu$ converges if and only if the sequence (h_ν) is an $\mathfrak{m}$-adic zero sequence.* □

Thus every $u \notin \mathfrak{m}$ is a unit. Formal power series cannot be evaluated at points $p \neq 0$. For the origin the value $f(0) \in k \cong k[[x_1, \ldots, x_n]]/\mathfrak{m}$ is given by the constant term.

Definition 10.2.4. Let $>$ be a **local monomial order** on $k[x_1, \ldots, x_n]$, i.e., $1 > x_i \ \forall i$. The leading term of a non-zero power series $f = \sum_{\alpha \in \mathbb{N}^n} f_\alpha x^\alpha$ with respect to $>$ is the term

$$\mathrm{Lt}(f) = f_\beta x^\beta,$$

where $x^\beta = \max\{x^\alpha \mid f_\alpha \neq 0\}$. This is well-defined because x^β is one of the finitely many generators of the monomial ideal $(\{x^\alpha \mid f_\alpha \neq 0\}) \subset k[x_1, \ldots, x_n]$ since $>$ is a local monomial order. We set $\mathrm{Lt}(0) = 0$.

Let $P = k[[x_1, \ldots, x_n]]$ denote the power series ring.

Theorem 10.2.5 (Grauert division). *Let $>$ be a local monomial order, and let $f_1, \ldots, f_r$ in P be non-zero power series. For every $f \in P$ there exists unique power series $g_1, \ldots, g_r$ in P and a remainder $h \in P$ such that the following holds:*

1) $f = g_1 f_1 + \cdots + g_r f_r + h$ and
2 a) No term of $g_i \mathrm{Lt}(f_i)$ is divisible by $\mathrm{Lt}(f_j)$ for some $j < i$.
 b) No term of h is divisible by an $\mathrm{Lt}(f_i)$.

Proof. Uniqueness follows as before. The non-zero leading terms of the $\mathrm{Lt}(g_i f_i) = \mathrm{Lt}(g_i) \mathrm{Lt}(f_i)$ and $\mathrm{Lt}(h)$ have different monomial parts. So they do not cancel. For the existence, we note that the result is trivially true if $f_1, \ldots, f_r$ are monomials. Thus there exists a unique expression

$$f = f^{(0)} = g_1^{(0)} \mathrm{Lt}(f_1) + \cdots + g_r^{(0)} \mathrm{Lt}(f_r) + h^{(0)}$$

satisfying condition 2a) and 2b).
Define

$$f^{(1)} = f^{(0)} - (g_1^{(0)} f_1 + \cdots + g_r^{(0)} f_r + h^{(0)})$$

and write similarly

$$f^{(1)} = g_1^{(1)} \mathrm{Lt}(f_1) + \cdots + g_r^{(1)} \mathrm{Lt}(f_r) + h^{(1)}.$$

10.2 Formal power series and completions

Iterating we obtain sequences $(f^{(\nu)}), (g_1^{(\nu)}), \ldots, (g_r^{(\nu)})$ and $(h^{(\nu)})$ of power series. Define

$$g_i = \sum_{\nu=0}^{\infty} g_i^{(\nu)} \text{ and } h = \sum_{\nu=0}^{\infty} h^{(\nu)}.$$

The existence follows if we can prove that the sequences are zero sequences in the $\mathfrak{m}$-adic topology. It suffices to prove that $(f^{(\nu)})$ is an $\mathfrak{m}$-adic zero sequence.

Clearly, we have

$$\text{Lt}(f^{(0)}) > \text{Lt}(f^{(1)}) > \ldots > \text{Lt}(f^{(\nu)}) > \ldots$$

This does not imply that $f^{(\nu)}$ is an $\mathfrak{m}$-adic zero sequence. However, if $>$ is a weight order $>_w$ with strictly negative weights $(w_1, \ldots, w_n)$ then $\lim_{\nu \to \infty} \text{Lt}(f^{(k)}) = 0$ implies $\lim_{\nu \to \infty} f^{(\nu)} = 0$.

To complete the proof we observe that the procedure only depends on knowing the leading terms $\text{Lt}(f_i)$ and use the following fact:

Claim. *There exists a weight order $>_w$ with strictly negative weights such that*

$$\text{Lt}_{>_w}(f_i) = \text{Lt}_>(f_i)$$

coincides for the finitely many power series $f_1, \ldots, f_r$.

The main step towards the proof has already been done in Exercise 1.2.19. However, since power series have infinitely many terms and Exercise 1.2.19 deals only with finitely many terms we need one more argument (which is left to the reader as Exercise 10.2.14) to deduce the claim from Exercise 1.2.19. □

Remark. In the case of $k = \mathbb{C}$ perturbing the local order to a weight order is also a key to the theorem of Grauert which says that if $f_1, \ldots, f_r \in \mathbb{C}[[x_1, \ldots, x_n]]$ and f are convergent power series then $g_1, \ldots, g_r$ and h are convergent series as well, see [32, 10].

Definition 10.2.6. Let $I \subset k[[x_1, \ldots, x_n]]$ be an ideal. Then

$$\text{Lt}(I) = (\{\text{Lt}(f) \mid f \in I\})$$

is called the **ideal of leading terms** of I. It is finitely generated, since it is a monomial ideal.

Corollary 10.2.7. *If $f_1, \ldots, f_r \in I \subset k[[x_1, \ldots, x_n]]$ are elements such that*

$$(\text{Lt}(f_1), \ldots, \text{Lt}(f_r)) = \text{Lt}(I)$$

then $I = (f_1, \ldots, f_r)$. In particular, $k[[x_1, \ldots, x_n]]$ is Noetherian. □

Corollary 10.2.8. *The monomials $x^\alpha \notin \text{Lt}(I)$ represent k-linearly independent elements of $k[[x_1, \ldots, x_n]]/I$ which are dense in the $\mathfrak{m}$-adic topology. These elements represent a basis if $\dim_k k[[x_1, \ldots, x_n]]/I < \infty$.* □

The definition of a Gröbner basis and a version of Buchberger's criterion work as before.

Definition 10.2.9. A power series $f \in k[[x_1,\ldots,x_n]]$ is called x_1-**general** if
$$f(x_1,0,\ldots,0) \in k[[x_1]]$$
is non-zero.

Example 10.2.10. Let $>$ be a local monomial order. If $f \in k[[x_1,\ldots,x_n]]$ is a power series with $\mathrm{Lt}(f) = ax_1^m$, then f is x_1-general. Conversely, for an x_1-general power series f there exists a local monomial order such that $\mathrm{Lt}(f) = ax_1^m$.

Theorem 10.2.11 (Weierstrass preparation theorem). *Let f be an x_1-general power series in $k[[x_1,\ldots,x_n]]$. Then there exists a unit $u \in k[[x_1,\ldots,x_n]]$ and a monic polynomial $p \in k[[x_2,\ldots,x_n]][x_1] \subset k[[x_1,\ldots,x_n]]$ such that*
$$f = up.$$
In particular, $k[[x_2,\ldots,x_n]] \subset k[[x_2,\ldots,x_n]]/(f)$ is an integral ring extension.

Remark. The original Weierstrass preparation theorem is the case when $k = \mathbb{C}$ and when f is a convergent power series. In that case, u and the coefficients of p are convergent power series as well.

Proof. Let $>$ be a local monomial order such that $\mathrm{Lt}(f) = ax_1^m$. Grauert division of x_1^m by f yields an expression
$$x_1^m = gf + r,$$
where $r \in k[[x_2,\ldots,x_n]][x_1]$ is a polynomial of degree less than m in x_1. Since $x_1^m = \mathrm{Lt}(gf) = \mathrm{Lt}(g)\mathrm{Lt}(f)$ we have $\mathrm{Lt}(g) = a^{-1} \in k$. Hence g is a unit in $k[[x_1,\ldots,x_n]]$ and
$$f = u(x_1^m - r) \quad \text{with } u = g^{-1}$$
is the desired expression. Since p is monic, we see that
$$k[[x_2,\ldots,x_n]] \subset k[[x_2,\ldots,x_n]][x_1]/(p) = k[[x_1,\ldots,x_n]]/(f)$$
is an integral ring extension. □

Lemma 10.2.12. *Let $I = (f) \subsetneq k[[x_1,\ldots,x_n]]$ be a non-zero ideal in a power series ring over an infinite field k. Then for a suitable tuple $(a_2,\ldots,a_n) \in \mathbb{A}^{n-1}(k)$ the substitution*
$$\varphi\colon k[[x_1,\ldots,x_n]] \to k[[x_1,\ldots,x_n]], x_1 \mapsto x_1, x_i \mapsto x_i + a_ix_1$$
maps f to an x_1-general power series $\varphi(f)$.

Proof. Write $f = f_m + f_{m+1} + \ldots$ as a series of homogeneous forms with $f_m \neq 0$. Choose a point $(1,a_2,\ldots,a_n) \in \mathbb{A}^n(k)$ with $f_m(1,a_2,\ldots,a_n) \neq 0$. Then as in Lemma

10.3 A lower bound on intersection multiplicities

1.4.6 the substitution automorphism φ maps f_m and f to x_1-general elements $\varphi(f_m)$ and $\varphi(f)$. □

Corollary 10.2.13. *The power series ring has dimension* $\dim k[[x_1,\ldots,x_n]] = n$.

Proof. The inequality $\dim k[[x_1,\ldots,x_n]] \geq n$ is clear because

$$(0) \subsetneq (x_1) \subsetneq \ldots \subsetneq (x_1,\ldots,x_n) = \mathfrak{m}$$

is a chain of prime ideals of length n. To show equality we consider a minimal non-zero prime ideal $\mathfrak{p}_1$. This is a principal ideal generated by an irreducible element f. Write $f = f_m + f_{m+1} + \cdots$ as a series of homogeneous polynomials with $f_m \neq 0$. Pick a point $(1, a_2,\ldots, a_n) \in \mathbb{A}^n(L)$ with $f_m(1, a_2,\ldots, a_n) \neq 0$ for a **finite** field extension $L \supset k$. Then $k[[x_1,\ldots,x_n]] \subset L[[x_1,\ldots,x_n]]$ and $k[[x_1,\ldots,x_n]]/(f) \subset L[[x_1,\ldots,x_n]]/(f)$ are integral ring extensions. To see that $g = \sum g_\alpha x^\alpha \in L[[x_1,\ldots,x_n]]$ is integral over $k[[x_1,\ldots,x_n]]$ we consider a basis $b_1,\ldots,b_d$ of L as a k-vector space and write each coefficient

$$g_\alpha = \sum_{i=1}^{d} g_{\alpha,i} b_i$$

with $g_{\alpha,i} \in k$. Then $g_i = \sum_\alpha g_{\alpha,i} x^\alpha \in k[[x_1,\ldots,x_n]]$ and $g = \sum_{i=1}^{d} g_i b_i$ is integral over $k[[x_1,\ldots,x_n]]$ as a finite sum of products of integral elements. Now, consider the automorphism

$$\varphi: L[[x_1,\ldots,x_n]] \to L[[x_1,\ldots,x_n]], x_1 \mapsto x_1, x_i \mapsto x_i + a_i x_1$$

and a Weierstrass polynomial p for $\varphi(f)$. Then

$$\dim k[[x_1,\ldots,x_n]]/(f) = \dim L[[x_1,\ldots,x_n]]/(f) = \dim L[[x_1,\ldots,x_n]]/(\varphi(f))$$
$$= \dim L[[x_2,\ldots,x_n]][x_1]/(p) = \dim L[[x_2,\ldots,x_n]] = n - 1$$

by Theorem 6.4.6 and induction. Thus $\dim k[[x_1,\ldots,x_n]] = n$. □

Exercise 10.2.14. Complete the last step in the proof of Theorem 10.2.5.

10.3 A lower bound on intersection multiplicities

Definition 10.3.1. We call the local monomial order $k[x_1,\ldots,x_n]$ defined by

$$x^\alpha >_{\text{ldrlex}} x^\beta \Leftrightarrow \deg x^\alpha < \deg x^\beta \text{ or}$$
$$\deg x^\alpha = \deg x^\beta \text{ and } x^\alpha >_{\text{rdlex}} x^\beta$$

the **local degree reverse lexicographic order**.

If we have no further information, this is often the best choice for a local monomial order. We are now ready to prove the following statement.

Theorem 9.3.4. *Let $f, g \in K[x, y]$ be polynomials without a common factor which vanish at the origin $o \in \mathbb{A}^2$. Then*

$$i(f, g; o) \geq \mathrm{mult}_o(f) \, \mathrm{mult}_o(g)$$

and equality holds if and only if $V(f)$ and $V(g)$ have no common tangent line at o.

Proof. We work with the local reverse lexicographic order on $K[x, y]$. Interchanging f and g if necessary we can achieve $\mathrm{mult}_o(f) = m \leq \mathrm{mult}_o(g) = n$. So $f = f_m + \cdots + f_d$ and $g = g_n + \cdots + g_e$. We first assume that $V(f_m)$ and $V(g_n)$ have no common factor. After a linear change of coordinates and adjusting of the leading coefficient we may assume that $\mathrm{Lt}(f) = x^m$ and after we replace g by an $g_1 = \lambda(g - hf)$ with $\lambda \in K^*$ that $\mathrm{Lt}(g_1) = x^{a_1} y^{b_1}$ with $a_1 + b_1 = n$ and $a_1 < m$.

Taking the remainder of $x^{m-a_1} g_1 - y^{b_1} f$ leads to a new Gröbner basis element g_2 with leading term $\mathrm{Lt}(g_2) = x^{a_2} y^{b_2}$ with $a_2 < a_1$ whose degree is $a_2 + b_2 = m + b_1$. After finitely many steps our stair must reach the y-axes with a monomial y^{b_r}. If f_m and g_n have no common factor then the new leading terms always have degree $a_{\nu+1} + b_{\nu+1} = a_{\nu-1} + b_\nu$, i.e., they lie on the corresponding diagonal.

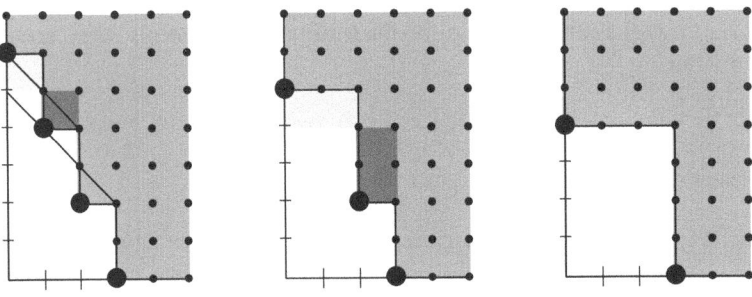

4 Gröbner basis elements.   3 Gröbner basis elements.   2 Gröbner basis elements.

In the diagrams above we indicate what happens if instead of the second to last leading term corresponding to (a_{r-1}, b_{r-1}) we already get at this step a leading term on the y-axis. This point would have coordinate $(0, a_{r-1} + b_{r-1})$. The dark grey area would be removed and the light grey area would be added. An elementary argument shows that the area under the stair does not change. Repeating this geometric transformation of the stair until we reach a rectangle proves $i(f, g; o) = m \cdot n$ if f_m and g_n have no common factor.

10.4 Mora division

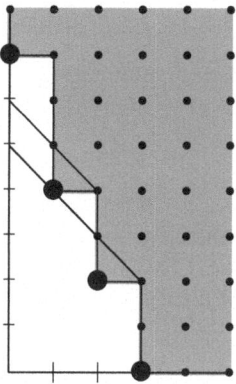

4 Gröbner basis elements. Two curves of multiplicity 3 and 4 with one common tangent.

On the other hand, if f_m and g_n have a common factor then the stair for f_m and g_n ends before it reaches the y-axis. Since the stair for f and g reaches the y-axis, we get a strictly larger area than $m \cdot n$ in the case when f_m and g_n have a common factor. □

10.4 Mora division

The proof of Grauert's division theorem does not yield an algorithm because the iteration usually does not terminate. For ideals of $k[x_1, \ldots, x_n]_{(x_1, \ldots, x_n)} \subset k[[x_1, \ldots, x_n]]$ there exists an algorithm to compute a Gröbner basis [61]. Without loss of generality we may assume that an ideal $I \subset k[x_1, \ldots, x_n]_{(x_1, \ldots, x_n)}$ is generated by elements of $k[x_1, \ldots, x_n]$, since the denominators are units in $k[x_1, \ldots, x_n]_{(x_1, \ldots, x_n)}$.

Theorem 10.4.1 (Mora division). *Let $>$ be a local monomial order and let $f_1, \ldots, f_r$ be elements of $k[x_1, \ldots, x_n]$. For each further element $g \in k[x_1, \ldots, x_n]$ there exists an element $u \in k[x_1, \ldots, x_n]$ with $u(0) = 1$, elements $g_1, \ldots, g_r \in k[x_1, \ldots, x_n]$ and a remainder $h \in k[x_1, \ldots, x_n]$ such that the following holds:*

1) $ug = g_1 f_1 + \cdots + g_r f_r + h$.
2) a) $\mathrm{Lt}(g) \geq \mathrm{Lt}(g_i f_i)$ *whenever both sides are non-zero.*
   b) *If $h \neq 0$ then $\mathrm{Lt}(h)$ is not divisible by any $\mathrm{Lt}(f_i)$.*

Definition 10.4.2. Let $>$ be a monomial order. The **ecart** of a non-zero polynomial f in $k[x_1, \ldots, x_n]$ is
$$\mathrm{ecart}(f) = \deg f - \deg \mathrm{Lt}(f).$$

Algorithm 10.4.3 (Mora division).
Input. A local monomial order $>$, polynomials $f_1, \ldots, f_r$ and g in $k[x_1, \ldots, x_n]$.
Output. A remainder h of a Mora division of g by $f_1, \ldots, f_r$.

1. Set $h := g$ and $D := \{f_1, \ldots, f_r\}$.

2. **while** ($h \neq 0$ **and** $D(h) := \{f \in D \mid \mathrm{Lt}(f) \text{ divides } \mathrm{Lt}(h)\} \neq \emptyset$) **do**
   - Choose $f \in D(h)$ with ecart(f) minimal.
   - **if** ecart(f) > ecart(h) **then** $D := D \cup \{h\}$.
   - $h := h - \frac{\mathrm{Lt}(h)}{\mathrm{Lt}(f)} f$.

3. return h.

Termination of Mora's algorithm. We write h_ν and D_ν for the value of h and D after ν iterations of the while loop. Let x_0 be a further variable. After ν iterations the while loop continues if and only if $\mathrm{Lt}(h_\nu) \in (\{\mathrm{Lt}(f) \mid f \in D_\nu\} \subset k[x_1, \ldots, x_n]$ and h_ν is added to D_ν if and only if

$$x_0^{\mathrm{ecart}(h_\nu)} \mathrm{Lt}(h_\nu) \notin I_\nu := (\{x_0^{\mathrm{ecart}(f)} \mathrm{Lt}(f) \mid f \in D_\nu\}) \subset k[x_0, x_1, \ldots, x_n].$$

Since the chain of monomial ideals

$$I_0 \subset I_1 \subset \ldots \subset I_\nu \subset \ldots \subset k[x_0, \ldots, x_n]$$

becomes stationary, there exists an N such that

$$D_N = D_{N+1} = D_{N+2} = \ldots$$

no longer increases.

After this point we homogenize h_N and the elements of D_N with x_0. The homogenization of $f \in D_N$, i.e.,

$$f^h = x_0^{\deg f} f(x_1/x_0, \ldots, x_n/x_0)$$

has leading term $\mathrm{Lt}(f^h) = x_0^{\mathrm{ecart}(f)} \mathrm{Lt}(f)$ with respect to the monomial order $>_g$ on $k[x_0, \ldots, x_n]$ defined by

$$x_0^a x^\alpha >_g x_0^b x^\beta \iff \deg x_0^a x^\alpha > \deg x_0^b x^\beta \text{ or}$$
$$\deg x_0^a x^\alpha = \deg x_0^b x^\beta \text{ and } x^\alpha > x^\beta.$$

Since D_N does not change after this point, we get a sequence

$$(h_\nu^h)_{\nu \geq N}$$

of homogeneous elements of the same degree with leading terms

$$\mathrm{Lt}(h_N^h) = x_0^{\mathrm{ecart}(h_N)} \mathrm{Lt}(h_N) >_g \mathrm{Lt}(h_{N+1}^h) >_g \ldots.$$

After finitely many further steps the algorithm stops with an $h_M = 0$ or an h_M with $\mathrm{Lt}(h_M) \notin (\{\mathrm{Lt}(f) \mid f \in D_N\})$ since $>_g$ is a global monomial order on $k[x_0, \ldots, x_n]$.

Correctness of the output. Recursively, starting with $u_0 = 1$, $g_i^{(0)} = 0$ and $h_0 = g$ suppose that we already have expressions

10.5 Differentiation and the tangent space

$$u_\ell g = g_1^{(\ell)} f_1 + \ldots + g_r^{(\ell)} f_r + h_\ell \quad \text{with } u_\ell(0) = 1$$

for $\ell = 0, \ldots, \nu - 1$. If the test condition for the ν-th iteration of the while loop is fulfilled, choose a polynomial $f = f^{(\nu)} \in D(h)$ as in the algorithm and set

$$h_\nu = h_{\nu-1} - m_\nu f^{(\nu)}, \text{ where } m_\nu = \frac{\text{Lt}(h_{\nu-1})}{\text{Lt}(f^{(\nu)})}.$$

There are two possibilities.

i) $f^{(\nu)}$ is one of $f_1, \ldots, f_r$ or
ii) $f^{(\nu)}$ is one of $h_1, \ldots, h_{\nu-1}$.

Thus substituting $h_{\nu-1} = h_\nu + m_\nu f^{(\nu)}$ into the expression for $u_{\nu-1} g$, we obtain the desired expression for $u_\nu g$ with

i) $u_\nu = u_{\nu-1}$,
ii) $u_\nu = u_{\nu-1} + m_\nu u_\ell$ for ℓ with $f^{(\nu)} = h_\ell$ and $g_j^{(\nu)} = g_j^{(\nu-1)} + m_\nu g_j^{(\ell)}$ for $j \in \{1, \ldots, r\}$.

In both cases we have $u_\nu(0) = u_{\nu-1}(0) = 1$. In case ii) this follows from

$$\text{Lt}(h_\ell) > \text{Lt}(h_\nu) = \text{Lt}(m_\nu h_\ell) = m_\nu \text{Lt}(h_\ell).$$

Hence $1 > m_\nu$ and $u_\nu(0) = u_{k-1}(0) + 0 u_\ell(0) = 1$.

The final expression satisfies condition 2a) because the leading terms of the h_ν decrease in each round of the while loop. Finally, condition 2b) is satisfied due to the stopping condition of the while loop. □

Example. Consider $g = x$ and $f_1 = x - x^2$ in $k[x]$. Mora division proceeds as follows.

$$h_0 = x, D_0 = \{x - x^2\}, 1 \cdot g = 0 \cdot f_1 + x,$$
$$f^{(1)} = x - x^2, m_1 = 1, D_1 = \{x - x^2, x\}, 1 \cdot g = 1 \cdot f_1 + x^2,$$
$$f^{(2)} = x, m_2 = x, D_2 = D_1, (1 - x) \cdot g = 1 \cdot f_1 + 0.$$

Greuel and Pfister [36] treat Mora's algorithm for mixed orders, i.e., weight orders with some negative and some positive weights. This allows us to do Gröbner basis computations in localizations at arbitrary primes of affine k-algebras, see [62].

10.5 Differentiation and the tangent space

Let k be an arbitrary field. Differentiation in $k[x]$ can be defined without analysis. For $f = \sum_{n \in \mathbb{N}} a_n x^n$ we define the derivative

$$f' = \sum_{n \in \mathbb{N}} n a_n x^{n-1}.$$

The usual differentiation rules hold with one exception if $\text{char } k = p > 0$.

Proposition 10.5.1. Let $f, g \in k[x]$ be polynomials. Then

1) $(f + g)' = f' + g'$,
2) $(fg)' = f'g + fg'$,
3) if char $k = 0$, then $f' = 0$ if and only if f is a constant polynomial,
4) if char $k = p > 0$, then $f' = 0 \iff f \in k[x^p]$.

Proof. 1) is clear. By 1) it suffices to prove 2) for monomials:

$$(x^{n+m})' = (n+m)x^{n+m-1} = nx^{n-1}x^m + mx^n x^{m-1}$$
$$= (x^n)' x^m + x^n (x^m)'.$$

3) and 4) are clear from the formula because $(x^{np})' = npx^{np-1} = 0$ if char $k = p > 0$, while $(x^m)' = mx^{m-1} \neq 0$ if p does not divide m. □

Remark. In the case of a finite field or an algebraically closed field of char $k = p$, we have

$$f \in k[x^p] \iff f = g^p \text{ for some } g \in k[x]$$

because the map $k \to k, a \mapsto a^p$ is surjective.

For multivariate polynomials $f \in k[x_1, \ldots, x_n]$ partial derivatives $\frac{\partial f}{\partial x_i}$ are defined analogously. The gradient

$$\left(\frac{\partial f}{\partial x_1}, \ldots, \frac{\partial f}{\partial x_n}\right)$$

of f is identically zero in char $k = p$ if and only if f lies in $k[x_1^p, \ldots, x_n^p]$.

Definition 10.5.2. Let $f \in k[x_1, \ldots, x_n]$. We define the **differential** of f at a point $p = (a_1, \ldots, a_n) \in \mathbb{A}^n$ as

$$d_p f = \sum_{i=0}^{n} \frac{\partial f}{\partial x_i}(p)(x_i - a_i).$$

In other words, $d_p f$ is the linear part in the Taylor expansion

$$f = f(p) + d_p f + \text{ terms of higher order in the } x_i - a_i$$

of f.

For a hypersurface $H \subset \mathbb{A}^n$ with $I(H) = (f)$ we define the **tangent space** of H at a point $p \in H$ as the linear subspace

$$T_p H = V(d_p f).$$

Definition 10.5.3. Let $A \subset \mathbb{A}^n$ be an algebraic set. The tangent space of A at a point $p \in A$ is defined by

$$T_p(A) = V(\{d_p f \mid f \in I(A)\}).$$

The local dimension of A at p is defined as

10.5 Differentiation and the tangent space

$$\dim_p A = \max\{\dim C \mid C \text{ is an irreducible component}$$
$$\text{of } A \text{ passing through } p\}.$$

The algebraic set A is **smooth** at p if $\dim T_p A = \dim_p A$.

Proposition 10.5.4 (Jacobian criterion). *Let $A \subset \mathbb{A}^n$ be an algebraic set and let $f_1, \ldots, f_r$ in $I(A)$ polynomials vanishing on A. Then*

$$n - \operatorname{rank}(\frac{\partial f_i}{\partial x_j}(p)) \geq \dim_p A$$

and A is smooth at p if equality holds.

Remark. If $1 \leq i_1 < \ldots < i_k \leq r$, $1 \leq j_1 < \ldots < j_k \leq n$ correspond to the indices of a non-vanishing minor of the Jacobian matrix $(\frac{\partial f_i}{\partial x_j}(p))$, then in the case of $K = \mathbb{R}$ or $\mathbb{C}$ the implicit function theorem says that one can solve the system of equations $f_{i_1} = \ldots = f_{i_k} = 0$ locally.

One can express $x_{j_1}, \ldots, x_{j_k}$ as differentiable or holomorphic functions of the x_j's with $j \notin \{j_1, \ldots, j_k\}$ respectively, and every solution of $f_{i_1} = \ldots = f_{i_k} = 0$ near p arises as a point on the corresponding graph.

Example. Consider the curve $C = V(f)$ defined by $f = 2y^3 + y^2 - y + x^2 + x - xy$. We have $\frac{\partial f}{\partial y}(p) = -1 \neq 0$ for $p = (a, b) = (0, 0)$.

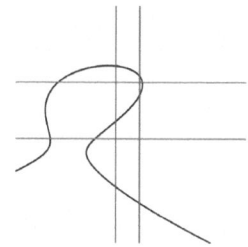

Hence there exists an $\epsilon > 0$, a $\delta > 0$ and a differentiable function $g: (-\delta + a, a + \delta) \to \mathbb{R}$ with $g(a) = b$ such that the intersection $C(\mathbb{R}) \cap ((-\delta+a, a+\delta) \times (-\epsilon+b, b+\epsilon))$ coincides with the graph

$$\Gamma = \{(x, g(x)) \mid x \in (-\delta + a, a + \delta)\}$$

of g.

Proof of the Jacobian criterion. We have

$$n - \operatorname{rank}(\frac{\partial f_i}{\partial x_j}(p)) \geq \dim T_p A \geq \dim_p A.$$

The first inequality is true by the definition of $T_p A$. It could be strict since we did not assume that $f_1, \ldots, f_r$ generate $I(A)$. The second inequality holds in a much more general setting, which we state below. $\square$

Theorem 10.5.5 (Krull's principal ideal theorem). *Let R be a Noetherian ring. Every minimal prime $\mathfrak{p}$ of a principal ideal $(f) \subset R$ has height*

$$\operatorname{height}(\mathfrak{p}) \leq 1.$$

Equality holds if f is a non-zero divisor. More generally, if $\mathfrak{p}$ is a minimal prime of an ideal $(f_1, \ldots, f_c) \subset R$ generated by c elements, then height($\mathfrak{p}$) $\leq c$.

For a proof, see [18] or [6].

Corollary 10.5.6. *Let $(R, \mathfrak{m}, k)$ be a Noetherian local ring. Then*

$$\dim_k \mathfrak{m}/\mathfrak{m}^2 \geq \dim R.$$

Proof. By Nakayama's lemma, the ideal $\mathfrak{m}$ is generated by $c = \dim_k \mathfrak{m}/\mathfrak{m}^2$ elements. Since $\mathfrak{m}$ is the unique maximal ideal of R we obtain

$$\dim R = \text{height}(\mathfrak{m}) \leq c$$

from the principal ideal theorem. $\square$

Definition 10.5.7. A **regular local** ring is a Noetherian local ring $(R, \mathfrak{m}, k)$ with

$$\dim_k \mathfrak{m}/\mathfrak{m}^2 = \dim R.$$

Proposition 10.5.8. *A point $p \in A$ of an algebraic set $A \subset \mathbb{A}^n$ is a smooth point of A if and only if $\mathcal{O}_{A,p}$ is a regular local ring.*

Proof. Since $n - \mathfrak{m}_{A,p}/\mathfrak{m}_{A,p}^2$ is the codimension of $T_p(A)$ we have $\dim T_p A = \dim A_p$ if and only if $\mathcal{O}_{A,p}$ is a regular local ring. $\square$

The K-vector space $\mathfrak{m}_{A,p}/\mathfrak{m}_{A,p}^2$ can be interpreted as the vector space of linear functions on $T_p(A)$ regarded as a K-vector space with origin p. Thus the dual vector space $(\mathfrak{m}_{A,p}/\mathfrak{m}_{A,p}^2)^* \cong T_p(A)$ is called the **Zariski tangent space** of A at p. Points $p \in A$ where A is not smooth are called **singular points of** A.

Example. Let $H \subset \mathbb{A}^n$ be a hypersurface and $(f) = I(A)$ be its ideal in $K[x_1, \ldots, x_n]$. Then the set of singular points is

$$H_{\text{sing}} = V(f, \frac{\partial f}{\partial x_1}, \ldots, \frac{\partial f}{\partial x_n}).$$

Notice that $(f) = (f, \frac{\partial f}{\partial x_1}, \ldots, \frac{\partial f}{\partial x_n})$ holds if and only if $\frac{\partial f}{\partial x_1} = 0, \ldots, \frac{\partial f}{\partial x_n} = 0$ since the partial derivative $\frac{\partial f}{\partial x_i}$ has smaller degree in x_i than f. Thus $(f) = (f, \frac{\partial f}{\partial x_1}, \ldots, \frac{\partial f}{\partial x_n})$ implies that $\text{char } K = p$ and $f \in K[x_1^p, \ldots, x_n^p]$. Since K is algebraically closed this gives $f = g^p$, contradicting the fact that (f) is a radical ideal. This leads to the following result.

Proposition 10.5.9. *The set of smooth points of a hypersurface $H \subset \mathbb{A}^n$ is a Zariski open dense subset of H.* $\square$

Theorem 10.5.10 (Generic smoothness). *Let $A \subset \mathbb{A}^n$ be an affine variety. Then the set of smooth points of A is a non-empty Zariski open dense subset of A.*

10.5 Differentiation and the tangent space

Proof. One can show that every variety is birational to a hypersurface H. In the case of char $K = 0$, this follows from the existence of a primitive element [82] for the field extensions $K(x_{n-d+1},\ldots,x_n) \subset K(A)$, where $A \to \mathbb{A}^d$ is a suitable linear projection. For arbitrary characteristic one has to express $K(A)$ as a separable field extension of a purely transcendental extension $K(y_1,\ldots,y_d)$ of K, see [18, 88]. For points p in an open set $U \subset A$ which is isomorphic to an open set of H we have

$$\mathcal{O}_{A,p} \cong \mathcal{O}_{H,p}$$

and the result follows from Proposition 10.5.9. □

At a singular point of an algebraic set $p \in A \subset \mathbb{A}^n$ the tangent space $T_p A$ is only a very rough approximation of A near p. The tangent cone, as defined below, is a better approximation.

Definition 10.5.11. For $I \subset k[x_1,\ldots,x_n]$ the **ideal of initial forms** of I is

$$J = (\{f_m \mid f_m \text{ is the smallest degree part of an element } f = f_m + \ldots + f_d \in I\}).$$

If $I = I(A) \subset K[x_1,\ldots,x_n]$ is the ideal of an affine algebraic set $A \subset \mathbb{A}^n$ we call $V(J)$ the **tangent cone** of A at p.

The ring $K[x_1,\ldots,x_n]/J$ is isomorphic to the **associated graded ring**

$$gr_{\mathfrak{m}} R = R/\mathfrak{m} \oplus \mathfrak{m}/\mathfrak{m}^2 \oplus \mathfrak{m}^2/\mathfrak{m}^3 \oplus \ldots = \bigoplus_{k=0}^{\infty} \mathfrak{m}^k/\mathfrak{m}^{k+1}$$

of $R = \mathcal{O}_{A,o}$ with respect to the maximal ideal $\mathfrak{m} = \mathfrak{m}_{A,o}$.

Algorithm 10.5.12 (Mora's tangent cone algorithm).
Input. Generators of the ideal I of an affine algebraic set $A \subset \mathbb{A}^n$.
Output. Generators of the ideal of initial forms of I at o.

1. Choose a local monomial order $>$ which refines the degree:

$$\deg x^\alpha < \deg x^\beta \implies x^\alpha > x^\beta.$$

2. Compute a Gröbner basis G of I with respect to the monomial order $>$ using Mora's algorithm.
3. Return the initial forms f_m of all $f = f_m + \cdots + f_d \in G$.

Hierarchy of approximations. Let $R = \mathcal{O}_{A,p}$ be the local ring of an algebraic set. We have introduced its $\mathfrak{m} = \mathfrak{m}_{A,p}$-adic completion $\widehat{\mathcal{O}}_{A,p}$, the associated graded ring $gr_{\mathfrak{m}} R$ and the Zariski tangent space $T_p A = (\mathfrak{m}/\mathfrak{m}^2)^*$.

For two local rings $\mathcal{O}_{A,p}$ and $\mathcal{O}_{B,q}$ we have the following implications.

142   10 Local rings and power series

$$\mathcal{O}_{A,p} \cong \mathcal{O}_{B,q} \implies \widehat{\mathcal{O}}_{A,p} \cong \widehat{\mathcal{O}}_{B,q},$$
$$\widehat{\mathcal{O}}_{A,p} \cong \widehat{\mathcal{O}}_{B,q} \implies \mathrm{gr}_{\mathfrak{m}_{A,p}}\mathcal{O}_{A,p} \cong \mathrm{gr}_{\mathfrak{m}_{B,q}}\mathcal{O}_{B,q},$$
$$\mathrm{gr}_{\mathfrak{m}_{A,p}}\mathcal{O}_{A,p} \cong \mathrm{gr}_{\mathfrak{m}_{B,q}}\mathcal{O}_{B,q} \implies T_p A \cong T_q B.$$

In general none of these implications is an equivalence.

Example 10.5.13. Consider $A = V(y^2 - x^2 - x^3)$ and $B = V(y^2 - x^2)$ at the origin o for $K = \mathbb{C}$. Then

$$\widehat{\mathcal{O}}_{A,o} \cong \widehat{\mathcal{O}}_{B,o}$$

are isomorphic via the ring homomorphism induced by the substitution

$$\mathbb{C}[[x,y]] \to \mathbb{C}[[x,y]], (x,y) \mapsto (x, y\sqrt{1+x}).$$

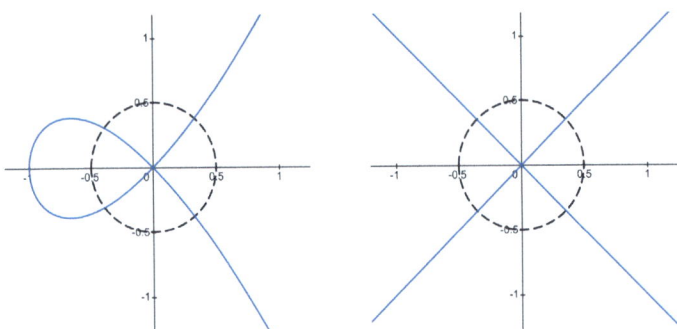

Indeed, we have that

$$\sqrt{1+x} = 1 + \frac{x}{2} - \frac{x^2}{8} + \ldots = \sum_{k=0}^{\infty} \binom{\frac{1}{2}}{k} x^k \in \mathbb{C}[[x]]$$

and its square $1 + x$ are units.

Definition 10.5.14. If $\widehat{\mathcal{O}}_{A,p} \cong \widehat{\mathcal{O}}_{B,q}$, then (A, p) and (B, q) are called **analytically isomorphic**.

Example 10.5.15. A point p of a plane curve C is called an **ordinary double point** or **node** if $\widehat{\mathcal{O}}_{C,p}$ is isomorphic to $K[[x,y]]/(xy)$. The point p is called a **cusp** if $\widehat{\mathcal{O}}_{C,p} \cong K[[x,y]]/(y^2 - x^3)$.

Exercise 10.5.16. Let $X^a \subset \mathbb{A}^4$ be the algebraic set defined by the polynomials

$$x_2^3 - x_1^2 x_3 + x_1 x_2 x_4 - x_1 x_3 x_4 - x_2 x_4^2 - x_1 x_2,$$
$$x_1 x_2^2 - x_1 x_3^2 + 2 x_2 x_3 x_4 - x_3^2 x_4 - x_2 x_3,$$
$$x_1^3 - x_1 x_2 x_3 + x_2^2 x_4 + x_1 x_4^2 - x_4^3 - x_1 x_4,$$
$$x_1^2 x_3 - x_2 x_3^2 + x_1 x_2 x_4 + 2 x_3 x_4^2 - x_3 x_4.$$

10.6 Discrete valuation rings

Prove that they form a Gröbner basis with respect to $>_{\text{ldrlex}}$. Thus the tangent cone at the origin is defined by

$$(x_1x_2, x_2x_3, x_1x_4, x_3x_4) = (x_1, x_3) \cap (x_2, x_4).$$

The tangent cone consists of two planes in $\mathbb{A}^4$ which intersect in a point.

Exercise 10.5.17. Let $X^a \subset \mathbb{A}^4$ be the algebraic set from Exercise 10.5.16 and let $X \subset \mathbb{P}^4$ be its algebraic closure. Let Y_0 and Y_1 be the planes defined by $J_0 = (x_1 + x_2, x_3 + x_4)$ and $J_1 = (x_1 + x_2 - x_0, x_3 + x_4 - x_0)$, respectively. Verify

1. $\deg X = 5$,
2. $X \cap Y_i$ consist of 4 respectively 5 points for $i = 0, 1$,
3. $\deg(I(X) + J_i)$ has values 6 and 5 for $i = 0, 1$,
4. $S/(I(X) + J_0)$ has the associated prime (x_1, x_2, x_3, x_4) three times in its filtration.

Conclude that the following definition of an intersection multiplicity for varieties of complementary dimension

$$i(X, Y; p) = \dim_K \mathcal{O}_{\mathbb{P}^4, p}/(I(X) + I(Y))\mathcal{O}_{\mathbb{P}^4, p}$$

does not always satisfy

$$\deg X \cdot \deg Y = \sum_{p \in X \cap Y} i(X, Y; p),$$

the desired Bézout formula. At the point $o \subset \mathbb{A}^4 \subset \mathbb{P}^4$ the intersection multiplicity should be 2 instead of 3: A general plane Y intersects the tangent cone in two points, but for Y_0 we have that

$$\mathcal{O}_{\mathbb{A}^4, o}/(x_1x_2, x_2x_3, x_1x_4, x_3x_4, x_1 + x_2, x_3 + x_4) \cong K[x_2, x_4]/(x_2^2, x_2x_4, x_4^2)$$

is a three-dimensional K-vector space.

It was one of Gröbner's research topics to find the right definition of intersection multiplicities [38]. This was finally solved by Serre [80].

Another problem when extending Theorem 9.4.1 to two arbitrary varieties $X, Y \subset \mathbb{P}^n$ is that $X \cap Y$ might have components of dimension larger than $\dim X + \dim Y - n$. How to measure their contribution to a Bézout formula is the topic of excess intersection theory [27, 24].

10.6 Discrete valuation rings

Definition 10.6.1. Let L be a field. A **discrete valuation** on L is a surjective map

$$v : L \setminus \{0\} \to \mathbb{Z}$$

such that for all $a, b \in L \setminus \{0\}$ the following holds
1. $v(ab) = v(a) + v(b)$,
2. $v(a + b) \geq \min\{v(a), v(b)\}$ if $a + b \neq 0$.

Note that the first condition says that $(L \setminus \{0\}, \cdot) \to (\mathbb{Z}, +)$ is a group homomorphism. In particular, we have $v(1) = 0$. By convention, we set $v(0) = \infty$. The set
$$R = \{a \in L \mid v(a) \geq 0\}$$
is a subring of L which is called the **valuation ring** of v. The subset of non-units in R
$$\mathfrak{m} = \{a \in L \mid v(a) > 0\}$$
is an ideal. It is the unique maximal ideal of R. Hence $(R, \mathfrak{m})$ is a local ring.

Definition 10.6.2. A **discrete valuation ring** (DVR) R is an integral domain such that R is the valuation ring of a discrete valuation v on its quotient field $L = Q(R)$.

Example. The formal power series ring $R = k[[t]]$ in one variable over a field k is a discrete valuation ring. Indeed, the quotient field of R is
$$L = k((t)) = \{\sum_{n=N}^{\infty} a_n t^n \mid N \in \mathbb{Z}\},$$
the ring of **formal Laurent series**, and
$$v(\sum_{n=N}^{\infty} a_n t^n) = \min\{n \mid a_n \neq 0\}$$
for a non-zero Laurent series defines a valuation on L with valuation ring $k[[t]]$. Following the notion for power series in one complex variable, we say that $f \in k[[t]]$ has a zero of order n if $v(f) = n$ and $f \in k((t))$ with $n = v(f) < 0$ has a pole of order $-n$.

Proposition 10.6.3 (Characterization of DVR's). *Let R be a ring. The following are equivalent:*

1) R is a DVR,
2) R is a Noetherian regular local ring of Krull dimension 1.

Proof. Suppose R is a DVR. Let $t \in R$ be an element with $v(t) = 1$. Every element $f \in R$ with $v(f) = n$ is of the form $f = ut^n$ with u a unit in R. In particular, t is a generator of $\mathfrak{m}$, and the only other ideals $I \neq 0$ are of the form $I = (t^n) = \mathfrak{m}^n$ with $n = \min\{v(f) \mid f \in I\}$. Hence $(0) \subset \mathfrak{m}$ is the only chain of prime ideals in R and R is a PID. So R is Noetherian and a regular local ring of Krull dimension 1, because $\mathfrak{m}$ is generated by a single element. Hence $\mathfrak{m}/\mathfrak{m}^2$ is 1-dimensional by Nakayama's lemma.

Conversely, let R be a Noetherian regular local ring of Krull dimension 1. By Nakayama's lemma the maximal ideal $\mathfrak{m}$ is a principal ideal, say $\mathfrak{m} = (t)$. Hence the

10.6 Discrete valuation rings

powers $\mathfrak{m}^n = (t^n)$ are principal ideals as well. Let $f \in R$ be a non-zero element. Since $\bigcap_{\nu=1}^{\infty} \mathfrak{m}^\nu = (0)$ by Krull's intersection theorem 10.1.5, the number

$$n = \max\{\nu \mid f \in \mathfrak{m}^\nu\}$$

is the maximum of finitely many integers and $f = ut^n$ for a unit $u \in R$. We set $v(f) = n$. Hence R is a domain and $v(f_1 f_2) = v(f_1) + v(f_2)$. We extend v to a map

$$v: Q(R) \setminus \{0\} \to \mathbb{Z} \quad \text{by} \quad v\left(\frac{f_1}{f_2}\right) = v(f_1) - v(f_2).$$

Then v is a discrete valuation on $Q(R)$ and R is its valuation ring. □

Corollary 10.6.4. *Let $p \in C$ be a smooth point of an irreducible curve. Then $\mathcal{O}_{C,p}$ is a discrete valuation ring.*

Notation. We denote the valuation of $K(C)$ corresponding to $\mathcal{O}_{C,p}$ for a smooth point p in C by v_p.

Remark 10.6.5. In the case of a smooth projective curve C one can show [42, Corollary I.6.6] that the map

$$p \mapsto v_p$$

induces a bijection between the points of C and the discrete valuations

$$v: K(C) \setminus \{0\} \to \mathbb{Z}$$

of the function field $K(C)$ with $v(a) = 0$ for all $a \in K \setminus \{0\}$.

Chapter 11
Products and morphisms of projective varieties

In this chapter we introduce some basic constructions of projective geometry and prove some fundamental results. After introducing of the Segre products we define morphisms between projective varieties. Section 11.3 gives the basic dimension bounds for projective varieties and their intersections.

After introducing the Veronese embeddings we prove in Section 11.5 that projective morphisms are closed maps. In particular, the image of a projective algebraic set under a morphism is always closed in the Zariski topology. So they are better behaved under morphisms than affine varieties. Finally we prove the semi-continuity of the fiber dimension for projective morphisms and prove that the general fiber of a dominant projective morphism $f: X \to Y$ between varieties has dimension $\dim X - \dim Y$ with a Gröbner basis argument. The proof actual gives that the Hilbert functions of the fibers are constant over a non-empty open subset $U \subset Y$. A similar result is true for homogeneous ideals I generated by homogeneous forms in $\mathbb{Z}[x_0, \ldots, x_n]$. The ideals $I_\mathbb{Q} \subset \mathbb{Q}[x_0, \ldots, x_n]$ and $I_p \subset (\mathbb{Z}/p)[x_0, \ldots, x_n]$ have the same Hilbert function for all but finitely prime numbers p. This is important for Computer Algebra experiments in projective geometry.

11.1 The Segre product

For two affine algebraic sets $A \subset \mathbb{A}^n$ and $B \subset \mathbb{A}^m$ the product

$$A \times B \subset \mathbb{A}^n \times \mathbb{A}^m = \mathbb{A}^{n+m}$$

is simply the algebraic set defined by

$$(\mathrm{I}(A) \cup \mathrm{I}(B)) \subset K[x_1, \ldots, x_n, y_1, \ldots, y_m],$$

where $\mathrm{I}(A) \subset K[x_1, \ldots, x_n]$ and $\mathrm{I}(B) \subset K[y_1, \ldots, y_m]$ are the vanishing ideals of A and B, respectively.

For projective algebraic sets the definition of a product is not so clear. To start we have to give $\mathbb{P}^n \times \mathbb{P}^m$ the structure of an algebraic set. One uses the **Segre embedding**.

Define
$$\sigma_{n,m}: \mathbb{P}^n \times \mathbb{P}^m \to \mathbb{P}^N \text{ with } N = (n+1)(m+1) - 1$$
by
$$((a_0 : \ldots : a_m), (b_0 : \ldots : b_n)) \mapsto (a_0 b_0 : \ldots : a_i b_j : \ldots : a_m b_n).$$

This is a well-defined map. For any pair of points at least one component $a_i b_j$ is different from 0. We will use variables
$$\mathbf{x} = x_0, \ldots, x_n, \ \mathbf{y} = y_0, \ldots, y_m \text{ and } \mathbf{z} = z_{00}, \ldots, z_{0m}, z_{10}, \ldots, z_{nm}$$
for the homogeneous coordinate rings of $\mathbb{P}^n, \mathbb{P}^m$ and $\mathbb{P}^N$, respectively. Moreover, we call a polynomial
$$f = \sum_{|\alpha|=d, |\beta|=e} f_{\alpha,\beta} x^\alpha y^\beta \in K[\mathbf{x}, \mathbf{y}]$$
bihomogeneous (in $\mathbf{x}$ and $\mathbf{y}$) of **bidegree** (d, e).

Proposition 11.1.1. *Let $\Sigma_{n,m} \subset \mathbb{P}^N$ be the projective algebraic set defined by the 2×2-minors of the $(n+1) \times (m+1)$-matrix (z_{ij}). Then*
$$\sigma_{n,m}: \mathbb{P}^n \times \mathbb{P}^m \to \Sigma_{m,n}$$
is a bijection which induces isomorphisms $U_i \times U_j \cong \Sigma_{n,m} \cap U_{ij}$ on the standard charts. Moreover $\Sigma_{n,m} \subset \mathbb{P}^N$ is irreducible, and the ideal of 2×2-minors coincides with the homogeneous ideal of $\Sigma_{m,m}$.

Proof. The minor
$$\det \begin{pmatrix} z_{i_1 j_1} & z_{i_1 j_2} \\ z_{i_2 j_1} & z_{i_2 j_2} \end{pmatrix}$$
vanishes on the image of $\sigma_{n,m}$ because
$$\det \begin{pmatrix} x_{i_1} y_{j_1} & x_{i_1} y_{j_2} \\ x_{i_2} y_{j_1} & x_{i_2} y_{j_2} \end{pmatrix} = 0.$$

Thus the image of $\sigma_{m,n}$ is contained in $\Sigma_{m,n}$.

The point $r = (1 : c_{01} : \ldots : c_{nm}) \in \Sigma_{n,m} \cap U_{00}$ satisfies
$$c_{ij} = c_{i0} c_{0j}.$$

Thus the pair of points
$$(p, q) = ((1 : c_{10} : \ldots, c_{n0}), (1 : c_{01} : \ldots : c_{0m})) \in U_0 \times U_0 \subset \mathbb{P}^n \times \mathbb{P}^m$$

11.1 The Segre product

is the unique preimage point of r and $\Sigma_{n,m} \cap U_{00} \cong U_0 \times U_0$. The same argument in other charts gives that $\sigma_{n,m} : \mathbb{P}^n \times \mathbb{P}^m \to \Sigma_{n,m}$ is bijective and gives isomorphisms $\Sigma_{n,m} \cap U_{ij} \cong U_i \times U_j$.

To prove that $\Sigma_{m,n}$ is irreducible and that the ideal J of 2×2-minors of (z_{ij}) is its homogeneous ideal, it suffices to prove that J is a prime ideal.

Consider the ring homomorphism

$$\varphi : K[\mathbf{z}] \to K[\mathbf{x},\mathbf{y}], z_{ij} \mapsto x_i y_j.$$

Clearly, $J \subset \ker \varphi$. To prove equality we consider a reverse lexicographic order $>_{\text{rdlex}}$ which refines the following order on the variables

$$\begin{array}{ccccc}
z_{00} & > & z_{01} & > \ldots > & z_{0m} \\
\vee & & \vee & & \vee \\
z_{10} & > & z_{11} & > \ldots > & z_{1m} \\
\vee & & \vee & & \vee \\
\vdots & & \vdots & & \vdots \\
\vee & & \vee & & \vee \\
z_{n0} & > & z_{n1} & > \ldots > & z_{nm}
\end{array}$$

We have

$$\mathrm{Lt}\left(\det \begin{pmatrix} z_{i_1 j_1} & z_{i_1 j_2} \\ z_{i_2 j_1} & z_{i_2 j_2} \end{pmatrix}\right) = -z_{i_2 j_1} z_{i_1 j_2}$$

whenever $i_1 < i_2$ and $j_1 < j_2$.

Thus the remainder of a monomial in $K[\mathbf{z}]$ divided by the 2×2-minors has the form

$$z_{i_1 j_1} z_{i_2 j_2} \cdots z_{i_d j_d} \text{ with } i_1 \leq i_2 \leq \ldots \leq i_d \text{ and } j_1 \leq j_2 \leq \ldots \leq j_d.$$

Since φ induces a bijection between such monomials and bihomogeneous monomials of bidegree (d,d), we conclude that the 2×2-minors form a Gröbner basis of $\ker \varphi$. In particular, $J = \ker \varphi$ and this is a prime ideal because $K[\mathbf{z}]/\ker \varphi$ is isomorphic to a subring of the domain $K[\mathbf{x},\mathbf{y}]$. □

Definition 11.1.2. We give $\mathbb{P}^n \times \mathbb{P}^m$ the structure of a projective variety by identifying $\mathbb{P}^n \times \mathbb{P}^m$ and $\Sigma_{n,m}$ via $\sigma_{n,m}$.

Example. We identify $\mathbb{P}^1 \times \mathbb{P}^1$ with the quadric

$$\Sigma_{1,1} = V(z_{00} z_{11} - z_{10} z_{01}) \subset \mathbb{P}^3.$$

The Zariski topology on $\mathbb{P}^n \times \mathbb{P}^m$ is finer than the product of the Zariski topologies of the factors. For example, if

$$f = \sum_{|\alpha|=d, |\beta|=e} f_{\alpha,\beta} x^\alpha y^\beta \in K[\mathbf{x},\mathbf{y}]$$

is a bihomogeneous polynomial of bidegree (d,e), then

$$V(f) = \{(a,b) \in \mathbb{P}^n \times \mathbb{P}^m \mid f(a,b) = 0\}$$

is a Zariski closed subset which for general f is not closed in the product topology. To see that $V(f)$ is an algebraic subset of $\mathbb{P}^n \times \mathbb{P}^m$ we argue as follows: Suppose $d \geq e$. Then multiplying f by all monomials $y^\beta \in K[\mathbf{y}]$ of degree $d-e$, we get $\binom{d-e+m}{m}$ polynomials fy^β of bidegree (d,d), each of which is the image of a polynomial in $F_\beta \in K[\mathbf{z}]$ of degree d. Then $V(f)$ coincides with the zero locus of $(\{F_\beta \mid |\beta| = d - e\}) + \ker\varphi$.

The zero locus $V(f)$ is called a **hypersurface of bidegree** (d,e) in $\mathbb{P}^n \times \mathbb{P}^m$.

Definition 11.1.3. Let $A \subset \mathbb{P}^n \times \mathbb{P}^m$ be a subset. The **bihomogeneous vanishing ideal** of A is

$$I(A) = (\{f \in K[\mathbf{x},\mathbf{y}] \mid f \text{ is bihomogeneous and } f(a,b) = 0 \text{ for all } (a,b) \in A\})$$

and $V(I(A)) = \overline{A}$ is its Zariski closure. For an algebraic subset $A \subset \mathbb{P}^n \times \mathbb{P}^m$ the bigraded ring $K[\mathbf{x},\mathbf{y}]/I(A)$ is called the **bihomogeneous coordinate ring** of A.

Remark. For $J \subset K[\mathbf{x},\mathbf{y}]$ a bihomogeneous ideal we have

$$I(V(J)) = ((\operatorname{rad}(J) : (x_0,\ldots,x_n)) : (y_0,\ldots,y_m).$$

□

We now can define the product $A \times B \subset \mathbb{P}^n \times \mathbb{P}^m \subset \mathbb{P}^N$ of two arbitrary projective algebraic sets $A \subset \mathbb{P}^n$ and $B \subset \mathbb{P}^m$. It is the algebraic set defined by the bihomogeneous polynomials $f_i \in I(A) \subset K[\mathbf{x}]$ of bidegree $(d_i, 0)$ and $g_j \in I(B) \subset K[\mathbf{y}]$ of bidegree $(0, e_j)$.

Exercise 11.1.4. Prove that the Segre product $\mathbb{P}^n \times \mathbb{P}^m \subset \mathbb{P}^N$ with $N = (n+1)(m+1) - 1$ has dimension $\dim \mathbb{P}^n \times \mathbb{P}^m = n + m$ and degree $\deg \mathbb{P}^n \times \mathbb{P}^m = \binom{n+m}{n}$.

Exercise 11.1.5. Compute the Hilbert polynomial of a curve $C \subset \mathbb{P}^1 \times \mathbb{P}^1 \subset \mathbb{P}^3$ of bidegree (a, b).

11.2 Morphisms

Definition 11.2.1. A **quasi-affine algebraic set** is a Zariski open subset of an affine algebraic set. Similarly, we have the notion of a **quasi-projective algebraic set**. Every quasi-affine algebraic set is also quasi-projective because $\mathbb{A}^n = \mathbb{P}^n \setminus V(x_0)$.

The product of two quasi-affine (quasi-projective) algebraic sets $A = A_1 \setminus A_2$ and $B = B_1 \setminus B_2$ is again quasi-affine (quasi-projective) since

$$A \times B = A_1 \times B_1 \setminus (A_2 \times B_1 \cup A_1 \times B_2).$$

11.2 Morphisms

Definition 11.2.2. For $A \subset \mathbb{P}^n$ a quasi-projective algebraic set we define the **ring of regular functions** $\mathcal{O}(A)$ as the ring of functions

$$f: A \to K$$

such that for every point $p \in A$ there exists an open neighborhood $U \subset A$ and homogeneous polynomials $g, h \in K[x_0, \ldots, x_n]$ of the same degree with $h(q) \neq 0$ for all $q \in U$ such that

$$f(q) = \frac{g(q)}{h(q)} \ \forall q \in U.$$

Definition 11.2.3. Let A be a quasi-projective algebraic set.

1. Let $B \subset \mathbb{A}^m$ be a quasi-affine algebraic set. A **morphism** $\varphi: A \to B$ is a map which is given by an m-tuple of regular functions $f_j \in \mathcal{O}(A)$, i.e.,

$$\varphi(p) = (f_1(p), \ldots, f_m(p)) \ \forall p \in A.$$

2. Let $B \subset \mathbb{P}^m$ be a quasi-projective algebraic set. A map $\varphi: A \to B$ is a **morphism** if it is locally given by regular functions, i.e., for each point $p \in A$ there exist an open neighborhood $U \subset A$ and regular functions $f_0, \ldots, f_m \in \mathcal{O}(U)$ such that

$$\varphi(q) = (f_0(q) : \ldots : f_m(q)) \ \forall q \in U.$$

Clearly, morphisms can be composed.

A morphism $\varphi: A \to B$ is an isomorphism if there exists a morphism $\psi: B \to A$ such that $\psi \circ \varphi = \mathrm{id}_A$ and $\varphi \circ \psi = \mathrm{id}_B$.

Examples 11.2.4. 1) Let $A \subset \mathbb{P}^n$ be a quasi-projective algebraic set, and let $f_0, \ldots, f_m$ be homogeneous polynomials in $K[x_0, \ldots, x_n]$ of the same degree d such that the intersection $V(f_0, \ldots, f_m) \cap A$ is empty. Then

$$\varphi: A \to \mathbb{P}^m, p \mapsto (f_0(p) : \ldots : f_m(p))$$

is a well-defined morphism. Indeed on the open set $U = A \cap (\mathbb{P}^n \setminus V(f_i))$ the map φ is given by the regular functions

$$(\frac{f_0}{f_i} : \ldots : \frac{f_m}{f_i}),$$

and these open sets cover A since $V(f_0, \ldots, f_m) \cap A = \emptyset$.

In particular, we see that the regular functions in $\mathcal{O}(U)$ which define φ on U might not exist globally.

2) For an explicit example, consider the morphism $\rho_d: \mathbb{P}^1 \to \mathbb{P}^d$ defined by

$$(t_0 : t_1) \mapsto (t_0^d : t_0^{d-1} t_1 : \ldots : t_1^d).$$

The image of ρ_d is called the **rational normal curve of degree** d. It has the homogeneous ideal generated by the 2×2-minors of the $2 \times d$-matrix

$$\begin{pmatrix} x_0 & x_1 & \ldots & x_{d-1} \\ x_1 & x_2 & \ldots & x_d \end{pmatrix}.$$

Remark. Morphisms $\varphi\colon A \to B$ between affine algebraic sets are easier to describe because they simply correspond to K-algebra homomorphisms $\varphi^*\colon K[B] \to K[A]$. Morphisms $\varphi\colon A \to B$ between projective algebraic sets have a more complicated description. However they are better behaved: We will see in Section 11.5 that the image of a projective algebraic set under a morphism is always an algebraic subset of the target. This was not the case for morphisms between affine algebraic sets.

Example 11.2.5. Consider $A = V(xy - z^2) \subset \mathbb{P}^2$. On the affine chart $U_{z=1}$ we saw that the projection

$$\mathbb{A}^2 \supset V(xy - 1) \to \mathbb{A}^1, (a,b) \mapsto a$$

is not surjective, because the origin o is not in the image.

The map

$$A \setminus \{(0:1:0)\} \to \mathbb{P}^1, (x:y:z) \mapsto (x:z)$$

extends to a surjective morphism $\pi\colon A \to \mathbb{P}^1$ because

$$(x:z) = (xy:yz) = (z^2:yz) = (z:y)$$

holds on $A \setminus V(yz)$. Thus the missing preimage point of $o = (0:1) \in \mathbb{A}^1 \subset \mathbb{P}^1$ is the point $p = (0:1:0)$ on the line $V(z)$ at infinity.

Proposition 11.2.6. *Let C be a smooth irreducible quasi-projective curve. A rational map $\varphi'\colon C \dashrightarrow \mathbb{P}^n$ extends to a morphism $\varphi\colon C \to \mathbb{P}^n$.*

Proof. Suppose that φ' is given by a tuple $f_0, \ldots, f_n$ of rational functions. There are two reasons why $(f_0(p) : \ldots : f_n(p))$ might be not defined at $p \in C$. One of the rational functions might have a pole at p or all rational functions might vanish at p.

Take $v = \min\{v_p(f_j) \mid j = 0, \ldots, n\}$ and let $t \in \mathfrak{m}_p \subset \mathcal{O}_{C,p}$ be a generator. Then we see that $(t^{-v} f_0 : \ldots : t^{-v} f_n)$ is defined at $p \in C$ and coincides with φ' where t has no zeroes or poles. □

Remark. The proposition is not true for a higher-dimensional source: The morphism

$$\mathbb{A}^2 \setminus \{o\} \to \mathbb{P}^1, p \mapsto (x(p):y(p))$$

has no extension to $\mathbb{A}^2$. Instead the closure of the graph is the blow-up of $o \in \mathbb{A}^2$, see Section 12.1.

Exercise 11.2.7. Consider the algebraic set $S(e,d) \subset \mathbb{P}^{d+e+1}$ for $d, e \geq 1$ defined by the 2×2-minors of the matrix

$$\begin{pmatrix} x_0 & x_1 & \ldots & x_{d-1} & y_0 & y_1 & \ldots & y_{e-1} \\ x_1 & x_2 & \ldots & x_d & y_1 & y_2 & \ldots & y_e \end{pmatrix},$$

where $x_0 \ldots, x_d, y_0, \ldots, y_e$ are the homogeneous coordinates on $\mathbb{P}^{d+e+1}$.

1) Prove that there exists a morphism $\pi\colon S(d,e) \to \mathbb{P}^1$ whose fibers are lines.

2) Let $\phi_1\colon \mathbb{P}^1 \to \mathbb{P}^d = V(y_0,\dots,y_e)$ and $\phi_2\colon \mathbb{P}^1 \to \mathbb{P}^e = V(x_0,\dots,x_d) \subset \mathbb{P}^{d+e+1}$ be the parametrization of the rational normal curve of degree d and e in disjoint linear subspaces $\mathbb{P}^d \cup \mathbb{P}^e \subset \mathbb{P}^{d+e+1}$. Prove

$$S(e,d) \cong \bigcup_{p \in \mathbb{P}^1} \overline{\phi_1(p)\phi_2(p)},$$

where $\overline{\phi_1(p)\phi_2(p)}$ denotes the line joining $\phi_1(p)$ and $\phi_2(p)$.

11.3 Dimension bounds

Let $A \subset \mathbb{P}^n$ be a projective variety. Let $\ell_0,\dots,\ell_r \in K[x_0,\dots,x_n]$ be $r+1$ linearly independent linear forms such that $L = V(\ell_0,\dots,\ell_r) \cong \mathbb{P}^{n-r-1}$ does not intersect A. Then

$$\pi_L\colon A \to \mathbb{P}^r, a \mapsto (\ell_0(a) : \dots : \ell_r(a))$$

is called the **linear projection from** L.

The condition $A \cap L = \emptyset$ is equivalent to $\mathrm{rad}(I(A) + (\ell_0,\dots,\ell_r)) = (x_0,\dots,x_n)$. If we choose coordinates on $\mathbb{P}^n$ such that $\ell_0 = x_{n-r},\dots,\ell_r = x_n$, then $A \cap L = \emptyset$ is equivalent to the condition that there are homogeneous equations $f_i \in I(A)$ with

$$f_i \equiv x_i^{d_i} \mod (x_{n-r},\dots,x_n) \text{ for } i = 0,\dots,n-r-1.$$

So the map

$$\varphi\colon K[x_{n-r},\dots,x_n] \to K[A] = K[x_0,\dots,x_n]/I(A)$$

induces an integral ring extension $K[A'] \hookrightarrow K[A]$, where $A' = V(\ker(\varphi))$.

Thus in this situation π_L induces a finite surjective map from A to $A' \subset \mathbb{P}^r$. In particular, we have $\dim A' = \dim A = \dim K[A'] - 1 \leq \dim \mathbb{P}^r = r$.

Corollary 11.3.1. *Let $A \subset \mathbb{P}^n$ be a projective algebraic set. If there exists a linear subspace $L \subset \mathbb{P}^n$ of dimension $n - r - 1$ with $A \cap L = \emptyset$, then $\dim A \leq r$.* □

Definition 11.3.2. Let $A \subset \mathbb{P}^n$ be a projective algebraic set. We call a linear projection $\pi_L\colon A \to \mathbb{P}^r$ with $L \cap A = \emptyset$ and $r = \dim A$ a **linear Noether normalization**.

Corollary 11.3.3. *Let $A \subset \mathbb{P}^n$ be a projective algebraic set of dimension $\dim A = r$. Then every linear subspace L of dimension $\dim L \geq n - r$ intersects A.*

Proof. If $L \cap A = \emptyset$, then $\dim A < r$. □

Theorem 11.3.4 (Dimension bound). *Let $X, Y \subset \mathbb{P}^n$ be projective algebraic sets. Then*

$$\dim X \cap Y \geq \dim X + \dim Y - n.$$

In particular, the intersection of algebraic sets of complementary dimensions is always non-empty.

Proof. Consider the projective space $\mathbb{P}^{2n+1}$ with coordinate ring $K[x_0,\ldots,x_n,y_0,\ldots,y_n]$ and the algebraic set $J(X,Y)$ defined by $(I(X) \cup I(Y))$, where $I(X) \subset K[x_0,\ldots,x_n]$ and $I(Y) \subset K[y_0,\ldots,y_n]$ denote the homogeneous ideals in disjoint sets of variables. The algebraic set $J(X,Y)$ is called the **join** of X and Y because it is the union of all lines joining a point of X with a point of Y. We have

$$X \subset \mathbb{P}^n \cong V(y_0,\ldots,y_n) \subset \mathbb{P}^{2n+1} \supset V(x_0,\ldots,x_n) \cong \mathbb{P}^n \supset Y.$$

Clearly,
$$\dim J(X,Y) = \dim X + \dim Y + 1,$$

as one can see by combining linear Noether normalizations of X and Y. The intersection $X \cap Y$ is isomorphic to $J(X,Y) \cap V(x_0 - y_0,\ldots,x_n - y_n)$. So it is the intersection of $J(X,Y)$ with a linear subspace of dimension n. Thus the intersection $X \cap Y \neq \emptyset$ if

$$n \geq 2n + 1 - (\dim X + \dim Y + 1) \Leftrightarrow \dim X + \dim Y - n \geq 0$$

by Corollary 11.3.3.

Suppose $\dim X \cap Y = e > 0$. Let $\ell_0,\ldots,\ell_e \subset K[x_0,\ldots,x_n]$ define a linear Noether normalization of $X \cap Y$. Then

$$J(X,Y) \cap L = \emptyset,$$

where $L = V(x_0 - y_0,\ldots,x_n - y_n,\ell_0,\ldots,\ell_e)$ is a linear space of dimension $2n+1-(n+1+e+1) = 2n+1-(n+1+e)-1$, and

$$\dim J(X,Y) \leq n + 1 + e$$

holds by Corollary 11.3.1. Thus

$$\dim X \cap Y = e \geq \dim J(X,Y) - n - 1 = \dim X + \dim Y - n.$$

$\square$

Remark. Let $X, Y \subset \mathbb{P}^n$ be projective subvarieties. Using Krull's principal ideal theorem 10.5.5, one can show that every component C of $X \cap Y$ has dimension $\dim C \geq \dim X + \dim Y - n$.

11.4 The Veronese embeddings

Definition 11.4.1. Let $n, d \geq 1$, $N = \binom{n+d}{n} - 1$ and let $m_0 = x_0^d, \ldots, m_N = x_n^d$ denote the monomials of degree d in $K[x_0,\ldots,x_n]$ in some order. The monomials define a morphism

$$\rho_{n,d} \colon \mathbb{P}^n \to \mathbb{P}^N, p \mapsto (m_0(p) : \ldots : m_N(p))$$

11.4 The Veronese embeddings

which turns out to be an embedding, i.e., an isomorphism to its image $V_{n,d} \subset \mathbb{P}^N$. The map $\rho_{n,d}$ is called the **Veronese** or **d-uple embedding** of $\mathbb{P}^n$.

Example. In the case of $n = 1$ the morphism $\rho_{1,d}$ embeds $\mathbb{P}^1 \hookrightarrow \mathbb{P}^d$ as the rational normal curve of degree d in $\mathbb{P}^d$. The image is defined by the 2×2-minors of

$$\begin{pmatrix} x_0 & x_1 & \ldots & x_{d-1} \\ x_1 & x_2 & \ldots & x_d \end{pmatrix}.$$

Example 11.4.2 (The Veronese surface). We discuss

$$\rho_{2,2}: \mathbb{P}^2 \to \mathbb{P}^5, (x : y : z) \mapsto (x^2 : xy : y^2 : xz : yz : z^2)$$

in some detail. We use homogeneous coordinates $w_0, \ldots, w_5$ on $\mathbb{P}^5$. Consider the symmetric matrix

$$\Delta = \begin{pmatrix} w_0 & w_1 & w_3 \\ w_1 & w_2 & w_4 \\ w_3 & w_4 & w_5 \end{pmatrix}$$

and let $V \subset \mathbb{P}^5$ be the algebraic set defined by the ideal I of 2×2-minors of Δ. Clearly, these minors vanish at all points of $\rho_{2,2}(\mathbb{P}^2)$. We show that $\rho_{2,2}$ induces an isomorphism of $\mathbb{P}^2$ with V by describing the inverse morphism ψ. Note that $V \subset U_0 \cup U_2 \cup U_5$ holds because

$$w_1^2, w_3^2, w_4^2 \in I + (w_0, w_2, w_5).$$

On $V \cap U_0$ the inverse map is given by

$$p \mapsto (w_0(p) : w_1(p) : w_3(p))$$

because $(x^2 : xy : xz) = (x : y : z)$ on $U_x = \{x \neq 0\} \subset \mathbb{P}^2$.

Similarly, the inverse map ψ is given on $V \cap U_2$ and $V \cap U_5$ by the second, respectively third, row of Δ. The maps coincide on $V \cap U_i \cap U_j$ for $i, j \in \{0, 2, 5\}$ since the 2×2-minors of Δ vanish on V. Thus these pieces glue to a well-defined morphism $\psi : V \to \mathbb{P}^2$ and $\mathbb{P}^2 \cong V$.

Remark. Notice that ψ is a morphism which cannot be defined globally by only one tuple of three homogeneous polynomials of the same degree.

Our next goal is to establish the following.

Claim. *The ideal I of 2×2 minors of Δ is the homogeneous ideal of V.*

Proof. Consider the ring homomorphism

$$\varphi: K[w_0, \ldots, w_5] \to K[x, y, z], w_0 \mapsto x^2, w_1 \mapsto xy, \ldots, w_5 \mapsto z^2.$$

Then $I \subset \ker(\varphi)$. To prove equality we consider a reverse lexicographic order where $w_0, \ldots, w_5$ are ordered such that $w_1, w_3, w_4 > w_0, w_2, w_5$. Then the leading terms of the minors are

$$w_1^2, w_1w_3, w_3^2, w_1w_4, w_3w_4, w_4^2$$

up to sign. Thus a remainder of the division by the minors is at most linear in w_1, w_3 and w_4, and there are precisely

$$\binom{d+2}{2} + 3\binom{d+1}{2} = 2d^2 + 3d + 1 = \binom{2d+2}{2}$$

different monomials of degree d occurring as remainders. Since φ is surjective, the homogeneous coordinate ring S_V has precisely that many elements in degree d. Thus $I = \ker(\varphi)$ and the minors form a Gröbner basis of $I(V)$. The ideal I is prime because

$$S_V = K[w_0, \ldots, w_5]/I \cong K[x^2, xy, y^2, xz, yz, xz] \subset K[x, y, z]$$

is isomorphic to a subring of a domain. Finally, we note that the Hilbert polynomial of V is

$$p_V(t) = \binom{2t+2}{2} = 4\frac{t^2}{2!} + 3t + 1, \text{ hence } \deg V = 4.$$

□

Theorem 11.4.3 (Veronese embedding). *The Veronese morphism*

$$\rho_{n,d} \colon \mathbb{P}^n \to \mathbb{P}^N, (x_0 : \ldots : x_n) \mapsto (x_0^d : x_0^{d-1}x_1 : \ldots : x_n^d),$$

where $N = \binom{n+d}{d} - 1$, *induces an isomorphism onto its image* $V_{n,d}$ *which is a subvariety of* $\mathbb{P}^N$ *of degree* $\deg V_{n,d} = d^n$.

Proof. The homogeneous coordinate ring of $\mathbb{P}^N$ has a variable y_α for each $\alpha \in \mathbb{N}^n$ with $|\alpha| = d$, and $\rho_{n,d}$ corresponds to the ring homomorphism

$$\varphi \colon K[y_\alpha's] \to K[x_0, \ldots, x_n], y_\alpha \mapsto x^\alpha.$$

We will show that $V_{n,d}$ coincides with the projective variety $V(\ker(\varphi)) \subset \mathbb{P}^N$. Some equations in $I = \ker(\varphi)$ can be obtained as follows: Consider the $\binom{n+d-1}{n} \times (n+1)$-matrix Δ with rows corresponding to monomials of x^β of degree $d-1$ and columns corresponding to the variables $x_0, \ldots, x_n$. The entries are $\Delta_{x^\beta, x_j} = y_\alpha$, where $x^\alpha = x^\beta x_j$. The 2×2-minors of Δ are contained in I.

The algebraic set $V_{n,d}$ is contained in the union of $n+1$ standard charts of $\mathbb{P}^N$: The intersection $V_{n,d} \cap V(y_{(d,0,\ldots,0)}, \ldots, y_{(0,\ldots,0,d)})$ is empty because

$$y_\alpha^d - y_{(d,0,\ldots,0)}^{\alpha_0} \cdot \ldots \cdot y_{(0,\ldots,0,d)}^{\alpha_n} \in I.$$

Thus

$$V_{n,d} \subset \widetilde{U}_0 \cup \ldots \cup \widetilde{U}_n$$

for $\widetilde{U}_j = \{y_{(0,\ldots,d,\ldots 0)} \neq 0\}$ corresponding to the monomial x_j^d.

11.4 The Veronese embeddings

The map $\rho_{n,d}$ restricts to an isomorphism between U_0 and $V_{n,d} \cap \tilde{U}_0$. Its inverse is given by the map

$$p \mapsto (y_{(d,0,\ldots,0)}(p) : y_{(d-1,1,\ldots,0)}(p) : \ldots : y_{(d-1,0,\ldots,1)}(p))$$

corresponding to the row $\Delta_{x_0^{d-1}}$. Similarly,

$$V_{n,d} \cap \tilde{U}_j \cong U_j$$

by the map defined by the row $\Delta_{x_j^{d-1}}$. These maps glue to a well-defined inverse morphism

$$\psi : V_{n,d} \to \mathbb{P}^n$$

since the 2×2-minors of Δ vanish on $V_{n,d}$. To compute the degree we compute the Hilbert polynomial. Since $K[y_\alpha \ 's]/I \cong K[x^\alpha \ 's] \subset K[x_0, \ldots, x_n]$, we obtain

$$p_{V_{n,d}}(t) = \binom{dt+n}{n} = d^n \frac{t^n}{n!} + \text{lower terms}.$$

□

Corollary 11.4.4. *Every quasi-projective algebraic set has a finite open covering by affine algebraic sets.*

Proof. Let

$$f = \sum_\alpha f_\alpha x^\alpha \in K[x_0, \ldots, x_n]$$

be a homogeneous polynomial of degree d. Consider the open set $U_f = \mathbb{P}^n \setminus V(f)$ and the corresponding hyperplane $H_f = V(\sum_\alpha f_\alpha y_\alpha) \subset \mathbb{P}^N$. Under the Veronese embedding the set U_f is isomorphic to the Zariski-closed subset of $\mathbb{A}^N$ since

$$U_f \cong V_{n,d} \cap (\mathbb{P}^N \setminus H_f) \subset \mathbb{P}^N \setminus H_f \cong \mathbb{A}^N.$$

Hence U_f is an affine variety.

Let $A = A_1 \setminus A_2$ be a quasi-projective set, where $A_2 \subset A_1 \subset \mathbb{P}^n$ are projective algebraic subsets. If $A_2 = V(f_1, \ldots, f_r)$, then

$$A = \bigcup_{j=1}^{r} (A_1 \cap U_{f_j})$$

is an open covering. Since $A_1 \cap U_{f_j}$ is a closed subset of the affine variety U_{f_j}, it is isomorphic to an affine algebraic set. □

Exercise 11.4.5. Let $X \subset \mathbb{P}^n$ be a projective variety of dimension r, and consider the image $X' = \rho_{n,d}(X)$ of X under the Veronese embedding $\rho_{n,d} : \mathbb{P}^n \hookrightarrow \mathbb{P}^N$ where $N = \binom{d+n}{n} - 1$. Prove: The Hilbert polynomials $p_X(t)$ and $p_{X'}(t)$ have the same constant term and $\deg X' = d^r \deg X$. In particular, the arithmetic genus of X and X' as defined in Remark 8.4.8 coincide.

Exercise 11.4.6. Consider the Veronese surface $V \subset \mathbb{P}^5$. Recall V is defined by the 2×2-minors of
$$\Delta = \begin{pmatrix} w_0 & w_1 & w_3 \\ w_1 & w_2 & w_4 \\ w_3 & w_4 & w_5 \end{pmatrix}.$$
Consider the point $p = (0:1:0:1:1:0) \in \mathbb{P}^5$ and compute the image $V' \subset \mathbb{P}^4$ of the projection of V from p. Show that $V' \cong V$ in the case of char $K \neq 2$. A famous result of Veronese says that if $X \subset \mathbb{P}^5$ is a smooth surface which can be projected isomorphically from a point $p \notin X$, then either X is already contained in a linear subspace $\mathbb{P}^4 \subset \mathbb{P}^5$ or X is projectively equivalent to the Veronese surface.

Exercise 11.4.7. Let $V_{n,d}$ denote the image of the Veronese embedding $\rho_{n,d} : \mathbb{P}^n \hookrightarrow \mathbb{P}^N$ where $N = \binom{d+n}{n} - 1$, and let Δ denote the $(n+1) \times \binom{n+d-1}{n}$-matrix of linear forms on $\mathbb{P}^N$ from the proof of Theorem 11.4.3. Prove: The homogeneous ideal of $V_{n,d}$ is generated by the 2×2 minors of Δ.

11.5 Morphisms from projective algebraic sets

Theorem 11.5.1. *Let A be a projective algebraic set and $\varphi : A \to B$ a morphism to a quasi-projective algebraic set. Then $\varphi(A) \subset B$ is a Zariski-closed subset.*

Corollary 11.5.2. *Let A be a projective variety. Every regular function $f : A \to K$ is constant.*

Proof of the Corollary. The regular function f defines a morphism $f : A \to \mathbb{A}^1 \subset \mathbb{P}^1$. The image is closed in $\mathbb{P}^1$, hence different from $\mathbb{A}^1$. Since it is also closed in $\mathbb{A}^1$, it is a finite union of points. Since A is irreducible, it is a single point. □

Remark. When $K = \mathbb{C}$, this corollary is similar to the maximum principle: If f is a holomorphic function on a compact complex connected manifold A, then $|f|$ attains its maximum, and hence f is constant.

Lemma 11.5.3. *Let $\varphi : A \to B$ a morphism between quasi-projective algebraic sets. Then the graph of φ is a closed subset of $A \times B$.*

Proof. To be a closed subset is a local property. Hence we may replace B by an open affine subset U and A by an open affine subset of $\varphi^{-1}(U)$ since every quasi-projective algebraic set has an open affine covering by Corollary 11.4.4. Thus we may assume that A and B are subsets of $\mathbb{A}^n$ and $\mathbb{A}^m$, respectively, and that φ is given by a tuple of polynomial functions $(f_1, \ldots, f_m)$. Then the graph of φ is defined by the ideal
$$(y_1 - f_1(x_1, \ldots, x_n), \ldots, y_m - f_m(x_1, \ldots, x_n))$$
on $A \times B$. □

11.5 Morphisms from projective algebraic sets

Passing to the graph reduces the proof of Theorem 11.5.1 to the proof of the following theorem.

Theorem 11.5.4 (Fundamental theorem of elimination). *Let A be a projective algebraic set and B a quasi-projective algebraic set. Then the projection onto the second factor $A \times B \to B$ is a closed map, i.e., it maps closed subsets of $A \times B$ to closed subsets of B.*

Proof. We may replace B by an open affine subset. Hence we may assume that $A \subset \mathbb{P}^n$ and $B \subset \mathbb{A}^m$ are closed subsets, and it suffices to prove that the projection $\mathbb{P}^n \times \mathbb{A}^m \to \mathbb{A}^m$ is closed.

Any algebraic subset $X \subset \mathbb{P}^n \times \mathbb{A}^m$ is defined by finitely many polynomials $f_1, \ldots, f_r$ in $K[x_0, \ldots, x_n, y_1, \ldots, y_m]$, where each f_i is homogeneous of some degree d_i in $x_0, \ldots, x_n$. By the projective Nullstellensatz 8.2.3, a point $q \in \mathbb{A}^m$ lies in the image of X if and only if the ideal

$$I(q) := (f_1(\mathbf{x}, q), \ldots, f_r(\mathbf{x}, q)) \subset K[\mathbf{x}]$$

does not contain any of the ideals $(x_0, \ldots, x_n)^d$ for $d \geq 1$.

Define

$$Y_d = \{q \in \mathbb{A}^m \mid I(q) \not\supset (x_0, \ldots, x_n)^d\}.$$

Then the image of X is

$$Y = \bigcap_{d=1}^{\infty} Y_d$$

and it suffices to prove that each Y_d is an algebraic subset of $\mathbb{A}^m$.

To obtain equations for Y_d we multiply each f_i by all monomials of degree $d - d_i$ in $\mathbf{x}$. Let T_d denote the resulting set of polynomials. Then $q \notin Y_d$ if and only if each monomial in $K[x_0, \ldots, x_n]_d$ of degree d is a linear combination of the polynomials $f(\mathbf{x}, q)$ with $f \in T_d$. Comparing coefficients we obtain a $\binom{d+n}{n} \times \sum_{i=1}^{r} \binom{d-d_i+n}{n}$-matrix M_d with entries in $K[y_1, \ldots, y_m]$ such that $q \in Y_d$ if and only if rank $M_d(q) < \binom{d+n}{n}$. Thus the $\binom{d+n}{n} \times \binom{d+n}{n}$-minors of M_d define Y_d. □

The proof of the theorem does not yield a practical algorithm to compute the image. Here is an approach which works frequently in practice.

Definition 11.5.5. Let I, J be ideals in a ring. Then the **saturation of I with respect to J** is

$$I : J^{\infty} = \bigcup_{N=1}^{\infty} (I : J^N).$$

To compute the saturation in Noetherian rings one can iterate

$$I_{k+1} = I_k : J$$

starting with $I_0 = I$ until $I_{N+1} = I_N$. Then $I_N = I : J^{\infty}$.

In the situation of $X \subset \mathbb{P}^n \times \mathbb{A}^m$ defined by $f_1, \ldots, f_r$ as above, we obtain equations of the image $Y \subset \mathbb{A}^m$ by taking the elements of degree 0 in $\mathbf{x}$ of

$$(f_1, \ldots, f_r) : (x_0, \ldots, x_n)^\infty.$$

The proof of Theorem 11.5.4 actually gives a stronger result.

Definition 11.5.6. A morphism $\varphi : A \to B$ is called a **projective morphism** if it is the composition of a closed embedding $\iota : A \to \mathbb{P}^n \times B$ with the projection onto B. Here a morphism $\iota : A \to C$ is called a **closed embedding** if ι induces an isomorphism $A \to \iota(A)$ and $\iota(A)$ is a Zariski-closed subset of C.

Theorem 11.5.7. *A projective morphism is a closed map.*

Proof. Indeed, in the proof of the fundamental theorem of elimination we replaced $A \subset \mathbb{P}^n$ with the graph of φ in $A \times B \subset \mathbb{P}^n \times B$. The map $\iota : A \to \mathbb{P}^n \times B$ to the graph of φ in $\mathbb{P}^n \times B$ is a closed embedding. In the remaining part of the proof all we used was that $A \cong \iota(A) \subset \mathbb{P}^n \times B$ is closed. Thus closed subsets of $X \subset A$ are also closed subsets of $\mathbb{P}^n \times B$, and our argument showed that the image Y of X under the projection to B is closed in B. □

11.6 Semi-continuity of the fiber dimension

Definition 11.6.1. Let $\varphi : X \to Y$ be a morphism. The **fiber** of φ over a point $q \in Y$ is $X_q := \varphi^{-1}(q) \subset X$. Since morphisms are continuous in the Zariski topology, the preimage of the point q is a Zariski-closed subset of X.

If φ is a projective morphism, say φ factors over a closed embedding $\iota : X \to \mathbb{P}^n \times Y$, then the situation is better: The fiber

$$X_q \subset \mathbb{P}^n \times \{q\} \cong \mathbb{P}^n$$

is a projective algebraic set for every $q \in Y$.

11.6 Semi-continuity of the fiber dimension

Example. On the right there is an illustration of the surface

$$X = V(y^2z - x^2(t^2z - x)) \subset \mathbb{P}^2 \times \mathbb{A}^1$$
$$\downarrow \varphi$$
$$Y = \mathbb{A}^1$$

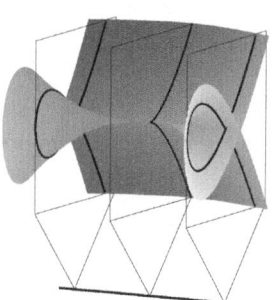

together with three fibers of the projection onto the t-axis.

Theorem 11.6.2 (Semi-continuity of the fiber dimension). *Let $\varphi \colon X \to Y$ be a projective morphism and let $r \geq -1$ be an integer. Then the set*

$$U_r = \{q \in Y \mid \dim X_q \leq r\}$$

is Zariski-open in Y.

The fiber dimension of special points can be larger than the fiber dimension for more general points.

Proof. The set

$$U_{-1} = \{q \in Y \mid \dim X_q \leq -1\} = \{q \in Y \mid X_q = \emptyset\}$$

is open in Y because it is the complement of the closed subset $\varphi(X) \subset Y$.

Suppose $\dim X_q = r \geq 0$. We assume that φ factors over $\mathbb{P}^n \times Y$. Consider a linear space $L \subset \mathbb{P}^n$ of dimension $n - r - 1$ with $X_q \cap L = \emptyset$ and

$$Z = X \cap (L \times Y) \subset \mathbb{P}^n \times Y.$$

Fibers X_q with $\dim X_q > r$ intersect Z by Corollary 11.3.3. Thus $U = Y \setminus \varphi(Z)$ is an open neighborhood of $q \in U_r$. □

In the case of a surjective projective morphism between varieties, the result can be strengthened.

Theorem 11.6.3 (Dimension of the general fiber). *Let $\varphi \colon X \to Y$ be a surjective projective morphism between varieties. Then*

$$\dim X_q \geq \dim X - \dim Y,$$

and equality holds for q in a non-empty open subset U of Y.

Proof. We may assume that Y is affine and that $X \subset \mathbb{P}^n \times Y$ is a closed subset. Consider the function fields
$$K(Y) \subset K(X).$$
We have
$$\operatorname{trdeg}_K K(X) = \operatorname{trdeg}_{K(Y)} K(X) + \operatorname{trdeg}_K K(Y).$$
Let $I \subset K[Y][x_0, \ldots, x_n]$ be the ideal of $X \subset \mathbb{P}^n \times Y$. Consider the ideal
$$J \subset K(Y)[x_0, \ldots, x_n]$$
generated by I. Then J corresponds to a variety $V(J)$ defined over the function field $K(Y)$ of dimension
$$\dim V(J) = \operatorname{trdeg}_{K(Y)} K(X) = \dim X - \dim Y.$$
Here we regard $V(J)$ as a subset of $\mathbb{P}^n(L)$, where L denotes the algebraic closure of $K(Y)$, and use that $\operatorname{trdeg}_{K(Y)} K(X) = \operatorname{trdeg}_L L(X)$ holds. We compute a normalized Gröbner basis of J, i.e., one where the leading coefficients of all Gröbner basis elements are 1. To do this, we have to divide by finitely many polynomial functions in $K[Y]$. Let $f \in K[Y]$ be the product of these polynomials and $U_f = Y \setminus V(f)$ the corresponding non-empty open subset. We claim that for a point $q \in U_f$ the ideal
$$I_q = (\{g(x, q) | g \in I\}) \subset K[x_0, \ldots, x_n]$$
defines an algebraic set of dimension $\dim X - \dim Y$.

Indeed, the computation of the Gröbner basis of I_q follows the same steps as the computation for $J = (I)$. We simply have to substitute q into the rational functions in $K(Y)$ which are the coefficients. Since each coefficient has a representation as a fraction with a power of f in the denominator, the coefficients can be evaluated in q. Thus J and I_q have the same ideal of leading terms. Hence $K(Y)[x_0, \ldots, x_n]/J$ and $K[x_0, \ldots, x_n]/I_q$ have the same Hilbert polynomial. In particular,
$$\dim X_q = \dim V(J) = \operatorname{trdeg}_{K(Y)} K(X) = \dim X - \dim Y$$
for all $q \in Y$. Since Y is irreducible, U_f is dense in Y. So we obtain
$$\dim X_q \geq \dim X - \dim Y$$
from the semi-continuity of the fiber dimension 11.6.2. □

An example of a morphism between varieties where a special fiber has a larger dimension is the blow-up which we discuss in Section 12.1.

As a corollary of the proof we note

Corollary 11.6.4. *Let Y be an affine variety and let $I \subset K[Y][x_0, \ldots, x_n]$ be an ideal which is homogeneous in $x_0, \ldots, x_n$. Then there exists a non-empty open subset $U \subset Y$ such that the ideals*

11.6 Semi-continuity of the fiber dimension

$$I_q = (\{g(x,q) \mid g \in I\}) \subset K[x_0, \ldots, x_n]$$

have the same Hilbert function for all $q \in U$.

Theorem 11.6.5 (Gröbner basis over prime fields). *Let $f_1, \ldots, f_r \in \mathbb{Z}[x_0, \ldots, x_n]$ be homogeneous polynomials. Let $I_\mathbb{Q} \subset \mathbb{Q}[x_0, \ldots, x_n]$ and $I_p \subset \mathbb{F}_p[x_0, \ldots, x_n]$ for p a prime number denote the ideals generated by $f_1, \ldots, f_r$ in these rings. Then for all but finitely many primes the ideals of leading terms*

$$\mathrm{Lt}(I_p) \text{ and } \mathrm{Lt}(I_\mathbb{Q})$$

are generated by the same monomials. In particular, their Hilbert polynomials coincide.

Proof. We compute a normalized Gröbner basis of $I_\mathbb{Q}$. In this process we have to divide by finitely many leading coefficients, and the Gröbner basis of the ideal I_p where p does not divide any of the leading coefficients is obtained by mapping the coefficients $\frac{a}{b} \in \mathbb{Q}$ to $ab^{-1} \in \mathbb{F}_p$. □

Remark. Notice that a Gröbner basis over $\mathbb{Q}$ can have very large coefficients: In adding or multiplying two rational numbers

$$\frac{a}{b} + \frac{c}{d} = \frac{ad+bc}{bd} \quad \text{or} \quad \frac{a}{b} \cdot \frac{c}{d} = \frac{ac}{bd},$$

one often obtains numbers with twice the number of digits in the numerator and denominator.

By passing to a finite prime field, this effect is avoided. If we are only interested, say, in the degree and the dimension of the $V(I_\mathbb{Q})$, then the result does not change for almost all primes. This is frequently used in experiments in algebraic geometry.

Exercise 11.6.6. Let $\varphi \colon X \to Y$ be a projective morphism. Then

$$A_r = \{q \in Y \mid \dim X_q \geq r\} \subset Y$$

is a Zariski-closed subset of Y. Suppose that X and Y are varieties and that φ is surjective. Prove that

$$\dim A_r + r < \dim X$$

for r with $r > \dim X - \dim Y$.

Exercise 11.6.7. Let $X \subset \mathbb{P}^n$ be a projective variety. The **secant variety** of X is

$$\mathrm{Sec}(X) = \overline{\bigcup_{p,q \in X^*, p \neq q} \overline{pq}} \subset \mathbb{P}^n,$$

where $\overline{pq}$ is the line spanned by the two points p and q and $X^* = X \setminus X_{\mathrm{sing}}$ is the open set of smooth points. Prove the following.

1) The secant variety has dimension $\dim \mathrm{Sec}(X) \leq 2 \dim X + 1$.

2) If X is smooth and p a point not contained in $\mathrm{Sec}(X)$, then the projection

$$\pi_p : \mathbb{P}^n \dashrightarrow \mathbb{P}^{n-1}$$

from p induces an isomorphism $X \cong \pi_p(X)$.

Conclude that every irreducible smooth projective curve can be embedded into $\mathbb{P}^3$.

Exercise 11.6.8. Make the following experiment with Macaulay2 repeatedly. Choose a finite prime field $k = \mathbb{F}$ and three random forms $f_0, f_1, f_2 \in k[x_0, x_1]$ of degree d. Compute the degree and arithmetic genus of the image $C \subset \mathbb{P}^2$ of

$$\mathbb{P}^1 \dashrightarrow \mathbb{P}^2, p \mapsto (f_0(p) : f_1(p) : f_2(p)).$$

How many singular points does C have? What kind of singularities?

Make a conjecture! Can you prove it?

Exercise 11.6.9. For a variety $X \subset \mathbb{P}^n$ we denote by $\mathrm{Sec}_k(X) \subset \mathbb{P}^n$ the **k-th secant variety of** X, i.e., the closure of the union of the linear span of any k-tuple of points on X.

Consider the $(k+1) \times (n-k+1)$ Hankel matrix

$$H_{k,n} = \begin{pmatrix} x_0 & x_1 & \cdots & x_{n-k} \\ x_1 & x_2 & \cdots & x_{n-k+1} \\ \vdots & \vdots & & \vdots \\ x_k & x_{k+1} & \cdots & x_n \end{pmatrix} = (x_{i+j})_{\substack{i=0,\ldots,k \\ j=0,\ldots,n-k}}$$

of linear forms on $\mathbb{P}^n$. Prove: The ideal of $(k+1) \times (k+1)$-minors of $H_{k,n}$ define the k-th secant variety $\mathrm{Sec}_k(R)$ of the rational normal curve $R = V_{1,n} \subset \mathbb{P}^n$.

Chapter 12
Resolution of curve singularities

In this chapter we introduce the blow-up of the origin o in $\sigma\colon X \to \mathbb{A}^2$ and more generally of a surface in smooth points. σ replaces o with the projective tangent space $\mathbb{P}(T_p\mathbb{A}^2) \cong \mathbb{P}^1$. Repeated blow-ups can transform a singular plane curve to a smooth curve. In Section 12.2 we introduce the quadratic transformation

$$q\colon \mathbb{P}^2 \dashrightarrow \mathbb{P}^2, (x:y:z) \mapsto (\frac{1}{x}:\frac{1}{y};\frac{1}{z})$$

and prove that the existence of a Cremona resolution implies the existence of a resolution of singularities by repeated blow-up for plane curves. Here a Cremona resolution is a sequence of quadratic transformations in suitable coordinates which transforms a plane curve into a curve with only ordinary singularities. The existence of a Cremona resolution is postponed until Chapter 15, since we need Bertini's theorem.

12.1 The blow-up

Definition 12.1.1. Let $X \subset \mathbb{P}^1 \times \mathbb{A}^2$ be defined by

$$\det \begin{pmatrix} z_0 & z_1 \\ x & y \end{pmatrix} \in K[z_0, z_1, x, y]$$

and let $\sigma\colon X \to \mathbb{A}^2$ denote the projection onto the second component. σ is called the **blow-up** of $\mathbb{A}^2$ at the origin o.

X is covered by two affine charts $U_j = X \cap (U_{z_j} \times \mathbb{A}^2)$ which are both isomorphic to $\mathbb{A}^2$.

$$K[U_0] \cong K[z, x, y]/(y - xz) \cong K[x, z]$$

and the map $\sigma|_{U_0}\colon U_0 \to \mathbb{A}^2$ is given by $(x, z) \mapsto (x, xz)$. Similarly

$K[U_1] \cong K[w, y]$ and $\sigma|_{U_1} : U_1 \to \mathbb{A}^2, (w, y) \mapsto (wy, y)$.

The fiber of σ over $o = (0, 0) \in \mathbb{A}^2$ is $E = \mathbb{P}^1 \times \{o\} \cong \mathbb{P}^1$. The curve E is called the **exceptional curve of** σ. Outside E the map σ restricts to an isomorphism $X \setminus E \cong \mathbb{A}^2 \setminus \{o\}$. $X \setminus E \subset \mathbb{P}^1 \times (\mathbb{A}^2 \setminus \{o\})$ is isomorphic to the graph of the morphism

$$\mathbb{A}^2 \setminus \{o\}, (x, y) \mapsto (x : y),$$

and $X \subset \mathbb{P}^1 \times \mathbb{A}^2$ is its Zariski closure.

In other words, we may think of

$$X = V(\det \begin{pmatrix} z_0 & z_1 \\ x & y \end{pmatrix}) \subset \mathbb{P}^1 \times \mathbb{A}^2$$

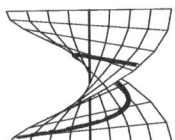

as obtained from $\mathbb{A}^2$ by replacing the origin o by the projective space $E \cong \mathbb{P}^1$ of lines through o.

Definition 12.1.2. Let $C \subset \mathbb{A}^2$ be a plane curve. The **strict transform** of C is the closure $C' = \overline{\sigma^{-1}(C \setminus \{o\})} \subset X$. The **total transform** is $\sigma^{-1}(C)$.

Proposition 12.1.3. *Let $C = V(f)$ be a curve of multiplicity m at the origin. Then the strict transform $C' \subset X$ intersects E in precisely m points counted with multiplicities.*

Proof. Suppose $f = f_m + \ldots + f_d \in K[x, y]$ with f_j homogeneous of degree j. The total transform of C in the chart U_0 is defined by

$$f(x, xz) = x^m(f_m(1, z) + x f_{m+1}(1, z) + \ldots + x^{d-m} f_d(1, z)) = 0.$$

The exceptional curve E is defined by $x = 0$ on U_0. So the strict transform C' is defined by
$$f_m(1, z) + x f_{m+1}(1, z) + \ldots + x^{d-m} f_d(1, z) = 0.$$

The intersection points of $E \cap C'$ contained in U_0 are defined by $V(f_m(1, z), x)$. Let $f_m = \prod_{i=1}^r \ell_i^{e_i}$ be the factorization of f_m into distinct linear factors. The intersection multiplicity at the point $p_j = (a_j : b_j) \in \mathbb{P}^1 = E$ corresponding to the tangent line $V(\ell_j)$ with $\ell_j = b_j x - a_j y$ is

$$i(C', E; p_j) = e_j$$

since the factors ℓ_i for $i \neq j$ are units in $\mathcal{O}_{X, p_j}$. Hence $\sum_{i=1}^r e_i = m$ implies the result. □

Corollary 12.1.4. *If o is an ordinary m-fold point, then E and C' have transversal intersections and C' is non-singular at the intersection points.* □

12.1 The blow-up

Since X is covered by charts isomorphic to $\mathbb{A}^2$ we can iterate this process.

Example 12.1.5. Consider $C = V(y^3 - x^5)$. The strict transform of C is contained in the chart U_0 where the total transform is defined by $x^3(z^3 - x^2)$.

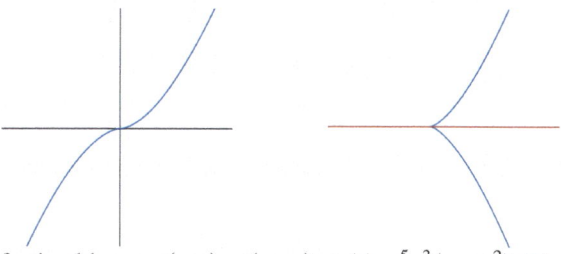

The further blow-up $(x, z) = (uz, u)$ yields $u^5 z^3 (u - z^2)$. Blowing-up the intersection point of the second exceptional curve $E_2 = \{u = 0\}$ with C'' via $(u, z) = (wz, z)$ yields the local equation $w^5 z^9 (w - z)$, and all curves intersect transversally.

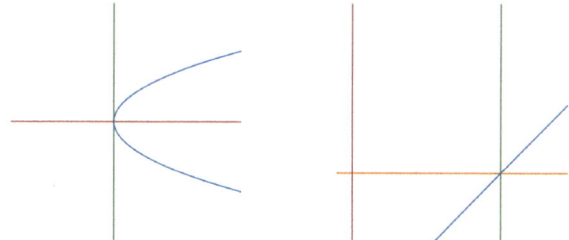

Remark 12.1.6. The real points of $C = V(y^3 - x^5)$ form the graph of a differentiable function $C(\mathbb{R}) = \{(x, y) \in \mathbb{R}^2 \mid y = x^{\frac{5}{3}}\}$. The singularity is more apparent if we look at the complex solutions: Intersect $C \subset \mathbb{C}^2$ with a small sphere

$$S_\epsilon^3 = \{(x, y) \in \mathbb{C}^2 \mid |x|^2 + |y|^2 = \epsilon^2\}$$

of radius ϵ. The intersection $C \cap S_\epsilon^3$ is a knot. In this case it is a torus knot on the torus

$$T = \{(x, y) \mid |x| = \epsilon_1 \text{ and } |y| = \epsilon_2\},$$

where $(\epsilon_1, \epsilon_2) \in C(\mathbb{R})$ is the real point with positive coordinates of distance ϵ from the origin, because $|x|^5 = |y|^3$ holds on C.

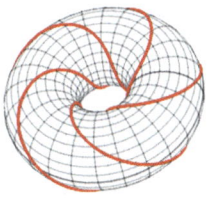

The winding numbers are 3 and 5. For more on the fascinating story about the links of plane curve singularities, see Brieskorn-Knörrer [11].

Theorem 12.1.7 (Resolution of singularities). *Let $C \subset \mathbb{A}^2$ (or $\mathbb{P}^2$) be a plane algebraic curve. Then there exists a sequence*

$$X_r \xrightarrow{\sigma_r} X_{r-1} \longrightarrow \ldots \longrightarrow X_1 \xrightarrow{\sigma_1} \mathbb{P}^2$$

of blow-ups such that the strict transform $C^{(r)}$ of C in X_r is a non-singular curve.

The main difficulty in proving this theorem is to prove that some numerical invariant improves along the process of blow-ups. In the example above, such an invariant was the multiplicity of the singular points. However, in general, a more subtle invariant is needed.

Example 12.1.8. Consider $y^2 - x^4 + x^6 = 0$. Substituting $(x, y) = (x, xz)$ leads to the strict transform $z^2 - x^2 + x^4 = 0$, which still has a double point at the origin. A second blow-up $(x, z) = (uz, z)$ gives the strict transform $u^2 - 1 + u^4 z^2 = 0$, which actually is now a smooth curve.

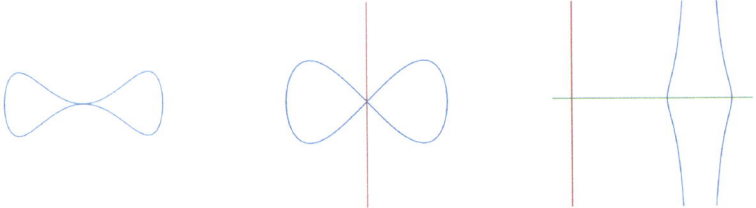

Example 12.1.9. The blow-up of a point $p \in \mathbb{A}^2 \cong U_2 \subset \mathbb{P}^2$ glues with the U_0, U_1 to the blow-up $\mathbb{P}^2(p)$ of $\mathbb{P}^2$ at p. This is a projective surface which one can describe explicitly as follows: Consider the rational map

$$\mathbb{P}^2 \dashrightarrow \mathbb{P}^4, (x : y : z) \mapsto (x^2 : xy : y^2 : xz : yz).$$

The image is the **cubic scroll** defined by

$$\operatorname{rank} \begin{pmatrix} w_0 & w_1 & w_3 \\ w_1 & w_2 & w_4 \end{pmatrix} < 2.$$

It is the projection of the Veronese surface $V_{2,2} \subset \mathbb{P}^5$ from the point

$$\rho_{2,2}((0 : 0 : 1)) = (0 : \ldots : 0 : 1).$$

12.2 Blow-up of smooth projective surfaces

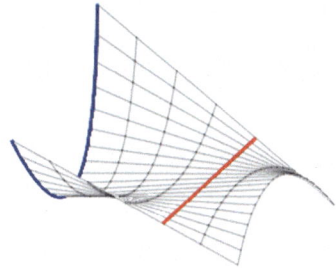

Note that we can identify the exceptional line

$$E = \mathbb{P}^1 = V(w_0, w_1, w_2)$$

with the projective tangent space $E \cong \mathbb{P}(T_p\mathbb{P}^2)$. The lines through p correspond to lines defined by the three linear forms

$$\begin{pmatrix} \lambda_0 & \lambda_1 \end{pmatrix} \begin{pmatrix} w_0 & w_1 & w_3 \\ w_1 & w_2 & w_4 \end{pmatrix}$$

of a generalized row of the 2×3 matrix.

12.2 Blow-up of smooth projective surfaces

More generally one can define the blow-up $X(p)$ of a smooth projective surface $X \subset \mathbb{P}^n$ at a point p as the image of the composition of

$$X \hookrightarrow \mathbb{P}^n \hookrightarrow \mathbb{P}^{\binom{n+2}{2}-1} \dashrightarrow \mathbb{P}^{\binom{n+2}{2}-2}$$

of the 2-uple embedding $\rho_{n,2}$ with the projection π from the image point. The image is again a smooth projective surface $X(p)$ with an exceptional curve $E \cong \mathbb{P}(T_p X) \cong \mathbb{P}^1$ and

$$X(p) \setminus E \cong X \setminus p.$$

Remarks 12.2.1. 1) Frequently one can project $X \subset \mathbb{P}^n$ directly from p without the Veronese re-embedding. However, this does not work if X has a 3-secant line which passes through p, e.g., if X contains a line which passes through p.
2) If p, q are distinct points on a smooth surface X and $q' \in X(p)$ and $p' \in X(q)$ are the pre-images then one can show that $X(p)(q') \cong X(q)(p')$. We denote this iterated blow-up by $X(p,q)$.
3) More generally given a set of finitely many distinct points $\Gamma = \{p_1, \ldots, p_r\}$ on a smooth surface $X \subset \mathbb{P}^n$ we denote by $X(p_1, \ldots, p_r)$ the blow-up of these points in some order. One can compute a projective model of $X(p_1, \ldots, p_r)$ by considering

homogeneous elements $f_0, \ldots, f_N$ of the homogeneous ideal $I(\Gamma) \subset K[x_0, \ldots x_n]$ of degree d which represent a basis of $I(\Gamma)_d/I(X)_d$. For d sufficiently large $X(p_1, \ldots, p_r)$ is isomorphic the closure of the image of X under the rational map

$$X \dashrightarrow \mathbb{P}^N, p \mapsto (f_0(p) : \ldots : f_N(p)).$$

A second place where the blow-up plays a crucial role is in the description of birational maps between smooth surfaces.

Theorem 12.2.2 (Castelnuovo, Zariski [87]).

1. *Let $\varphi \colon Z \to X$ be a birational morphism between smooth projective surfaces. Then there exists a sequence of blow-ups*

$$X_r \xrightarrow{\sigma_r} X_{r-1} \longrightarrow \ldots \longrightarrow X_1 \xrightarrow{\sigma_1} X$$

such that $Z \cong X_r$.
2. *Every birational map $Y \dashrightarrow X$ between smooth projective surfaces can be factored into birational morphisms from a smooth projective surface Z as follows:*

$$\begin{array}{ccc} & Z & \\ \swarrow & & \searrow \\ Y & & X, \end{array}$$

where both morphisms are sequences of blow-ups.

Example 12.2.3 (The birational projection of $\mathbb{P}^1 \times \mathbb{P}^1 \dashrightarrow \mathbb{P}^2$). Consider a point $p = (a, b) \in \mathbb{P}^1 \times \mathbb{P}^1 \subset \mathbb{P}^3$ and the rational map

$$\pi_p : \mathbb{P}^1 \times \mathbb{P}^1 \dashrightarrow \mathbb{P}^2$$

which maps a point $q \in \mathbb{P}^1 \times \mathbb{P}^1$ to the line $\overline{pq} \in \mathbb{P}^2$ where we identify $\mathbb{P}^2$ with the set of lines in $\mathbb{P}^3$ through p. Its factorization is

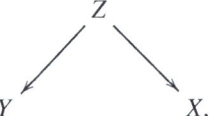

where $Z \to \mathbb{P}^1 \times \mathbb{P}^1$ is the blow-up of $\mathbb{P}^1 \times \mathbb{P}^1$ in p and $Z \to \mathbb{P}^2$ collapses the strict transforms of the lines $\mathbb{P}^1 \times \{b\}$ and $\{a\} \times \mathbb{P}^1$ to two points $p_1, p_2 \in \mathbb{P}^2$. The exceptional curve E over p is mapped to the line $\overline{p_1 p_2}$.

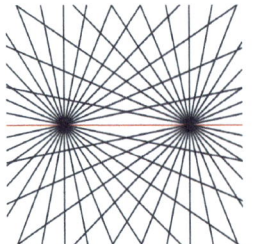

$$\begin{aligned} \mathbb{P}^2 \setminus \overline{p_1 p_2} &= \mathbb{A}^2 = \mathbb{A}^1 \times \mathbb{A}^1 \\ &\cong (\mathbb{P}^1 \setminus \{a\}) \times (\mathbb{P}^1 \setminus \{b\}) \\ &\subset \mathbb{P}^1 \times \mathbb{P}^1 \setminus \{(a,b)\} \end{aligned}$$

12.2 Blow-up of smooth projective surfaces

Definition 12.2.4. The birational map

$$q\colon \mathbb{P}^2 \dashrightarrow \mathbb{P}^2, \quad (x:y:z) \mapsto (\tfrac{1}{x}:\tfrac{1}{y}:\tfrac{1}{z}) = (yz:xz:yz)$$

is called the **quadratic transformation**. The map is not defined at the **fundamental points** $p_0 = (1:0:0), p_1 = (0:1:0)$ and $p_2 = (0:0:1)$.

To visualize the map it is convenient to choose coordinates such that all three fundamental points lie in an affine chart. We use convex coordinates from Exercise 8.1.5.

Proposition 12.2.5 (Graph of the quadratic transformation). *The graph of the quadratic transformation is isomorphic to the blow-up of $\mathbb{P}^2$ in the 3 fundamental points. The projection onto the second factor is again a blow-up of three points, the strict transforms of the 3 fundamental lines $L_{ij} = \overline{p_i p_j}$ are the exceptional curves of the second projection.*

Proof. The graph G is defined as the closure of the graph of the morphism $U \to \mathbb{P}^2$ which represents q. We use coordinates $(x:y:z)$ and $(u:v:w)$ on $\mathbb{P}^2 \times \mathbb{P}^2$. The graph is defined by

$$\operatorname{rank} \begin{pmatrix} yz & xz & xy \\ u & v & w \end{pmatrix} < 2$$

outside the fundamental points. However, over the fundamental points we need additional equations. If J denotes the ideal generated by the 2×2-minors of the matrix above. Then

$$I = J : (xy, xz, yz) = (vy - ux, ux - wz)$$

gives the defining ideal of G. Now restricted to the open set $G_{21} = G \cap \{z = 1, v = 1\}$, we obtain

$$K[G_{21}] \cong K[x, y, u, w]/(y - ux, w - ux) \cong K[x, u]$$

and the projection onto the first factor is

$$\mathbb{A}_2 \cong G_{21} \rightarrow U_2 \cong \mathbb{A}^2, (x,u) \mapsto (x,ux).$$

This is the chart of a blow-up. Since the sets G_{ij} for $i \neq j$ cover G and the equations of G are symmetric under the permutation $u \leftrightarrow x, v \leftrightarrow y, w \leftrightarrow z$ the proposition follows. □

Proposition 12.2.6. *Let C be a reduced plane curve of degree d which has multiplicity r_0, r_1, r_2 in the fundamental points p_0, p_1, p_2 of a quadratic transformation q. If none of the fundamental lines are tangent to C at a fundamental point, then the strict transform $q(C)$ has degree $2d - r_0 - r_1 - r_2$ and three new singular points with multiplicity $d - r_1 - r_2, d - r_0 - r_2$ and $d - r_0 - r_1$.* □

Example 12.2.7. Consider the curve

$$C = V(12x^4 - 44x^3y + 20x^2y^2 + 12xy^3 - 9x^3 - 30x^2y + 23xy^2 - 4y^3).$$

The strict transform of C under the quadratic transformation is

$$q(C) = V(4x^4 - 23x^3y + 30x^2y^2 + 9xy^3 - 12x^3 - 20x^2y + 44xy^2 - 12y^3).$$

Below are plots of these curves in the convex coordinates of Exercise 8.1.5.

Note that the circle on one side is the strict transform of the line at infinity of the visible chart on the other side. The original curve has a non-ordinary triple point at the origin because

$$-9x^3 - 30x^2y + 23xy^2 - 4y^3 = -(x+4y)(3x-y)^2.$$

The strict transform $q(C)$ has an ordinary triple point.

Theorem 12.2.8 (Cremona resolution). *Every irreducible plane curve can be transformed by a sequence of quadratic transformations into a plane curve with only ordinary singularities.*

The Cremona resolution allows us to deduce the existence of a resolution of singularities.

Proof of Theorem 12.1.7. Let C be a plane curve. We first assume that C is irreducible and consider a Cremona resolution of C. Now instead of performing the first quadratic transformation, we only blow-up the fundamental points.

12.2 Blow-up of smooth projective surfaces

The fundamental points of the second quadratic transformation give points on the blown-up surface, which we blow-up in the second step. Iterating this process we end up with a strict transform of C on an iterated blow-up of $\mathbb{P}^2$ which has only ordinary singular points. Blowing-up all those points we arrive at a smooth curve C'.

Now assume $C = C_1 \cup \ldots \cup C_s$ is reducible. We can perform this process in parallel for each component C_i arriving at an iterated blow-up of $\mathbb{P}^2$ where all strict transforms C'_i are smooth. Thus the only singularities of $C' = C'_1 \cup \ldots \cup C'_s$ are points where two components intersect.

We leave it as Exercise 12.2.9 to prove that two smooth curves which intersect with multiplicity $i(C_1, C_2; p) = m$ at a point p get separated after m blow-ups. □

Ingredients of the proof for the Cremona resolution. The proof of the existence of a Cremona resolution needs two results.

1. We need Bertini's theorem 14.1.2, which will allow us to find good fundamental points for the quadratic transformations.
2. We have to introduce two invariants of plane curves which improve under a suitable chosen quadratic transformation. The first invariant is the number of non-ordinary singularities of C. The second invariant is the difference

$$\binom{d-1}{2} - \sum_{p \in C} \binom{r_p}{2},$$

where r_p denotes the multiplicity of C at p. This difference is non-negative by Theorem 15.1.1.

We will return to this after the proof of Bertini's theorem.

Exercise 12.2.9. Let $f, g \in K[x, y]$ denote the defining equations of two affine curves which are smooth at $o \in \mathbb{A}^2$. Prove: An iterated blow-up over o separates the strict transforms after precisely m steps if and only if $i(f, g; p) = m$.

Exercise 12.2.10. Compute the equation of C and $q(C)$ in the affine charts corresponding to the convex coordinate plotted in Example 12.2.7 and make your own plots.

Chapter 13
Families of varieties

A phenomenon unique to algebraic geometry is that algebraic sets occur naturally in families, which themselves carry the structure of an algebraic set. The main point is that we can vary the coefficients of the defining equations.

We start by identifying the projective spaces of homogeneous polynomials of degree d up to scalars with the set of projective hypersurfaces of degree d.

We then study the linear systems $L(d; r_1 p_1, \ldots, r_s p_s)$ of plane curves of degree d with assigned base points p_i of multiplicity r_i. These linear systems play a major role in the parametrization of a rational curve in Chapter 15 and in the proof of the Riemann-Roch theorem in Chapter 16.

In Section 13.3 we define the Grassmannians and discuss $\mathbb{G}(2,4)$ in some detail. Section 13.4 takes a brief look at the Hilbert scheme. We take the ad hoc definition of subschemes of $\mathbb{A}^n$ to be in bijection with arbitrary ideals of $K[x_1, \ldots, x_n]$, and projective subschemes of $\mathbb{P}^n$ as saturated homogeneous ideals of $K[x_0, \ldots, x_n]$.

To give some flesh to the Hilbert scheme, we discuss in the next section $\text{Hilb}_{3t+1}(\mathbb{P}^3)$, following Piene and Schlessinger [71]. We identify two components of $\text{Hilb}_{3t+1}(\mathbb{P}^3)$ with a Macaulay2 computation.

In this last section we interpret the association $I \rightsquigarrow \text{Lt}(I)$ as the limits of a one-parameter family in the Hilbert scheme induced by a diagonal coordinate change. Our main reason to talk vaguely about the Hilbert scheme is to give this interpretation of the ideal leading terms.

13.1 The family of hypersurfaces

The family of hypersurfaces $X \subset \mathbb{P}^n$ of degree d is a Zariski open subspace of a projective space.

Proposition 13.1.1. *Let* $L(n,d) = K[x_0, \ldots, x_n]_d$ *denote the K-vector space of homogeneous polynomials of degree d. Then the $\binom{d+n}{n} - 1$-dimensional projective space $\mathbb{P}(L(n,d))$ contains a Zariski open subset which is in bijection with the set of irreducible hypersurfaces of degree d.*

Proof. The equation f of a hypersurface $X = V(f)$ is uniquely determined up to a scalar at least in case when f has no multiple factors. By Example 13.1.2 the set

$$\{X \subset \mathbb{P}^n \mid X \text{ is an irreducible hypersurface of degree } d\}$$

is in bijection with a Zariski open subset of $\mathbb{P}(L(n,d))$. □

In the following we will consider the projective space of all equations $\mathbb{P}(L(n,d))$ for simplicity, since the set of reducible or not square-free polynomials has a complicated structure.

Example 13.1.2. Let $d = d_1 + d_2$. The set of reducible hypersurfaces

$$\{[f] \in \mathbb{P}(L(n,d)) \mid f = f_1 f_2 \text{ with } \deg f_i = d_i\}$$

is the image of the morphism

$$\mathbb{P}(L(n,d_1)) \times \mathbb{P}(L(n,d_2)) \to \mathbb{P}(L(n,d)), ([f_1],[f_2]) \mapsto [f_1 f_2].$$

Hence it is a projective variety. If $d_1 \neq d_2$, then it is a birational linear projection from the Segre embedding of $\mathbb{P}(L(n,d_1)) \times \mathbb{P}(L(n,d_2))$, hence of large degree.

Example 13.1.3. The map

$$\mathbb{P}(L(n,1)) \to \mathbb{P}(L(n,d)), [\ell] \mapsto [\ell^d]$$

can be identified with the Veronese embedding

$$\rho_{n,d} : \mathbb{P}^n \to \mathbb{P}^{\binom{d+n}{n}-1}.$$

Example 13.1.4 (Plane conics). In the special case of plane conics we have the following: Write the equation of a plane conic in the form

$$q(x,y,z) = (x \ y \ z) \begin{pmatrix} w_0 & w_1 & w_3 \\ w_1 & w_2 & w_4 \\ w_3 & w_4 & w_5 \end{pmatrix} \begin{pmatrix} x \\ y \\ z \end{pmatrix}$$

and identify $\mathbb{P}^5 = \mathbb{P}(L(2,2))$. Then the Veronese surface $V_{2,2} \subset \mathbb{P}^5$ corresponds to the squares of linear forms, i.e., to **double lines**, and

$$V(\det \begin{pmatrix} w_0 & w_1 & w_3 \\ w_1 & w_2 & w_4 \\ w_3 & w_4 & w_5 \end{pmatrix}) \subset \mathbb{P}^5$$

corresponds to the set of reducible conics, i.e., to **pairs of lines**.

Definition 13.1.5 (Linear systems of hypersurfaces). A **linear system** of hypersurfaces is a projective space $\mathbb{P}(L) \subset \mathbb{P}(L(n,d))$ for a linear subspace $L \subset L(n,d)$. We speak of a **pencil** if $\mathbb{P}(L) \cong \mathbb{P}^1$. A **net** or **web** is a linear system of dimension 2 and 3 respectively.

13.2 Linear systems of plane curves

Example 13.1.4 continued. A pencil of conics contains, counted with multiplicities, precisely three reducible conics unless all conics are reducible, because

$$\deg \det \begin{pmatrix} w_0 & w_1 & w_3 \\ w_1 & w_2 & w_4 \\ w_3 & w_4 & w_5 \end{pmatrix} = 3.$$

A general net of conics contains no double lines because a general $\mathbb{P}^2 \subset \mathbb{P}^5$ does not intersect the Veronese surface $V_{2,2} \subset \mathbb{P}^5$.

13.2 Linear systems of plane curves

In the following we study linear systems of plane curves and abbreviate our notation:

$$L(d) = L(2, d) \ (= K[x, y, z]_d).$$

Definition 13.2.1. Let $\mathbb{P}(L) \subset \mathbb{P}(L(d))$ be a linear system of plane curves. A point $p \in \mathbb{P}^2$ is called a **base point** of $\mathbb{P}(L)$ if $p \in V(f)$ for all $f \in L(d)$.

Let $p_1, \ldots, p_s \in \mathbb{P}^2$ be distinct points and let $r_1, \ldots, r_s$ be positive integers. Then we set

$$L(d; r_1 p_1, \ldots, r_s p_s) := \{ f \in L(d) \mid f \text{ has multiplicity} \geq r_i \text{ at } p_i \ \forall i \}.$$

The space $\mathbb{P}(L(d; r_1 p_1, \ldots, r_s p_s))$ is called the linear system of plane curves of degree d with **assigned base points** p_i of **multiplicity** r_i.

Proposition 13.2.2. *Let $p_1, \ldots, p_s \in \mathbb{P}^2$ be distinct points and let $r_1, \ldots, r_s$ be positive integers. Then*

$$\dim_K L(d; r_1 p_1, \ldots, r_s p_s) \geq \binom{d+2}{2} - \sum_{i=1}^{s} \binom{r_i + 1}{2}$$

and equality holds if $d \geq (\sum_{i=1}^{s} r_i) - 1$.

Proof. Since $L(d; r_1 p_1, \ldots, r_s p_s) = \bigcap_{i=1}^{s} L(d; r_i p_i)$ it suffices to prove that

$$L(d; rp) \subset L(d)$$

has codimension $\binom{r+1}{2}$ for the first statement. If $p = (0 : 0 : 1)$, then $f \in L(d; rp)$ if and only if in the affine equation

$$f(x_1, x_2, 1) = \sum_{|\alpha| \leq d} f_\alpha x^\alpha$$

the coefficients f_α vanish for $|\alpha| \leq r$. These are $\binom{r+1}{2}$ coefficients.

The second statement is proved by induction on $\sum r_i$. The key step is to prove that $L(d; r_1p_1, r_2p_2, \ldots, r_sp_s) \subset L(d; (r_1-1)p_1, r_2p_2, \ldots, r_sp_s)$ has the maximal possible codimension r_1 when $d \geq (\sum_{i=1}^{s} r_i) - 1$. We leave this as Exercise 13.2.7. □

Example 13.2.3. The inequality might be strict if the points lie in a special position. For example, if $p_1, \ldots, p_4$ lie on a line we have

$$\dim \mathbb{P}(L(2; p_1, \ldots, p_4)) = 2.$$

In all other cases $\mathbb{P}(L(2; p_1, \ldots, p_4))$ is a pencil as expected, since $\binom{2+2}{2} - 4 \cdot 1 - 1 = 1$.

$L(d; r_1p_1, \ldots, r_sp_s) \subset L(d)$ is defined by a linear system of equations whose coefficients are polynomials in the coordinates $(a_i : b_i : c_i)$ of p_i. Thus there exists an open subset

$$U \subset \mathbb{P}^2 \times \ldots \times \mathbb{P}^2$$

of the product of s copies of $\mathbb{P}^2$ such that $\dim_K L(d; r_1p_1, \ldots, r_sp_s)$ takes its minimal value for all tuples $(p_1, \ldots, p_s) \in U$.

The minimal value can be larger than $\binom{d+2}{2} - \sum \binom{r_i+1}{2}$.

Example 13.2.4. $\dim L(4; 2p_1, \ldots, 2p_5) \geq 1$ although $\binom{4+2}{2} - 5 \cdot 3 = 0$. The reason is that $L(2; p_1, \ldots, p_5) \geq 1$ and the equation q of a conic through the five points yields a non-zero quartic $q^2 \in L(4; 2p_1, \ldots, 2p_5)$.

It is ongoing research to characterize those multiplicities $r_1, \ldots, r_s$ for which

$$\dim L(d; r_1p_1, \ldots, r_sp_s) = \binom{d+2}{2} - \sum_{i=1}^{s} \binom{r_i+1}{2}$$

holds for a general collection of points $p_1, \ldots, p_s$. See [15] for some recent developments in this line of research.

For simple points we have

Proposition 13.2.5. *Let $p_1, \ldots, p_s$ be a general tuple of points in $\mathbb{P}^2$. Then*

$$\dim L(d; p_1, \ldots, p_s) = \binom{d+2}{2} - s$$

as long as the right-hand side is non-negative.

Proof. We have to prove that the Zariski-open subset $U \subset \mathbb{P}^2 \times \ldots \times \mathbb{P}^2$ where equality holds is non-empty. Suppose $\dim_K L(d; p_1, \ldots, p_{s-1}) \neq 0$. Choose a non-zero $f \in L(d; p_1, \ldots, p_{s-1})$ and a point $p_s \notin V(f)$. Then $L(d; p_1, \ldots, p_s) \subset L(d; p_1, \ldots, p_{s-1})$ has (the maximal possible) codimension 1, and $U \neq \emptyset$ follows by induction on s. □

Exercise 13.2.6. Let $p_1, \ldots, p_4$ be four distinct points in $\mathbb{P}^2$. Prove: The vector space $L(2; p_1, \ldots, p_4)$ has dimension 2 if and only if the four points do not lie on a line.

13.3 The Grassmannian

Exercise 13.2.7. Let $p_1, \ldots, p_s$ be s distinct points in $\mathbb{P}^2$, and let $r_1, \ldots r_s$ be non-negative integers. Then

$$\dim L(d; r_1 p_1, \ldots, r_s p_s) = \binom{d+1}{2} - \sum_{i=1}^{s} \binom{r_i + 1}{2}$$

if $d \geq (\sum_{i=1}^{r} r_i) - 1$.

Exercise 13.2.8. The projective closure $\overline{C} \subset \mathbb{P}^2(\mathbb{C})$ of every circle

$$C = \{(x, y) \in \mathbb{A}^2(\mathbb{R}) \mid (x-a)^2 + (y-b)^2 = r^2\}$$

passes through the two **circle points** $(1 : \pm i : 0)$ on the line $V(z)$ at infinity. Prove: Any three non-collinear points p_1, p_2, p_3 in $\mathbb{A}^2(\mathbb{R})$ are contained in a unique circle.

Exercise 13.2.9. A parabola $C \subset \mathbb{A}^2$ is a conic with a smooth projective closure $\overline{C} \subset \mathbb{P}^2$ which intersects the line at infinity in a single point. Let $p_1, \ldots, p_4$ be four points in $\mathbb{A}^2$ of which no three are collinear. Prove: There exists a parabola passing through $p_1, \ldots, p_4$ if and only if the points do not form a parallelogram.

Exercise 13.2.10. Consider a conic $C \subset \mathbb{P}^2$ and six different points $p_1, \ldots, p_6$ on C. Prove Pascal's theorem: The opposite sides of the hexagon $L_{12} = \overline{p_1 p_2}, L_{23} = \overline{p_2 p_3}, \ldots, L_{56} = \overline{p_5 p_6}, L_{61} = \overline{p_6 p_1}$ intersect in three points $q_1 = L_{12} \cap L_{45}, q_2 = L_{23} \cap L_{56}, q_3 = L_{34} \cap L_{61}$ which lie on a line.

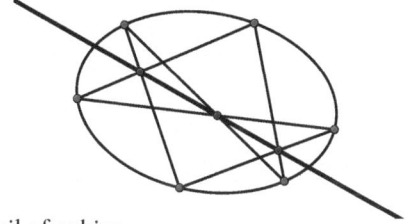

Hint: Consider the pencil of cubics

$$V(t_0 f + t_1 g) \subset \mathbb{P}^2 \text{ with } (t_0 : t_1) \in \mathbb{P}^1$$

where $f = \ell_{12} \ell_{34} \ell_{56}$ and $g = \ell_{23} \ell_{45} \ell_{61}$ are products of the equations ℓ_{ij} of the L_{ij}'s.

13.3 The Grassmannian

We now turn to the description of families of varieties of larger codimension. The first interesting case is perhaps the family of lines in $\mathbb{P}^3$ or, equivalently, two-dimensional subvector spaces $W \subset K^4$.

Definition 13.3.1. Let $1 \leq d < n$ be two integers. As a set we define the **Grassmannian** as

$$\mathbb{G}(d,n) = \{W \subset K^n \mid W \text{ is a subvector space of dimension } d\}.$$

If $M = \{A = (a_{ij}) \in K^{d \times n} \mid \text{rank } A = d\} \subset \mathbb{A}^{dn}$ denotes the quasi-affine variety of $d \times n$-matrices of maximal rank d, then we can identify

$$\mathbb{G}(d,n) = M/\text{GL}(d,K)$$

with the set of orbits under the action

$$\text{GL}(d,n) \times M \to M, (B,A) \mapsto BA.$$

Indeed, the map
$$M \to \mathbb{G}(d,n)$$
which maps the matrix
$$A = \begin{pmatrix} a_{11} & a_{12} & \cdots & a_{1n} \\ \vdots & \vdots & & \vdots \\ a_{d1} & a_{d2} & \cdots & a_{dn} \end{pmatrix}$$

to the subspace $W \subset K^n$ spanned by the rows of A is surjective. The fibers correspond to the choices of a basis of W, i.e., to points in the orbit $\text{GL}(d,K)A \subset M$.

To give $\mathbb{G}(d,n)$ the structure of a projective variety we consider the **Plücker embedding**. For a subset

$$I = \{i_1 < i_2 < \ldots < i_d\} \subset \{1,\ldots,n\}$$

of d elements we denote by A_I the $d \times d$-submatrix of A with columns i_k for $k = 1,\ldots,d$.

Consider the map

$$\gamma : \mathbb{G}(d,n) \to \mathbb{P}^{\binom{n}{d}-1}, \quad [A] \mapsto [\det A_I]$$

induced by the **Plücker coordinates** of A, i.e., by all $d \times d$-minors of A. This induces a well-defined map because the $d \times d$ minors of A and BA differ by the common factor $\det B \in K^*$ since $\det(BA)_I = \det B \det A_I$ and at least one minor is non-zero because rank $A = d$.

In algebraic terms we have a variable p_I in the homogeneous coordinate ring of $\mathbb{P}^{\binom{n}{d}-1}$, and we define $\mathbb{G}(d,n) \subset \mathbb{P}^{\binom{n}{d}-1}$ as the projective variety defined by the ideal $\ker(\gamma^*)$ of the ring homomorphism

$$\gamma^* : K[p_I] \to K[a_{ij}], \quad p_I \mapsto \det A_I.$$

Proposition 13.3.2. *The affine charts $U_I = \{p_I \neq 0\}$ of $\mathbb{P}^{\binom{n}{d}-1}$ intersect the Grassmannian in affine varieties $\mathbb{G}(d,n) \cap U_I \cong \mathbb{A}^{d(n-d)}$. In particular, $\mathbb{G}(d,n)$ is a smooth projective variety of dimension $d(n-d)$.*

13.3 The Grassmannian

Proof. We consider $U_I \cap \mathbb{G}(d,n)$ for $I = \{1,\ldots,d\}$. The points of $\gamma^{-1}(U_I)$ are represented by matrices A' with $\det A'_I \neq 0$. Thus we have a distinguished representative $A = (A'_I)^{-1}A'$ of shape

$$A = \begin{pmatrix} 1 & & 0 & a_{1,d+1} & \cdots & a_{1n} \\ & \ddots & & \vdots & & \vdots \\ 0 & & 1 & a_{d,d+1} & \cdots & a_{dn} \end{pmatrix}.$$

On this chart we have

$$p_{(\{1,\ldots,d\}\setminus\{i\})\cup\{j\}} = (-1)^{d-i}a_{ij} \text{ for } j > d.$$

Every Plücker coordinate $\det A_J$ is a polynomial in the a_{ij} with $j > d$. Thus interpreting these a_{ij}'s in terms of the Plücker coordinates $p_{(\{1,\ldots,d\}\setminus\{i\})\cup\{j\}}$ above and homogenizing with respect to $p_{\{1,\ldots,d\}}$ we obtain elements of $\ker(\gamma^*)$ which show that $\mathbb{G}(d,n) \cap U_{\{1,\ldots,d\}}$ is isomorphic to $\mathbb{A}^{d(n-d)}$. The arguments in other charts are analogous. In particular, we see that $\mathbb{G}(n,d)$ is covered by $\binom{n}{d}$ charts which are all needed to cover $\mathbb{G}(d,n)$. □

Example 13.3.3. The Grassmannian $\mathbb{G}(2,4) \subset \mathbb{P}^5$ is a hypersurface. It is actually a quadric. In terms of coordinates $p_{12},\ldots,p_{34}$ the ideal is generated by the **Plücker quadric**

$$p_{12}p_{34} - p_{13}p_{24} + p_{14}p_{23}.$$

We can see that this equation is satisfied for the minors of the 2×4-matrix

$$A = \begin{pmatrix} a_{11} & a_{12} & a_{13} & a_{14} \\ a_{21} & a_{22} & a_{23} & a_{24} \end{pmatrix}$$

by expanding the determinant

$$0 = \det \begin{pmatrix} a_{11} & a_{12} & a_{13} & a_{14} \\ a_{21} & a_{22} & a_{23} & a_{24} \\ a_{11} & a_{12} & a_{13} & a_{14} \\ a_{21} & a_{22} & a_{23} & a_{24} \end{pmatrix}$$

with respect to the first two rows. Thus $2(\det A_{12} \det A_{34} - \det A_{13} \det A_{24} + \det A_{14} \det A_{23}) = 0 \in \mathbb{Z}[a_{ij}]$.

$\mathbb{P}^n = \mathbb{G}(1, n+1)$ has a stratification by affine strata:

$$\mathbb{P}^n = \mathbb{A}^n \cup \mathbb{P}^{n-1} = \mathbb{A}^n \cup \mathbb{A}^{n-1} \cup \ldots \cup \mathbb{A}^1 \cup \mathbb{A}^0.$$

A similar stratification exists for all Grassmannians. We describe this for the case $\mathbb{G}(2,4)$. The Grassmannian $\mathbb{G}(2,4)$ is the disjoint union of the following six affine spaces

$$S_{12} = \left\{\begin{pmatrix} 1 & 0 & * & * \\ 0 & 1 & * & * \end{pmatrix}\right\} \cong \mathbb{A}^4$$

$$S_{13} = \left\{\begin{pmatrix} 1 & * & 0 & * \\ 0 & 0 & 1 & * \end{pmatrix}\right\} \cong \mathbb{A}^3$$

$$S_{14} = \left\{\begin{pmatrix} 1 & * & * & 0 \\ 0 & 0 & 0 & 1 \end{pmatrix}\right\} \cong \mathbb{A}^2 \qquad S_{23} = \left\{\begin{pmatrix} 0 & 1 & 0 & * \\ 0 & 0 & 1 & * \end{pmatrix}\right\} \cong \mathbb{A}^2$$

$$S_{24} = \left\{\begin{pmatrix} 0 & 1 & * & 0 \\ 0 & 0 & 0 & 1 \end{pmatrix}\right\} \cong \mathbb{A}^1$$

$$S_{34} = \left\{\begin{pmatrix} 0 & 0 & 1 & 0 \\ 0 & 0 & 0 & 1 \end{pmatrix}\right\} \cong \mathbb{A}^0$$

More generally, in which strata a point $[A] \in \mathbb{G}(d, n)$ lies depends on the row echelon form of the matrix $A \in K^{d \times n}$. The closure of the strata

$$\overline{S_{i_1 \ldots i_d}}$$

is called a **Schubert variety**. Schubert varieties are intensely studied objects of algebraic geometry. The closure of the strata can also be characterized by how the corresponding linear subspace intersect the subspaces of a complete flag of linear subspaces.

We illustrate this in the case of $\mathbb{G}(2, 4)$. Consider the flag

$$p_0 \subset L_0 \subset P_0 \subset \mathbb{P}^3$$

of the point, line and plane defined by

$$V(x_0, x_1, x_2) \subset V(x_0, x_1) \subset V(x_0) \subset \mathbb{P}^3.$$

$$\overline{S_{12}} = \mathbb{G}(2, 4)$$

$$\overline{S_{13}} = \{L \in \mathbb{G}(2, 4) \mid L \cap L_0 \neq \emptyset\}$$

$$\overline{S_{14}} = \{L \mid p_0 \in L\} \qquad \overline{S_{23}} = \{L \mid L \subset P_0\}$$

$$\overline{S_{24}} = \{L \mid p_0 \in L \subset P_0\} \cong \mathbb{P}^1$$

$$\overline{S_{34}} = \{L_0\}$$

Corollary 13.3.4. *The set of lines L in the affine three space is the quasi-projective variety*

$$\{L \subset \mathbb{A}^3 \mid L \text{ is a line}\} = \mathbb{G}(2, 4) \setminus \overline{S_{23}},$$

where $\overline{S_{23}} = \mathbb{P}^2$ is the space of lines contained in the plane at infinity of $\mathbb{P}^3$.

Remark 13.3.5. Hermann Schubert (1848 –1911) developed a general machinery to solve enumerative problems. For example: How many lines intersect four given lines in $\mathbb{P}^3$? The answer is two.

Here is how Schubert would argue. Take $L_1, \ldots, L_4$ into special position such that L_1 and L_2 intersect at a point p_{12} and similarly L_3 and L_4 intersect at a point p_{34}. Then L_1 and L_2 span a plane P_{12} and L_3 and L_4 span a plane P_{34}. The line $\overline{p_{12}p_{34}}$ spanned by the intersection points and the intersection line $P_{12} \cap P_{34}$ are the only two lines which intersect all four lines.

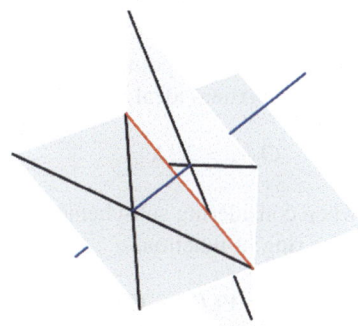

One of Schubert's most famous results is that 5 general conics are tangent to precisely 3264 smooth conics. To put Schubert calculus on a solid foundation was Hilbert's 15th problem.

Exercise 13.3.6. Prove that there are precisely two lines in $\mathbb{P}^3$ which intersect four given general lines. Hint: Prove that three general skew lines lie on a quadric surface $Q \cong \mathbb{P}^1 \times \mathbb{P}^1 \subset \mathbb{P}^3$.

13.4 A glance at the Hilbert scheme

Subschemes of $\mathbb{A}^n$. We have frequently seen that non-radical ideals occur naturally.

Example. Consider the ideal $(x^2, xy) \subset K[x, y]$. It arises if we intersect the algebraic set $V(xz, yz)$ with the hyperplane $V(z - x)$. On the algebra side we have

$$K[x, y, z]/(xz, yz, z - x) \cong K[x, y]/(x^2, xy)$$

corresponding to this intersection.

It is thus natural to extend our algebra-geometry dictionary beyond radical ideals as follows:

$$\{\text{radical ideals of } K[x_1,\ldots,x_n]\} \stackrel{1:1}{\longleftrightarrow} \{\text{algebraic subsets of} \mathbb{A}^n\}$$

$$\{\text{arbitrary ideals of } K[x_1,\ldots,x_n]\} \stackrel{1:1}{\longleftrightarrow} \{\text{subschemes of } \mathbb{A}^n\}.$$

The **coordinate ring** of the **subscheme** $X \subset \mathbb{A}^n$ corresponding to the ideal $J \subset K[x_1,\ldots,x_n]$ is the ring $K[X] = K[x_1,\ldots,x_n]/J$.

A morphism $\varphi \colon X \to Y$ between a subscheme $X \subset \mathbb{A}^n$ and a subscheme $Y \subset \mathbb{A}^m$ is given by a K-algebra homomorphism

$$\varphi^* \colon K[Y] \to K[X].$$

The local ring $\mathcal{O}_{X,p}$ of a subscheme at a point $p \in V(J)$ can be defined as the localization in the corresponding maximal ideal $\mathfrak{m}$ of $K[X]$:

$$\mathcal{O}_{X,p} = K[X]_{\mathfrak{m}}.$$

The only disadvantage when considering subschemes is that $K[X]$ can no longer be regarded as a subring of the ring of functions

$$\{f \colon V(J) \to K\}.$$

In our example above with $K[X] = K[x,y]/(x^2,xy)$ the element $\bar{x} \in K[X]$ would correspond to the zero function because $(\bar{x})^2 = 0$. But $\bar{x} \neq 0 \in K[X]$. We allow non-trivial nilpotent elements in the coordinate ring of a subscheme. Subschemes which do not correspond to radical ideals are called **non-reduced**.

Subschemes of $\mathbb{P}^n$.

Definition 13.4.1. A subscheme Y of $\mathbb{P}^n$ is a collection of subschemes in the affine charts $U_i \cong \mathbb{A}^n$ that coincide in the intersections $U_i \cap U_j$.

To be more precise, if $Y_i \subset U_i$ denotes the subscheme in the i-th chart with coordinate ring $K[Y_i] = K[x_0,\ldots,x_{i-1},x_{i+1},\ldots,x_n]/J_i$ then the restriction $Y_i|_{U_i \cap U_j}$ to $U_i \cap U_j$ is the affine scheme with coordinate ring $K[Y_i][x_j^{-1}]$. We require that this ring coincides with $K[Y_j][x_i^{-1}]$ under the map induced by the coordinate change

$$K[x_0,\ldots,x_{i-1},x_j^{-1},x_{i+1},\ldots,x_n] \to K[x_0,\ldots,x_{j-1},x_i^{-1},x_{j+1},\ldots,x_n],$$

$$x_j \mapsto \frac{1}{x_i}, \quad \frac{1}{x_j} \mapsto x_i, \quad x_\ell \mapsto \frac{x_\ell}{x_i}.$$

Definition 13.4.2. A homogeneous ideal $I \subset K[x_0,\ldots,x_n]$ is **saturated** if it coincides with its **saturation**

$$I_{\text{sat}} = I : (x_0,\ldots,x_n)^\infty = \bigcup_{N=1}^{\infty} (I : (x_0,\ldots,x_n)^N).$$

13.4 A glance at the Hilbert scheme

One can show the following:

Proposition 13.4.3. *There is a bijection between the set of saturated proper homogeneous ideals of $K[x_0, \ldots, x_n]$ and the set of subschemes of $\mathbb{P}^n$.*

Remark. A homogeneous ideal I is saturated if and only if the irrelevant ideal $(x_0, \ldots, x_n)$ is not an associated prime ideal of I.

The graded ring $K[X] = K[x_0, \ldots, x_n]/I_{\text{sat}}$ is called the **homogeneous coordinate ring** of the subscheme $X \subset \mathbb{P}^n$ corresponding to the subscheme defined by I. The **Hilbert polynomial** $p_X \in \mathbb{Q}[t]$ is the Hilbert polynomial of $K[X]$.

A fundamental result in algebraic geometry, due to Grothendieck, is the following.

Theorem 13.4.4 (Grothendieck). *Let $p \in \mathbb{Q}[t]$ be a polynomial of degree $< n$. The set of subschemes*

$$\text{Hilb}_p(\mathbb{P}^n) = \{X \subset \mathbb{P}^n \mid p_X = p\}$$

carries the structure of a projective scheme.

These schemes were baptised **Hilbert schemes** by Grothendieck.

Example 13.4.5. Let $p(t) = \binom{t+n}{n} - \binom{t-d+n}{n} = d\frac{t^{n-1}}{(n-1)!} +$ lower terms $\in \mathbb{Q}[t]$ be the Hilbert polynomial of a hypersurface $X \subset \mathbb{P}^n$ of degree d. Then

$$\text{Hilb}_p(\mathbb{P}^n) \cong \mathbb{P}(L(d,n)).$$

By Proposition 13.1.1 we add to the quasi-projective space of subvarieties of degree d and dimension $n-1$ reducible algebraic sets and non-reduced schemes.

Some ideas of the proof.

Step 1. Let $p \in \mathbb{Q}[t]$. There exists an integer r such that the saturated ideal $I = I_X$ of any subscheme $X \subset \mathbb{P}^n$ with $p_X = p$ satisfies the following:

$$I_d \subset K[x_0, \ldots, x_n]_d \text{ has codimension } p(d) \text{ for all } d \geq r. \tag{†}$$

The key point of this step is that the bound r depends only on p. This is proved using the concept of Castelnuovo-Mumford regularity, which builds upon coherent sheaves and their cohomology [67]. As an additional property one obtains

$$I_{\geq d} = \bigoplus_{d' \geq d} I_{d'} = (I_d) \text{ for } d \geq r. \tag{‡}$$

Remark. Since the homogeneous ideals J and $J \cap (x_0, \ldots, x_n)^N$ have the same Hilbert polynomial the statements (†) and (‡) fail if we consider non-saturated ideals.

Step 2. For a given polynomial $p \in \mathbb{Q}[t]$ fix an integer $d \geq r$. We then have an injection

$$\{X \subset \mathbb{P}^n \mid p_X = p\} \hookrightarrow \mathbb{G}(\binom{n+d}{n} - p(d), \binom{d+n}{n})$$

into a Grassmannian by mapping the subscheme X to the degree d part of its homogeneous ideal I_X

$$X \mapsto (I_X)_d \subset K[x_0, \ldots, x_n]_d.$$

This is a codimension $p(d)$ subvector space, hence a point in the Grassmannian above.

To get equations of the Hilbert scheme, we note that a subspace W lies in the image if and only if the ideal (W) generated by W has Hilbert function values $h_{(W)}(d') = \binom{n+d'}{n} - p(d')$ for $d' \geq d$.

Representing W by a $(\binom{n+d}{d} - p(d)) \times \binom{n+d}{n}$-coefficient matrix, we see that for $d' = d + 1$ this gives the condition that a

$$(n+1)(\binom{n+d}{n} - p(d)) \times \binom{n+d+1}{n}\text{-matrix}$$

in these coefficients has rank precisely $\binom{n+d+1}{n} - p(d+1)$. Thus the corresponding minors give equations for the Hilbert scheme.

We leave it as an open question to explain why the rank of any W satisfying these equations cannot be smaller, and why these equations define the Hilbert scheme correctly from a conceptual point of view, see [40].

The key concept for the latter is the notion of a representable functor [39].

13.5 Hilb$_{3t+1}(\mathbb{P}^3)$

To give the vague ideas of Section 13.4 some concrete content we describe Hilb$_{3t+1}(\mathbb{P}^3)$ following the work of Piene-Schlessinger [71]. The polynomial $p(t) = 3t + 1$ is the Hilbert polynomial of the rational normal curve in $\mathbb{P}^3$. We give a description of the open part of this Hilbert scheme of all saturated ideals I whose ideal of leading terms with respect to $>_{\text{rdlex}}$ is $\text{Lt}(I) = (x_0^2, x_0 x_1, x_1^2)$. If I is such an ideal, then its equations take the form

$$x_0^2 + a_1 x_0 x_2 + a_2 x_1 x_2 + a_3 x_0 x_3 + a_4 x_1 x_3 + a_5 x_2^2 + a_6 x_2 x_3 + a_7 x_3^2$$
$$x_0 x_1 + b_1 x_0 x_2 + b_2 x_1 x_2 + b_3 x_0 x_3 + b_4 x_1 x_3 + b_5 x_2^2 + b_6 x_2 x_3 + b_7 x_3^2$$
$$x_1^2 + c_1 x_0 x_2 + c_2 x_1 x_2 + c_3 x_0 x_3 + c_4 x_1 x_3 + c_5 x_2^2 + c_6 x_2 x_3 + c_7 x_3^2$$

with coefficients $a_1, \ldots, c_7$. The Buchberger test gives $2 \cdot 10$ equations for the coefficients obtained from the 10 possible different remainders of the division by $x_0^2, x_0 x_1, x_1^2$ in degree 3. The Macaulay2 computation of Section B.3 shows that the the ideal is generated by only nine equations

$$a_5 + a_2 b_1 - a_1 b_2 + b_2^2 - a_2 c_2, \ldots, c_7 + b_3^2 - a_3 c_3 + b_4 c_3 - b_3 c_4,$$

which tell us that $a_5, \ldots, c_7$ are quadratic polynomials in $a_1, \ldots, c_4$.

To phrase this differently: The open subset of $\text{Hilb}_{3t+1}(\mathbb{P}^3)$ corresponding to subschemes whose ideal of leading terms is $\text{Lt}(I) = (x_0^2, x_0 x_1, x_1^2)$ is isomorphic to $\mathbb{A}^{12}$ and the best way to describe the equations between the coefficients is to say that the three generators of I can be written as the 2×2-minors of the matrix

$$\begin{pmatrix} -x_1 - x_2 b_1 - x_3 b_3 & -x_2 c_1 - x_3 c_3 \\ x_0 + x_2 a_1 + x_3 a_3 - x_2 b_2 - x_3 b_4 & -x_1 + x_2 b_1 + x_3 b_3 - x_2 c_2 - x_3 c_4 \\ x_2 a_2 + x_3 a_4 & x_0 + x_2 b_2 + x_3 b_4 \end{pmatrix}.$$

Indeed, for example, the coefficient of x_2^2 in the minor with leading term x_0^2 is $a_5 = b_2(a_1 - b_2) - a_2(b_1 - c_2)$.

Thus $\text{Hilb}_{3t+1}(\mathbb{P}^3)$ has a 12-dimensional component in which the $\text{PGL}(4, K)$-orbit of the rational normal curve is dense. Note that

$$\dim \text{PGL}(4, K) - \dim \text{PGL}(2, K) = 15 - 3 = 12.$$

Actually, there are two components:

$$\text{Hilb}_{3t+1}(\mathbb{P}^3) = H_{12} \cup H_{15},$$

the component of dimension 12 from above and a further component of dimension 15. A general subscheme $X \in H_{15}$ is the union $X = E \cup \{p\}$ of a plane cubic curve E and a disjoint point. The choice of E depends on $3 + 9$ parameters, since we have to specify a plane in $\mathbb{P}^3$ and a cubic equation in that plane. The choice of p gives 3 further parameters. This fits with $15 = 3 + 9 + 3$.

To find the intersection of the two components we look at the open subset of the Hilbert scheme of ideals I with $\text{Lt}(I) = (x_0^2, x_0 x_1, x_0 x_2, x_1^3)$. This time we introduce $3 \cdot 7 + 10$ variables for the possible coefficients. The Buchberger test computation yields a system of equations for the coefficients. Its primary decomposition gives two smooth components isomorphic to $\mathbb{A}^{12}$ and $\mathbb{A}^{15}$ which intersect in a smooth subvariety isomorphic to $\mathbb{A}^{11}$. A general scheme corresponding to a point in the intersection consists of a nodal plane curve together with an embedded point at the node sticking out of the plane. An example of a scheme corresponding to a point in the intersection is given in Example 13.6.2 below.

13.6 Ideals of leading terms from a Hilbert scheme point of view

Let $I = I_{\text{sat}}$ be a saturated homogeneous ideal in $K[x_0, \ldots, x_n]$. The Hilbert scheme allows us to interpret the association

$$I \rightsquigarrow \text{Lt}(I)$$

in a more conceptual way. By Exercise 1.2.19 there exists a weight order $>_w$, which has the same leading terms as the given monomial order for the Gröbner basis of I. We may assume that the weights $(w_0, \ldots, w_n)$ are integers and all positive, since adding the vector $(1, \ldots, 1)$ does not change the ordering for monomials of a given degree. Consider the one-parameter subgroup

$$T = \left\{ \begin{pmatrix} t^{w_0} & & 0 \\ & \ddots & \\ 0 & & t^{w_n} \end{pmatrix} \mid t \in K^* \right\}$$

of the group of diagonal matrices in $GL(n+1, K)$. We ask how the ideal moves under the substitution $x_i \mapsto t^{-w_i} x_i$.

Let $f = \sum_{|\alpha|=d} f_\alpha x^\alpha \in I$ be a homogeneous element and let $f_\beta x^\beta = \mathrm{Lt}(f)$. Then

$$f^{(t)} = t^{<\beta, w>} f(t^{-w_0} x_0, \ldots, t^{-w_n} x_n)$$
$$= \sum_\alpha f_\alpha t^{<\beta-\alpha, w>} x^\alpha \in K[t, x_0, \ldots, x_n]$$

because $x^\beta >_w x^\alpha$ for all $\alpha \neq \beta$ with $f_\alpha \neq 0$. Consider a Gröbner basis $f_1, \ldots, f_r$ of I and the subscheme

$$Y \subset \mathbb{P}^n \times \mathbb{A}^1$$

defined by the ideal $J = (f_1^{(t)}, \ldots, f_r^{(t)}) \subset K[t, x_0, \ldots, x_n]$. The fibers Y_q for $q \in \mathbb{A}^1 \setminus \{o\}$ are all isomorphic to $Y_1 = X$, which is our original scheme. The special fiber Y_0 is defined by $\mathrm{Lt}(I)$. Thus

$$\lim_{t \to 0} Y_t = Y_0 \in \mathrm{Hilb}_{p_X}(\mathbb{P}^n).$$

Corollary 13.6.1. *Let $I = I_{sat} \subset K[x_0, \ldots, x_n]$ be a homogeneous saturated ideal, X the corresponding subscheme of $\mathbb{P}^n$ and p_X its Hilbert polynomial. The set of possible ideals of leading terms*

$$\{\mathrm{Lt}_>(I) \mid > \text{ is a global monomial order}\}$$

coincides with the set of monomial limits of X in $\mathrm{Hilb}_{p_X}(\mathbb{P}^n)$ along one-parameter subgroups of the torus

$$\left\{ \begin{pmatrix} a_0 & & 0 \\ & \ddots & \\ 0 & & a_n \end{pmatrix} \mid a_i \in K^* \right\} \subset GL(n+1, K).$$

□

Since $\mathrm{Hilb}_p(\mathbb{P}^n)$ is a projective scheme as a subscheme of a suitable Grassmannian, limits exist along arbitrary one-parameter families by Proposition 11.2.6.

13.6 Ideals of leading terms from a Hilbert scheme point of view

Example 13.6.2. Consider the rational normal curve in $\mathbb{P}^3$ defined by the 2×2-minors of the matrix

$$\begin{pmatrix} x_3 & x_0 & x_1 \\ x_0 x_1 + x_3 & x_2 \end{pmatrix}.$$

Substituting $x_0 \mapsto x_0, x_1 \mapsto tx_1, x_2 \mapsto tx_2, x_3 \mapsto tx_3$ and computing the Gröbner basis of the ideal I_t with respect to the lexicographic order we get

$x_0^2 - t^2 x_1 x_3 - t^2 x_3^2$	$-x_1$	$-x_2$	0	0
$x_0 x_1 - t\, x_2 x_3$	x_0	$tx_1 + tx_3$	$-x_2$	$-x_1^2 - x_1 x_3$
$x_0 x_2 - t\, x_1^2 - t\, x_1 x_3$	tx_3	x_0	x_1	$x_2 x_3$
$x_1^3 - x_2^2 x_3 + x_1^2 x_3$			t	x_0

Thus

$$I_0 = (x_0^2, x_0 x_1, x_0 x_2, x_1^3 - x_2^2 x_3 + x_1^2 x_3)$$
$$= (x_0, x_1^3 - x_2^2 x_3 + x_1^2 x_3) \cap (x_0^2, x_1, x_2)$$

is a nodal plane curve with an embedded point at the node $(0:0:0:1)$.

Remark 13.6.3. Like the Grassmannian, the Hilbert scheme allows a stratification. Fix a Hilbert polynomial p and a global monomial order, for example, $>_{\text{lex}}$ or $>_{\text{rdlex}}$. There are only finitely many monomial ideals J with Hilbert polynomial p which are generated in degree $\leq r$ for the bound r from the sketch of 13.4. Choose one of these monomial ideals J. Then the set

$$S_J = \{I \in \text{Hilb}_p(\mathbb{P}^n) \mid \text{Lt}(I) = J\}$$

of saturated ideals $I = I_{\text{sat}}$ with $\text{Lt}(I) = J$ are precisely those ideals which approach J under a suitable 1-parameter subgroup. The strata have a natural structure of affine schemes, which in principle can be computed similarly to the computation in Section 13.5.

A famous theorem of Hartshorne [41] says that the Hilbert schemes $\text{Hilb}_p(\mathbb{P}^n)$ are connected. The proof builds upon the idea of connecting these monomial ideals via a chain of 1-parameter families.

Chapter 14
Bertini's theorem and applications

In this chapter we prove Bertini's theorem, which in its simplest form says that a general hyperplane section of a smooth subvariety X is again smooth. Combined with Bézout's theorem 9.4.1, Bertini's theorem gives a geometric interpretation of the degree of a projective variety $X \subset \mathbb{P}^n$ of dimension r: $\deg X$ coincides with the number of intersection points $X \cap \mathbb{P}^{n-r}$ of X with a general linear subspace $\mathbb{P}^{n-r}$ of complementary dimension.

The space of hyperplanes $\check{X} = \{H \in \check{\mathbb{P}}^n \mid X \cap H \text{ is singular}\}$ is called the dual variety of X. In the case of plane curves $C \subset \mathbb{P}^2$ the dual curve $\check{C} \subset \check{\mathbb{P}}^2$ is the space of all tangent lines T_pC to C. If $\text{char}(K) = 0$ then $\check{\check{C}} = C$. There are classical formulas, due to Plücker, which relate the number bi-tangents and flexes to the number of nodes and cusps of C and $\check{C}$. These formulas will be proved in Exercise 15.3.5.

Section 14.2 contains a proof of the Riemann-Hurwitz formula for the topological genus of curves over $\mathbb{C}$ via a triangulation of the underlying oriented real 2-manifolds. Later in Theorem 16.2.9 we give a proof of this formula for arbitrary separable morphism between smooth curves.

In the last Section 14.3 we prove that intersection numbers of plane curves defined over $\mathbb{C}$ have a dynamical interpretation.

14.1 The dual variety

Definition 14.1.1. Let $\mathbb{P}^n$ be a projective space. Then the projective space of hyperplanes $H \subset \mathbb{P}^n$ is a called the **dual projective space**

$$\check{\mathbb{P}}^n = \{H \subset \mathbb{P}^n \mid H \text{ is a hyperplane}\}.$$

Remark. For a point $p \in \mathbb{P}^n$ the space of hyperplanes passing through p

$$H_p = \{H \in \check{\mathbb{P}} \mid p \in H\} \subset \check{\mathbb{P}}^n$$

is a hyperplane in $\check{\mathbb{P}}^n$, and any hyperplane in $\check{\mathbb{P}}^n$ arises this way: The subvariety

$$\mathbb{F} = V(a_0 x_0 + \ldots + a_n x_n) \subset \mathbb{P}^n \times \check{\mathbb{P}}^n$$

can be interpreted in two ways

$$\mathbb{F} = \{(p, H) \in \mathbb{P}^n \times \check{\mathbb{P}}^n \mid p \in H\} = \{(p, H) \in \mathbb{P}^n \times \check{\mathbb{P}}^n \mid H \in H_p\}.$$

The fibers of the projection $\mathbb{F} \to \check{\mathbb{P}}^n$ onto the second factor are hyperplanes in $\mathbb{P}^n$, and the fibers of the projection to the first factor $\mathbb{F} \to \mathbb{P}^n$ are hyperplanes in $\check{\mathbb{P}}^n$.

Theorem 14.1.2 (Bertini's theorem). *Let $X \subset \mathbb{P}^n$ be a projective variety of dimension d. Let X_{sing} denote its set of singular points. There exists a nonempty open subset $U \subset \check{\mathbb{P}}^n$ of hyperplanes such that $X \cap H$ is smooth outside $X_{sing} \cap H$ for every $H \in U$. In particular, if X is smooth, then $X \cap H$ is smooth as well for all $H \in U$.*

Proof. Consider the open set $X^* = X \setminus X_{sing}$ of smooth points of X and the variety

$$D^* = \{(p, H) \in X^* \times \check{\mathbb{P}}^n \mid T_p X \subset H\} \longrightarrow \check{\mathbb{P}}^n$$

$$\pi_1 \downarrow$$

$$X^*$$

with its two projections π_1 and π_2. A point $(p, H) \in D^*$ is a pair such that $X \cap H$ is singular in p or $X \subset H$.

The fiber of $\pi_1 \colon D^* \to X^*$ over a point $p \in X^*$ is a projective space of dimension $n - d - 1$

$$\{H \subset \mathbb{P}^n \mid H \supset T_p(X)\} \cong \mathbb{P}^{n-d-1}$$

because H is contained in the fiber if and only if H is defined by a linear combination of the $n - d$ equations of $T_p X \cong \mathbb{P}^d \subset \mathbb{P}^n$.

Thus $\dim D^* = d + n - d - 1 = n - 1$. We take

$$D = \overline{D^*} \subset X \times \check{\mathbb{P}}^n \subset \mathbb{P}^n \times \check{\mathbb{P}}^n.$$

Then $\dim D = \dim D^*$ and the projection $\pi_2(D) \subset \check{\mathbb{P}}^n$ is a Zariski closed subset of dimension

$$\dim \pi_2(D) \leq \dim D = n - 1.$$

Hence $U = \check{\mathbb{P}}^n \setminus \pi_2(D)$ is the desired open subset. □

Corollary 14.1.3 (Geometric interpretation of the degree). *Let $X \subset \mathbb{P}^n$ be a projective variety of dimension r. Then a general linear subspace $\mathbb{P}^{n-r} \subset \mathbb{P}^n$ intersects X in $\deg X$ many distinct points transversally.*

Proof. Let $H \subset \mathbb{P}^n$ be a general hyperplane. In particular, H does not contain any component of X_{sing}. Let $C_1 \cup \ldots \cup C_s = X \cap H$ be the irreducible components. Then

14.1 The dual variety

$$\deg X = \sum_{j=1}^{s} i(X, H; C_j) \deg C_j$$

holds by Bézout's theorem. By Bertini's theorem the intersection is smooth outside $X_{\text{sing}} \cap H$. In particular, the intersection is transversal at a general smooth point of each C_j, and the intersection multiplicity is 1. The result now follows by induction. A general complementary $\mathbb{P}^{n-r}$ is the intersection of r general hyperplanes $H_1 \cap \ldots \cap H_r$ such that H_i intersects each component of $X \cap H_1 \cap \ldots \cap H_{i-1}$ transversally. □

Definition 14.1.4. $\check{X} = \pi_2(D)$ is called the **dual variety** of X.

For $C \subset \mathbb{P}^2$ be an irreducible curve which is not a line, the dual variety is again a curve $\check{C} \subset \check{\mathbb{P}}^2$.

Theorem 14.1.5. *Let $C \subset \mathbb{P}^2$ be an irreducible curve over a field of characteristic 0. Then the double dual curve*

$$\check{\check{C}} = C$$

gives the original curve back.

Instead of a rigorous proof, see e.g. [28], we give a hand waving argument for curves and points defined over $\mathbb{R}$.

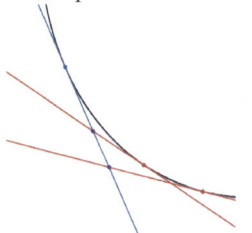

Consider two points p_1 and p_2 on C. The tangent lines $T_{p_1}C = \check{p}_1$ and $T_{p_2}C = \check{p}_2$ intersect at the point $\check{L}$ corresponding to the secant line $L = \check{p}_1\check{p}_2$. If p_2 approaches p_1 then the point $\check{L}$ approaches p_1 and L approaches $T_{\check{p}_1}\check{C}$. This says that the point corresponding to the line $T_{\check{p}_1}\check{C}$ coincides with p_1. □

Example 14.1.6. If $\text{char}(K) = p > 0$, it is possible that all tangent lines of an irreducible plane curve pass through a common point. Curves different from lines with this property are called **strange**.

Consider $C = V(x^p - yz^{p-1}) \subset \mathbb{P}^2$. In the affine chart $U_{z=1}$ this curve has the parametrization

$$\mathbb{A}^1 \to C \cap U_{z=1} \subset \mathbb{A}^2, t \mapsto q = (t, t^p)$$

and equation $f = x^p - y$. Since $d_q f = pt^{p-1}(x-t) - 1(y-t^p)$ the projective tangent lines are $T_q C = V(-y + t^p z)$. These lines all pass through the point $V(y, z) = (1 : 0 : 0)$. So the dual curve $\check{C} \subset \check{\mathbb{P}}^2$ coincides with $H_{(1:0:0)} \cong \mathbb{P}^1 \subset \check{\mathbb{P}}^2$, the set of lines passing through $(1 : 0 : 0)$. Hence $\check{\check{C}} \neq C$.

Notice that a strange curve can have at most one 'strange point', because the dual curve is a line.

Exercise 14.1.7. Consider the quartic curve $C = V(f)$ defined by

$$f = 8x^4 - 2x^2y^2 + 6y^4 + 60x^2yz - 37y^3z - 16x^2z^2 + 78y^2z^2 - 60yz^3 + 8z^4$$

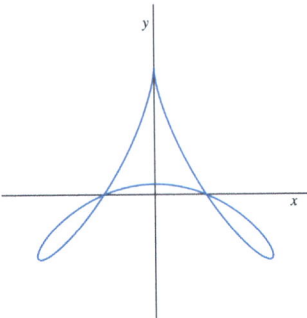

Verify with Macaulay2 that the dual curve is defined by

$$\check{C} = V(28462104u^4v - 3278600u^2v^3 - 682112v^5 + 14231052u^4w+$$
$$29374116u^2v^2w - 2990080v^4w + 6637962u^2vw^2 + 6389568v^3w^2-$$
$$2134405u^2w^3 - 1533376v^2w^3 - 1082504vw^4 - 106032w^5).$$

Hint: At a smooth point $p \in C$ the tangent line is defined by

$$T_pC = V\left(\frac{\partial f}{\partial x}(p)x + \frac{\partial f}{\partial y}(p)y + \frac{\partial f}{\partial z}(p)x\right).$$

Hence the 2×2 minors of

$$\begin{pmatrix} \frac{\partial f}{\partial x} & \frac{\partial f}{\partial y} & \frac{\partial f}{\partial z} \\ u & v & w \end{pmatrix}$$

vanish on the graph of the rational map $C \to \check{C}$ inside $\mathbb{P}^2 \times \check{\mathbb{P}}^2$.

The pictures below correspond to $\check{C}$ in the standard charts. Label the coordinates!

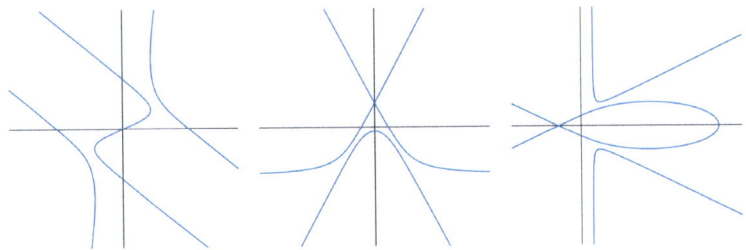

Observe that bitangents correspond to ordinary double points on the dual curve, and ordinary flexes to cusps. Recall that a smooth point p of a plane curve C is called a **flex** if the intersection multiplicity of C with the tangent line T_pC satisfies $i(C, T_pC; p) \geq 3$. The point p is an ordinary flex if equality holds. Where are these in the pictures?

14.2 The Riemann-Hurwitz formula

Notice that according Exercise 15.3.5 C has two bitangents and four ordinary flexes. The non-visible bitangent corresponds to a pair of complex conjugate points which lead to an isolated double point of $\check{C}$. Where is this point is the charts of $\check{C}$? The flexes and cusps are two pairs of complex conjugate points.

Compute a rational parametrization of C.

Exercise 14.1.8. Deduce Brianchon's theorem from Pascal's theorem 13.2.10 and projective duality: A hexagon in $\mathbb{P}^2$ is circumscribed to a smooth conic, if and only if the lines joining opposite vertices intersect in a point.

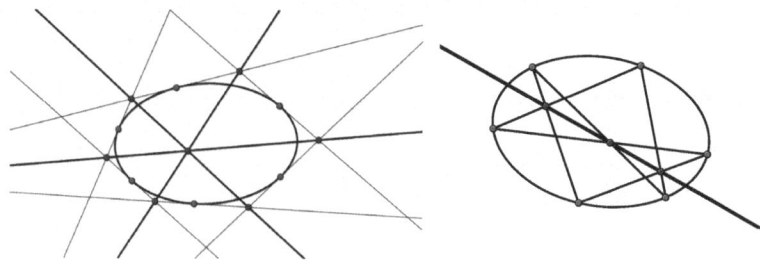

14.2 The Riemann-Hurwitz formula

Let $C \subset \mathbb{P}^n$ be an irreducible smooth projective curve, and let $f \in K(C)$ a non-constant rational function. The rational map

$$C \dashrightarrow \mathbb{P}^1, p \mapsto [1 : f(p)]$$

extends to a morphism $f : C \to \mathbb{P}^1$ by Proposition 11.2.6, which we denote by the same letter.

Definition 14.2.1. The **degree** of f is

$$\deg f = \sum_{p \in C : v_p(f) > 0} v_p(f),$$

the number of preimage points of $(1 : 0)$ counted with multiplicities.

Proposition 14.2.2. *Counted with multiplicities each fiber $f^{-1}(\lambda)$ of $\lambda \in \mathbb{P}^1$ has precisely $\deg f$ many points.*

Proof. Since rational functions are given by quotients of homogeneous polynomials of the same degree on the ambient $\mathbb{P}^n$ the number of poles $\sum_{p \in C : v_p(f) < 0} -v_p(f)$ coincides with the number of zeroes by Bézout's theorem. To see that the number of preimage points of $\lambda \in \mathbb{A}^1 = K$ coincides with $\deg f$, we note that f and $f - \lambda$ have the same poles. □

Remark. One can show that $\deg f$ also coincides with the degree of the field extension $[K(C) : K(f)]$. Note that $K(f) \cong K(\mathbb{P}^1)$.

More generally for a morphism $\varphi \colon C \to E$ between smooth projective curves the **degree** can be defined as

$$\deg \varphi = [K(C) : K(E)],$$

and this number coincides with the number of preimage points of any point $p \in E$ counted with multiplicities.

Remark 14.2.3. Let C be an irreducible smooth projective curve over $K = \mathbb{C}$. Then C is also a compact Riemann surface, which turns out to be connected.

One proof of the connectedness builds upon analytic continuation and monodromy from one complex variable theory and Galois theory from algebra. This is beyond the scope of this book.

The underlying compact two-dimensional differential or topological real manifolds are orientable and classified by their **topological genus** $g_{\text{top}} \in \mathbb{N}$.

The topological genus can be obtained from any triangulation: If C has a triangulation with c_0 vertices, c_1 edges and c_2 triangles, then the **topological Euler characteristic** satisfies

$$\chi_{\text{top}}(C) := c_0 - c_1 + c_2 \stackrel{!}{=} 2 - 2g_{\text{top}}.$$

Let $\varphi \colon C \to E$ be a non-constant morphism of smooth projective curves defined over $\mathbb{C}$. Let $p \in C$ a point and $q = \varphi(p) \in E$ its image. Let $s \in \mathfrak{m}_{C,p} \subset \mathcal{O}_{C,p}$ and $t \in \mathfrak{m}_{E,q} \subset \mathcal{O}_{E,q}$ be generators of the maximal ideals. Then $\varphi^*(t) = u s^r$ for some integer $r > 0$ and $u \in \mathcal{O}_{C,p}$ a unit.

Definition 14.2.4. With this notation the integer

$$e_p := r$$

is called the **ramification index** of φ at p. If $e_p > 1$ we call p a **ramification point** and $q = \varphi(p) \in E$ a **branch point** of φ.

$$R = \sum_{p \in C} (e_p - 1)$$

is called the **total ramification number** of φ. Note that the left-hand side is a finite sum, since the ramification points are isolated in C.

Theorem 14.2.5 (The Riemann-Hurwitz formula). *Let $\varphi \colon C \to E$ be a non-constant morphism of smooth irreducible projective curves defined over $K = \mathbb{C}$ of $d = \deg \varphi$. Let g_C and g_E denote the topological genus of C and E respectively, and let R be the total ramification number of φ. Then*

$$2 - 2g_C = d(2 - 2g_E) - R.$$

Proof. Consider a triangulation of the underlying real manifold of E with c_0 vertices, c_1 edges and c_2 triangles. We take the triangulation fine enough such that each triangle contains at most one branch point, which if present, is a vertex of the triangle. Moreover each triangle should have precisely d preimage triangles in C which are disjoint except for possible ramification points. So the preimages give a triangulation of C which has dc_2 triangles, dc_1 edges but only $dc_0 - R$ vertices because of the ramification. Hence

$$\begin{aligned} 2 - 2g_C = \chi_{\text{top}}(C) &= dc_0 - R - dc_1 + dc_2 \\ &= d\chi_{\text{top}}(E) - R = d(2 - 2g_E) - R. \end{aligned}$$

□

14.3 Dynamical interpretation of intersection numbers

We assume that $K = \mathbb{C}$. Let $f \in K[x, y, z]$ be a square-free polynomial of degree d, and let $g \in K[x, y, z]$ be a homogeneous polynomial of degree e which has no common factor with f. Then

$$d \cdot e = \sum_{p \in V(f,g)} i(f, g; p)$$

by Bézout's theorem. We will show that the intersection multiplicities can be interpreted dynamically.

As an application of Bertini's theorem we see that there exists a homogeneous polynomial g_1 of degree e such that $C = V(f)$ and $D = V(g_1)$ intersect transversally in $d \cdot e$ distinct points.

Indeed, consider the e-uple embedding

$$\rho_{2,e} \colon \mathbb{P}^2 \to \mathbb{P}^{\binom{e+2}{2}-1}.$$

Curves of degree e in $\mathbb{P}^2$ correspond to hyperplanes H in $\mathbb{P}^{\binom{e+2}{2}-1}$, and a general hyperplane H_1 intersects every component of $\rho_{2,e}(C)$ transversally in smooth points of $\rho_{2,e}(C)$.

Let g_1 be the polynomial corresponding to the equation of H_1 and consider the pencil of curves of degree e

$$D = V(t_0 g + t_1 g_1) \in \mathbb{P}^1 \times \mathbb{P}^2.$$

All but finitely many fibers D_λ over $\lambda \in \mathbb{P}^1$ intersect C in $d \cdot e$ distinct points. Consider now the curve

$$X' = D \cap (\mathbb{P}^1 \times C)$$

and the union X of components which dominate $\mathbb{P}^1$. Let $\sigma \colon Y \to X$ be a birational morphism from a smooth projective curve and let Y_0 be the preimage of $(1:0) \in \mathbb{P}^1$ under $f = \pi_1 \circ \sigma$, where π_1 denotes the projection onto the first factor of $\mathbb{P}^1 \times \mathbb{P}^2$. Each point of Y_0 maps to a point of $V(f,g)$ under π_2.

Let $q \in Y_0$ be a point and $s \in \mathfrak{m}_{Y,q} \subset \mathcal{O}_{Y,q}$ a local generator. The rational function $t = t_1/t_0 \in \mathcal{O}_{\mathbb{P}^1,(1:0)}$ pulls back to $h = us^r$ with $r = v_q(h)$ and $u \in \mathcal{O}_{Y,q}$ a unit. For a point $\lambda \in \mathbb{A}^1 = \mathbb{C}$ with $|\lambda|$ small there are precisely r preimage points in the holomorphic chart defined by s with absolute value approximately $(\frac{|\lambda|}{|u(q)|})^{1/r}$. For $\lambda \to 0$ the images of these points in C approach the image of $p \in C \cap D_0$ of q.

Let $p_1, \ldots, p_k$ denote the distinct points of $C \cap V(g)$. Let q_{ij} for $j = 1, \ldots d_i$ denote the distinct pre-images of p_i in Y and r_{ij} denote the ramification numbers as above. Then precisely $\sum_{j=1}^{d_i} r_{ij}$ images of the points in the fiber $f^{-1}(\lambda)$ approach p_i for $\lambda \to 0$.

Thus the dynamical interpretation of the intersection numbers is the following:

Claim. $i(f, g; p_i) = \sum_{j=1}^{d_i} r_{ij}$.

This identity fits with the fact that $\sum_{i=1}^k \sum_{j=1}^{d_i} r_{ij} = d \cdot e$ counts the number of points in the fibers of $Y \to \mathbb{P}^1$.

To prove this identity one can use that $i(f, g; p_i)$ can also be computed as the multiplicity of the resultant $\operatorname{Res}_x(f, g) \in K[y, z]$ at the point $(b_i : c_i)$ for $p_i = (a_i : b_i : c_i)$ if our coordinate system is chosen general enough, cf. Exercise 9.3.11. For example, the $(b_i : c_i)$'s should be pairwise distinct. The resultant $\operatorname{Res}_x(f, g_\lambda)$ has precisely $\sum_{j=1}^{d_i} r_{ij}$ zeroes which approach $(b_i : c_i)$ for $\lambda \to 0$. □

Chapter 15
The geometric genus of a plane curve

In this chapter we bound the number of singular points of an absolutely irreducible plane curve C in terms of the degree $d = \deg C$.

$$\binom{d-1}{2} - \sum_{p \in C} \binom{r_p}{2} \geq 0,$$

where r_p denotes the multiplicity of C at p. Curves which attain the bound are rational, i.e., have a rational parametrization. In Section 15.2 we prove, for an irreducible plane curve C, that there exists a sequence of quadratic transformations in suitable coordinates such that the strict transform of C has only ordinary singularities. This is needed for the proof of Riemann-Roch theorem in Section 16.3. The basic idea is to remove a non-ordinary singularity p of C by a quadratic transformation which has p as one of its fundamental points. If $\operatorname{char}(K) > 0$ the proof is more complicated, because p might be a "strange" point of C, i.e., every line through p might be tangent to C somewhere. In that case we need an additional quadratic transformation so that the image of p is no longer a strange point. The existence of such a transformation follows from Bertini's theorem. The procedure terminates because the difference above or the number of non-ordinary points decreases during the Cremona resolution process.

Finally we formulate a necessary and sufficient criterion for the rationality of a plane curve and define the geometric genus g of C as the difference above for a plane model of C with only ordinary singularities.

15.1 A bound on the number of singular points

Theorem 15.1.1. 1) *Let $C \subset \mathbb{P}^2$ be a plane curve of degree d. Let $r_p = \operatorname{mult}(C; p)$ denote the multiplicity of C at p. Then*

$$\sum_{p \in C} \binom{r_p}{2} \le \binom{d}{2}.$$

2) *If C is irreducible, then*

$$\sum_{p \in C} \binom{r_p}{2} \le \binom{d-1}{2}.$$

Remark. Both bounds are sharp: A general union of d lines has $\binom{d}{2}$ double points. The image of $\mathbb{P}^1$ under a general morphism to $\mathbb{P}^2$ defined by three homogeneous forms of degree d has $\binom{d-1}{2}$ double points.

Proof. Let $I(C) = (f)$. Then f is square-free and $C_{\text{sing}} = V(f, \frac{\partial f}{\partial x}, \frac{\partial f}{\partial y}, \frac{\partial f}{\partial z})$ is a finite set. In general coordinates f and $\frac{\partial f}{\partial x}$ have no common factor. If $p \in C$ is a point of multiplicity r_p, then $\frac{\partial f}{\partial x}$ has multiplicity $\ge r_p - 1$ at p. Thus by Bézout 9.3.1 and the bound on intersection multiplicities 9.3.4 we have

$$\sum_{p \in C} r_p(r_p - 1) \le d(d-1).$$

For the second case we may assume $d \ge 2$. Let $p_1, \ldots, p_s$ denote the singular points of C, and let $r_1, \ldots, r_s$ denote their multiplicity. The vector space

$$L(d - 1; (r_1 - 1)p_1, \ldots, (r_s - 1)p_s)$$

has dimension $\ge \binom{d+1}{2} - \sum_{i=1}^{s} \binom{r_i}{2}$ by Proposition 13.2.2, which is at least d by the first bound.

In particular, $t = \binom{d+1}{2} - \sum_{i=1}^{s} \binom{r_i}{2} - 1 \ge 1$. Choose t further points $q_1, \ldots, q_t$ on C. Then

$$L(d-1; (r_1 - 1)p_1, \ldots, (r_s - 1)p_s, q_1, \ldots, q_t)$$

contains a non-zero element g. The polynomials f and g have no common factor because f is irreducible and $\deg g < d$. So they intersect only in finitely many points and Bézout gives

$$d(d-1) \ge \sum_{i=1}^{s} r_i(r_i - 1) + t.$$

Since $t = \binom{d+1}{2} - \sum_{i=1}^{s} \binom{r_i}{2} - 1 \ge 1$ this inequality is equivalent to the assertion:

$$d(d-1) - \frac{(d+1)d}{2} + 1 = \frac{1}{2}(d^2 - 3d + 2).$$

□

Theorem 15.1.2. *Let $C \subset \mathbb{P}^2$ be an irreducible plane curve of degree d with points of multiplicity r_p. If*

$$\sum_{p \in C} \binom{r_p}{2} = \binom{d-1}{2}$$

then there exists a birational map $\mathbb{P}^1 \to C$.

Proof. With the notation of the proof above we consider now only $t-1$ additional $q_1, \ldots, q_{t-1}$. Then

$$\mathbb{P}(L(d-1; (r_1-1)p_1, \ldots, (r_s-1)p_s, q_1, \ldots, q_{t-1}) \cong \mathbb{P}^1$$

is a pencil. The dimension cannot be larger, because otherwise we could find a curve in the linear system which passes through 2 further points, too many for Bézout. So for every point $q \in C \setminus \{p_1, \ldots, p_s, q_1, \ldots, q_{t-1}\}$ there is a unique curve D in the pencil which passes through q. This defines a birational map $C \dashrightarrow \mathbb{P}^1$ whose inverse extends to a birational morphism $\mathbb{P}^1 \to C$. □

15.2 Existence of a Cremona resolution

We now can prove

Theorem 12.2.8. *Every irreducible plane curve can be transformed by a sequence of quadratic transformations into a plane curve with only ordinary singularities.*

Proof. Let C be an irreducible plane curve of degree d. Then by the bound on the number of singular points the difference

$$\binom{d-1}{2} - \sum_{p \in C} \binom{r_p}{2} \geq 0$$

is non-negative. We compute how this expression changes under a suitable quadratic transformation. Let $p_0 \in C$ be a singular point of multiplicity r. We would like to choose $p_1, p_2 \notin C$ such that the fundamental lines are not tangent to C and such that the fundamental lines intersect C in smooth points outside p_0. This is possible over a field of char$(K) = 0$ because not every line through p_0 is tangent to C by Theorem 14.1.5.

If char$(K) \neq 0$, then possibly every line through p_0 is a tangent line to C, and we might need an additional quadratic transformation, as we will explain at the end of the proof.

Assume now that we can find p_1 and p_2 satisfying the desired requirement for the fundamental triangle p_0, p_1, p_2. Then the strict transform $C' = q(C)$ has degree $d' = 2d - r$ and three new ordinary singular points of multiplicities d, $d-r$ and $d-r$. Since

$$\binom{2d-r-1}{2} - \binom{d}{2} - 2\binom{d-r}{2} = \binom{d-1}{2} - \binom{r}{2}$$

we have

$$\binom{d'-1}{2} - \sum_{p' \in C'} \binom{r'_p}{2} \leq \binom{d-1}{2} - \sum_{p \in C} \binom{r_p}{2},$$

because the contribution of the singular points outside the fundamental triangle is unchanged. The inequality is strict, if the strict transform of C in the blow-up of $\mathbb{P}^2$ at p_0 is not smooth at some points of the exceptional curve E.

Thus such quadratic transformations based at some non-ordinary point either decreases the difference above (in which case the number of non-ordinary points might increase), or decreases the number of non-ordinary points. Since both numbers are non-negative, we obtain a curve with only ordinary singularities after a finite number of transformations. This completes the proof over a field of characteristic zero.

If $\mathrm{char}(K) \neq 0$, we might need additional steps. Suppose that every line through p_0 is tangent to C somewhere. Then the dual curve $\check{C} \cong \mathbb{P}^1$ of C is the line dual to p_0. We claim that after a general quadratic transformation q based at points q_0, q_1, q_2 outside C the strict transform C' of C, which now has degree $2d$ and three more ordinary singular points of multiplicity d, the point $q(p_0) \in C'$ is no longer "strange".

A line through $q(p_0)$ corresponds to a conic D through p_0 and q_0, q_1, q_2. We choose such a nowhere tangent conic D through p_0 first and then $q_0, q_1, q_2 \in D$. Consider the blow-up of $\mathbb{P}^2$ at p_0, that is, the ruled surface $F \subset \mathbb{P}^4$ defined by

$$\mathrm{rank} \begin{pmatrix} w_0 & w_1 & w_3 \\ w_1 & w_2 & w_4 \end{pmatrix} < 2$$

from Example 12.1.9. A smooth conic through p_0, which is nowhere tangent to C, corresponds to a transversal hyperplane section of the strict transform $C'' \subset F \subset \mathbb{P}^4$ of C in F. Such a hyperplane exists by Bertini's theorem 14.1.2. Pick such D. Furthermore, pick a general line L transversal to $C \cup D$. The two intersection points q_1, q_2 of $D \cap L$ are not strange points of C, because C has at most one strange point.

So only finitely many lines through q_1, respectively q_2, are tangent to C, and we can choose one of the infinitely many points $q_0 \in D$ such that $\overline{q_0 q_1}$ and $\overline{q_0 q_2}$ are neither tangent nor pass through a singular point of C. The quadratic transformation based on q_0, q_1, q_2 satisfies the assertion that $q(p_0)$ is no longer a strange point because D is not tangent. The number of non-ordinary singular points and the difference

$$\binom{d-1}{2} - \sum_{p \in C} \binom{r_p}{2}$$

remain unchanged under q. Indeed

$$\binom{d-1}{2} = \binom{2d-1}{2} - 3\binom{d}{2}.$$

Thus, the induction goes through also for fields of characteristic $\neq 0$. □

15.3 Rational curves

Remark. The equality

$$\sum_{p \in C} \binom{r_p}{2} = \binom{d-1}{2}$$

is sufficient but not necessary for rationality.

Example. The curve $V(z^2y^3 - x^5)$ is rational but has only two singular points of multiplicity 2 and 3. So

$$\binom{4}{2} > \binom{2}{2} + \binom{3}{2}.$$

Note that in the blow-up of the affine curve $z^2 - x^5$ we get another double point $u^2 - x^3$ in the chart $(x, z) = (x, ux)$. Over the triple point $y^3 - x^5$ we find a further double point $w^3 - x^2$ under the transformation $(x, y) = (x, wx)$.

Taking these singular points into account we get equality

$$6 = 1 + 3 + 1 + 1.$$

Definition 15.3.1. Let $X_2 \to X_1 \to \mathbb{P}^2$ be the blow-up of a point p followed by a blow-up of a point q on the exceptional $E_1 \subset X_1$. Then we call a point $p_1 \in E_1$ an **infinitesimal near point** to p of **first order** and the points $p_2 \in E_2$ on the exceptional curve of $X_2 \to X_1$ **infinitesimal near point** of p of **second order**.

So we have an infinite tree of infinitesimal near points to every point $p \in \mathbb{P}^2$.

Theorem 15.3.2. *Let $C \subset \mathbb{P}^2$ be an irreducible curve of degree d. Then*

$$\binom{d-1}{2} \geq \sum_p \binom{r_p}{2},$$

where the sum runs over all points of $\mathbb{P}^2$ including infinitesimal near points, and r_p denotes the multiplicity of the strict transform at p. Equality holds if and only if C is birational to $\mathbb{P}^1$.

Definition 15.3.3. The difference $g = \binom{d-1}{2} - \sum_p \binom{r_p}{2}$, where the sum runs over all points including infinitesimal near points as above, is called the **geometric genus** of the plane curve C.

Proof. The bound follows by observing that geometric genus stays constant during a Cremona resolution. This also proves that curves of geometric genus 0 can be rationally parametrized. The converse direction, rational curves have geometric genus 0, follows from the Riemann-Roch theorem 16.3.1 below. □

Remark 15.3.4. Let $C \subset \mathbb{P}^r$ be an irreducible smooth projective curve with Hilbert polynomial p_C. Recall that
$$p_a(C) = 1 - p_C(0)$$
denotes the **arithmetic genus** of C by Definition 8.4.8.

Actually the geometric genus of a plane model of a smooth curve and the arithmetic genus coincide. This follows from the Riemann-Roch theorem 16.3.1 below.

For curves over $\mathbb{C}$ we also have the topological genus g_{top} defined via the formula
$$\chi_{\text{top}}(C) = 2 - 2g_{\text{top}}$$
from the topological Euler characteristic, which coincides with g and $p_a(C)$ as well.

Exercise 15.3.5 (Plücker formulas). Let $C \subset \mathbb{P}^2$ and $\check{C} \subset \check{\mathbb{P}}^2$ be a pair of irreducible curves over $\mathbb{C}$ dual to each other, which have only ordinary cusps and nodes. Let $d, \delta, \kappa, b, f, g$ and $\check{d}$ denote the degree of C, the number of ordinary double points, the number of ordinary cusps, the number of bi-tangents, the number of flexes, the geometric genus and the degree of the dual curve. Since nodes and cusps on $\check{C}$ correspond to bi-tangents and flexes on C, the dual curve has b nodes and f flexes. Prove the Plücker formulas

$$g = \binom{d-1}{2} - \delta - \kappa = \binom{\check{d}-1}{2} - b - f,$$
$$\check{d} = d(d-1) - 2\delta - 3\kappa,$$
$$d = \check{d}(\check{d}-1) - 2b - 3f.$$

Hint: Use the Riemann-Hurwitz formula for the geometric genus for the projection $\pi_p : C \to \mathbb{P}^1$ from a general point $p \in \mathbb{P}^2 \setminus C$.

Conclude
$$f = 3d(d-2) - 6\delta - 8\kappa$$
$$\kappa = 3\check{d}(\check{d}-2) - 6b - 8f$$

and

$$b = \frac{1}{2}[d(d-2)(d-3)(d+3) - (4d^2 - 4d - 20)\delta - (6d^2 - 6d - 27)\kappa + (2\delta + 3\kappa)^2].$$

Exercise 15.3.6. Let $F \in K[x_0, x_1, x_2]$ be the equation of the curve C from Exercise 15.3.5. For a point $a = (a_0 : a_1 : a_2)$ the form
$$P_a F = \sum_{i=0}^{2} a_i \frac{\partial F}{\partial x_i}$$

15.3 Rational curves

and its zero locus $V(P_a F)$ is called the **polar** of C at a. Prove: The polar intersects C in the double point with multiplicity 2, in the cusps with multiplicity 3 and in addition in the ramification points of $\pi_p : C \to \mathbb{P}^1$. This gives a geometric interpretation of the formula $\check{d} = d(d-1) - 2\delta - 3\kappa$.

The formula $f = 3d(d-2) - 6\delta - 8\kappa$ has the following geometric interpretation: Let

$$H = \det\left(\frac{\partial^2 F}{\partial x_i \partial x_j}\right)_{i,j=0,1,2}$$

denote the Hessian of F. Then H is a form of degree $3(d-2)$ which intersects C precisely in the flexes, in the double points and in the cusps with intersection multiplicities 1, 6 and 8 respectively.

Chapter 16
Riemann-Roch

Throughout this chapter C will denote an irreducible smooth projective curve.

In Section 16.1 we introduce divisors D on C and their Riemann-Roch spaces $L(D)$. We denote their dimension by $\ell(D)$. We explain their connection to maps $C \to \mathbb{P}^r$. In particular, we prove that $\varphi_D : C \to \mathbb{P}^{\ell(D)-1}$ defines an embedding if and only if $\ell(D - p - q) = \ell(D) - 2$ holds for any pair of points $p, q \in C$. Finally, we prove Riemann's inequality in the form

$$\ell(D) \geq \deg D + 1 - p_a(C),$$

where $p_a(C)$ denotes the arithmetic genus of a smooth projective model of $C \subset \mathbb{P}^r$. The key point of the proof in the case when D is effective is to consider a hypersurface F of sufficiently large degree e which passes through D and the residual divisor E with $\mathrm{div}(F) = E + D$. Forms G of degree e which pass through E give rise to elements $f = G/F \in L(D)$.

In Section 16.2 we introduce the $K(C)$-vector space $\Omega(C)$ of rational differential forms $\omega = g\,dh$ on C and their canonical divisors $W = \sum_{p \in C} v_p(\omega)p$ on C. A key result is that $\Omega(C)$ is a 1-dimensional $K(C)$-vector space. So any two canonical divisors are linearly equivalent. We compute the divisor class of a canonical divisor for a plane model C' of C with only ordinary singularities in terms of the hyperplane class H and the multiple point divisor E:

$$W \sim (d-3)H - E,$$

where $d = \deg C'$. In particular, we obtain $\deg W = 2g - 2$, where g is the geometric genus of C', and see that g depends only on C.

In Section 16.3 we give Brill and Noether's proof of the famous Riemann-Roch formula

$$\ell(D) - \ell(W - D) = \deg D + 1 - g$$

using the completeness of the adjoint systems, Theorem 16.3.4, from which we deduce the residue theorem. The proof of Theorem 16.3.4 which can be found in Fulton's book [26] relies on Noether's famous AF+BG theorem. We however prefer to refer to some basic facts on sheaf theory, which makes the proof much more transparent. When $K = \mathbb{C}$ the residue theorem follows from the residue theorem on compact Riemann surfaces, leading to a proof of Riemann-Roch without the completeness of the adjoint systems. However, our algorithm 16.3.7 to compute complete linear systems via plane models relies on Theorem 16.3.4.

To prove the equality of the geometric genus with the arithmetic genus, we need the completeness of the hypersurface system of large degree, Theorem 16.3.10, which follows immediately from the Serre vanishing of cohomology of coherent sheaves, Theorem A.2.6.

In Section 16.4 we prove that curves over $\mathbb{C}$ of genus $g \geq 2$ depend on $3g-3$ moduli by using Riemann's existence theorem 16.4.1, Riemann-Roch and the Riemann-Hurwitz formula for branch coverings.

Finally, in Section 16.5 we prove Clifford's theorem, and discuss Green's famous conjecture 16.5.15 about the connection between syzygies of a canonical curve and its the Clifford index.

16.1 Divisors

Definition 16.1.1. A **divisor** on C is a finite formal sum

$$D = \sum_{p \in C} n_p p$$

where all but finitely many coefficients $n_p \in \mathbb{Z}$ are zero. D is called **effective**, in symbols,

$$D \geq 0$$

if all $n_p \geq 0$.

$$\deg D = \sum_{p \in C} n_p$$

is called the **degree** of the divisor

Example 16.1.2. Let $\varphi \colon C \to \mathbb{P}^n$ be a morphism. Let F be a homogeneous form on $\mathbb{P}^n$ of degree d which does not vanish on $\varphi(C)$. Let $p \in C$ be a point. Then the **vanishing order** of F at p is defined as

$$v_p(F) := v_p(\varphi^*(\frac{F}{x^d}))$$

where x is a linear form on $\mathbb{P}^n$ which does not vanish at $\varphi(p)$. The **divisor cut out on C by F** is

$$\mathrm{div}(F) = \sum_{p \in C} v_p(F) p.$$

16.1 Divisors

If φ is an embedding, we can regard $C \subset \mathbb{P}^n$ as a subvariety and this divisor coincides with

$$C.F = \sum_{p \in C \cap V(F)} i(C, F; p)p,$$

the **intersection divisor** of C and F. By Bézout's theorem 9.4.1 this a divisor of $\deg(C.F) = \deg C \cdot \deg F$. We sometimes use the notation $C.F = C'.F$ when C is birational to a curve $C' \subset \mathbb{P}^n$ and F intersects C' only in smooth points. See Corollary A.4.2 for a generalization if F intersects C' in singular points.

Definition 16.1.3. If $f \in K(C)$ is a non-zero function then

$$(f) = \sum_{p \in C} v_p(f)p$$

is called the **principal divisor** of f.

$$(f)_0 = \sum_{p \in C \text{ with } v_p(f) > 0} v_p(f)p$$

and

$$(f)_\infty = \sum_{p \in C \text{ with } v_p(f) < 0} -v_p(f)p$$

are called the divisor **zeroes** or **poles** of f respectively and

$$(f) = (f)_0 - (f)_\infty.$$

Principal divisors have degree 0 by Proposition 14.2.2.

The set of divisors

$$\mathrm{Div}(C) = \{D \mid D \text{ is a divisor on } C\}$$

forms a group under addition, and the map

$$K(C)^* \to \mathrm{Div}(C), f \mapsto (f)$$

is a group homomorphism from the multiplicative group $K(C)^* = K(C) \setminus \{0\}$ to the additive group of divisors because $v_p(fg) = v_p(f) + v_p(g)$.

Definition 16.1.4. Let D be a divisor on C. Then

$$L(D) = \{f \in K(C) \mid (f) + D \geq 0\} := \{f \in K(C)^* \mid (f) + D \geq 0\} \cup \{0\}$$

is called the **Riemann-Roch space** of D. This is a K-vector space because

$$v_p(f + g) \geq \min\{v_p(f), v_p(g)\}.$$

We denote by

$$\ell(D) = \dim_K L(D)$$

its dimension, which turns out to be finite by Corollary 16.1.8.

If $D = \sum_{p \in C} n_p p$ is effective then $L(D)$ can be interpreted as the vector space of rational functions which have poles up to order n_p at p. In particular, functions $f \in L(D)$ are regular outside

$$\mathrm{supp}(D) = \{p \in C \mid n_p \neq 0\},$$

the **support** of D.

Definition 16.1.5. Two divisors D and E on C are called **linearly equivalent**, in symbols,

$$D \sim E$$

if $D - E$ is a principal divisor. The set

$$|D| = \{E \in \mathrm{Div}(C) \mid E \geq 0 \text{ and } E \sim D\}$$

is called the **complete linear system** of effective divisors linearly equivalent to D.

The set $|D|$ can be identified with a projective space because the map

$$\mathbb{P}(L(D)) \to |D|, [f] \mapsto (f) + D$$

is a bijection. Indeed, f_1, f_2 with $(f_1) = (f_2)$ differ by the scalar

$$\lambda = \frac{f_1}{f_2} \in K^*$$

because f_1/f_2 has neither zeroes nor poles, hence is a constant function by Corollary 11.5.2.

Remark 16.1.6. Let D and E be two linearly equivalent divisors on C, say $D - E = (f)$. Then multiplication by f on $K(C)$ maps the subspace $L(E)$ to the subspace $L(D)$. In particular, the dimension $\ell(D)$ depends only on the linear equivalence class of D.

Proposition 16.1.7. *Let D be a divisor and p be a point on C. Then*

1) $\ell(D) = 0$ if $\deg D < 0$. If $\deg D = 0$ then $\ell(D) > 0$ if and only if $D \sim 0$.
2) $\ell(D) \geq \ell(D - p)$
3) If $\ell(D) > 0$ then

$$\ell(D - p) \geq \ell(D) - 1$$

and equality holds for a general point $p \in C$.

Proof. 1) If $f \in L(D)$ is a non-zero function then

$$(f) + D \geq 0 \implies \deg D \geq 0$$

16.1 Divisors

since $\deg(f) = 0$. If $\deg D = 0$ then $D = -(f) = (\frac{1}{f})$ is a principal divisor.

2) $L(D-p)$ is a subspace of $L(D)$ because for an effective divisor E the divisor $E+p$ is effective as well.

3) A non-zero element $f \in L(D)$ lies in $L(D-p)$ if and only if one more coefficient of the power series expansion of $f \in Q(\widehat{\mathcal{O}_{C,p}}) \cong K((t))$ vanishes. So $L(D-p)$ has codimension at most one in $L(D)$. To see that the inclusion is strict when $L(D) \neq 0$ for a general point, we may take $p \notin \mathrm{supp}((f))$. □

Corollary 16.1.8. *Let D be a divisor on C. Then*

$$\ell(D) \leq \deg D + 1.$$

If E is a further divisor which is effective then

$$\ell(D-E) \geq \ell(D) - \deg E.$$

Proof. For $n = \deg D + 1$ points $p_1, \ldots, p_n$ in C the Riemann-Roch spaces

$$L(D) \supset L(D-p_1) \supset L(D-p_1-p_2) \supset \ldots \supset L(D-p_1-\ldots-p_n) = 0$$

form a chain of subspaces of codimension at most one at each step, ending in the zero space. The second inequality is proved similarly by subtracting points in the support of E to obtain a chain which ends in $L(D-E)$. □

Definition 16.1.9. An **r-dimensional linear system of divisors of degree** d, traditionally called a g_d^r on C, is a linear subspace $\mathbb{P}^r = \mathbb{P}(L) \subset \mathbb{P}(L(D)) = |D|$ of a complete linear system of divisors $|D|$ of divisors of degree d corresponding to an $(r+1)$-dimension linear subspace $L \subset L(D)$. A point $p \in C$ is a **base point** of $\mathbb{P}(L)$ if $D' - p$ is effective for all $D' \in \mathbb{P}(L)$. A linear system $\mathbb{P}(L)$ without base points is called **base point free**.

Proposition 16.1.10. *Let D be a divisor on C. Then $|D|$ is base point free if and only if*

$$\ell(D-p) = \ell(D) - 1 \text{ for all } p \in C.$$

□

Definition 16.1.11. Let

$$D^j = \sum_{p \in C} n_p^j p$$

for $j = 0, \ldots, r$ be $r+1$ divisors. Then

$$\min\{D^0, \ldots, D^r\} = \sum_{p \in C} \min\{n_p^0, \ldots n_p^r\} p$$

is called the **minimum** of $D^0, \ldots, D^r$.

Proposition 16.1.12. *Let $f_0, \ldots, f_r \in K(C)$ be rational function and let*

$$\varphi': C \dashrightarrow \mathbb{P}^r, p \mapsto (f_0(p) : \ldots : f_r(p))$$

be the corresponding rational map. Let $D = -\min\{(f_0), \ldots, (f_r)\}$ be the negative of the minimum of the principal divisors (f_j). Then $f_0, \ldots, f_r \in L(D)$ and the K-vector space $L \subset L(D)$ spanned by these functions gives a base point free linear system $\mathbb{P}(L) \subset |D|$.

Proof. By the definition of the minimum the divisor

$$(f_j) \geq \min\{(f_0), \ldots, (f_r)\}.$$

Thus $(f_j) + D \geq 0$ and $f_j \in L(D)$. To see that $\mathbb{P}(L)$ is base point free, we consider for $p \in C$ and $D = \sum_{p \in C} n_p p$ a function f_i with $v_p(f_i) = -n_p$. Then p is not in the support $(f_i) + D$, hence not a base point of $\mathbb{P}(L) \subset |D|$. □

Let $D = \sum n_p p$ be a divisor such that $|D|$ is a base point free linear system of dimension $r = \ell(D) - 1 \geq 1$. We denote by

$$\varphi_D : C \to \mathbb{P}^r$$

the morphism which extends the rational map defined by a basis $f_0, \ldots, f_r$ of $L(D)$. So φ_D is well-defined up to an automorphism of $\mathbb{P}^r$.

We now describe φ_D from a conceptual point view with divisors. For points $p \notin \mathrm{supp}(D)$ none of the functions $f_0, \ldots, f_r$ has a pole, and not all are zero. Thus

$$L(D) \to K, f \mapsto f(p)$$

is a non-zero linear form on $L(D)$ whose kernel coincides with $L(D - p)$. The subspace $L(D - p)$ determines the tuple $(f_0(p), \ldots, f_r(p))$ up to a scalar.

Proposition 16.1.13. *Let $|D|$ be a base point free complete linear system of divisors of dimension $\ell(D) = r + 1 \geq 2$. Identify $\mathbb{P}^r = \mathbb{P}(L(D)^*)$ with the projective space of codimension 1 subspace of $L(D)$. Then the extension φ_D is given by*

$$\varphi_D : C \to \mathbb{P}(L(D)^*), p \mapsto L(D - p) \subset L(D).$$

Proof. Write $D = \sum n_p p$ and let $f_0, \ldots, f_r$ be a basis of $L(D)$ as before. In terms of coordinates we can describe φ_D as follows. For a point $p \in C$ let $f_i \in L(D)$ be a rational function with $v_p(f_i) = -n_p$. Then

$$(\frac{f_0}{f_i}(p) : \ldots : \frac{f_r}{f_i}(p)) = (f_0(p) : \ldots : f_r(p)) \in \mathbb{P}^r$$

where both sides are defined. The left-hand side is defined at p. If $(f_i) = D' - D$ then multiplication by f_i maps the linear subspace $L(D' - p) \subset L(D')$ to the linear subspace $L(D - p) \subset L(D)$. □

Theorem 16.1.14. *Let $|D|$ be a complete linear system of divisors on an irreducible smooth projective curve C. Then*

16.1 Divisors

$$\ell(D - p - q) = \ell(D) - 2$$

holds for every pair of points $p, q \in C$ if and only if

$$\varphi_D \colon C \to \mathbb{P}^{\ell(D)-1}$$

is an embedding.

Proof. We assume that the condition on the dimensions is satisfied. Then $|D|$ is base point free. Since

$$L(D - p - q) = L(D - p) \cap L(D - q) \subset L(D) \text{ for points } p \neq q$$

φ_D gives a bijection $C \to \varphi_D(C)$ by Proposition 16.1.13 and $L(D-p-q) \subsetneq L(D-p)$. This does not imply that $C \to \varphi_D(C)$ is an isomorphism, as Example 2.2.5 3) shows. This might be just a birational morphism.

The condition $\ell(D-2p) = \ell(D)-2$ is needed to conclude that $\mathcal{O}_{C,p}$ is isomorphic to $\mathcal{O}_{\varphi_D(C),\varphi_D(p)}$. Let $f_0 \in L(D) \setminus L(D - p)$ and $f_1 \in L(D - p) \setminus L(D - 2p)$. Then $t = \frac{f_1}{f_0} \in K(C)$ is a function with $v_p(t) = 1$. Let $(R, \mathfrak{m})$ be the local ring of $\varphi_D(C)$ at $\varphi_D(p)$. Then

$$t \in \mathfrak{m} \subset R \subset \mathcal{O}_{C,p} \subset K(C)$$

is a generator of the maximal ideal $\mathfrak{m}_{C,p}$ of the DVR $\mathcal{O}_{C,p}$. We prove that t generates $\mathfrak{m}$ in R as well.

Let $g_1, \ldots, g_e \in \mathfrak{m}$ be generators, which map to a basis of $\mathfrak{m}/\mathfrak{m}^2$. Replacing $g_1, \ldots, g_e$ by suitable K-linear combinations we may assume that the $v_p(g_i)$ are pairwise different and sorted

$$1 \leq v_p(g_1) < v_p(g_2) < \ldots < v_p(g_e) = m.$$

Then a non-trivial K-linear combination $\sum_{i=1}^{e} \lambda_i g_i$ has valuation

$$v_p\left(\sum_{i=1}^{e} \lambda_i g_i\right) = v_p(g_j) \text{ where } j = \min\{i \mid \lambda_i \neq 0\}.$$

In particular, we see that any function $g \in R$ with $v_p(g) > m = v_p(g_e)$ lies in $\mathfrak{m}^2$. Consider now the power series expansions

$$g_i = \sum_{n=1}^{\infty} g_{i,n} t^n \in K[[t]] = \widehat{\mathcal{O}_{C,p}}.$$

Then $\tilde{g}_i = g_i - \sum_{n=1}^{m} g_{i,n} t^n \in R$ satisfies $v_p(\tilde{g}_i) > m$. Hence $\tilde{g}_i \in \mathfrak{m}^2$ and we conclude

$$(t) + \mathfrak{m}^2 = \mathfrak{m} \subset R.$$

Nakayama's lemma 10.1.3 implies $(t) = \mathfrak{m}$. Hence R is a discrete valuation ring by Proposition 10.6.3 and its valuation coincides with v_p. Thus $R = \mathcal{O}_{C,p}$ and $C \cong \varphi(C)$.

Conversely, if $\varphi_D : C \to \mathbb{P}^{\ell(d)-1}$ is an embedding then a hyperplane which does not pass through p induces an element of $L(D) \setminus L(D - p)$. A hyperplane which passes through p but not through q induces an element of $L(D - p) \setminus L(D - p - q)$ and a hyperplane which intersect C at p transversally induces an element of $L(D - p) \setminus L(D - 2p)$. □

Theorem 16.1.14 above makes it clear that it is important to be able to compute $\ell(D)$ for a divisor precisely. A first step in this direction is Riemann's inequality.

Theorem 16.1.15 (Riemann's inequality). *Let $C \subset \mathbb{P}^n$ be an irreducible smooth projective curve of arithmetic genus $p_a(C)$ and let D be a divisor on C. Then*

$$\ell(D) \geq \deg D + 1 - p_a(C).$$

Proof. We first assume that $D = \sum_{i=1}^{s} n_i p_i$ is an effective divisor whose support consists of s points. The homogeneous ideal of D viewed as a subscheme of C is $I(D) = J_{\text{sat}}$, the saturation of

$$J = \bigcap_{i=1}^{s} (I(p_i)^{n_i} + I(C)).$$

A homogeneous form $F \in I(D) \setminus I(C)$ defines a hypersurface which intersects C in the p_i's with multiplicities

$$i(C, F; p_i) \geq n_i.$$

Choose e large enough such that

1. the value of the Hilbert function of C coincides with the value of the Hilbert polynomial, i.e., $h_C(e) = p_C(e)$,
2. $I(D)_e \supsetneq I(C)_e$.

The second condition can be satisfied because the Hilbert polynomial of D is the constant polynomial $p_D(t) = \deg D$ while $p_C(t) = dt + 1 - p_a(C)$, where $d = \deg C$ grows linearly.

Fix $F_0 \in I(D)_e \setminus I(C)_e$. The residual scheme E defined by

$$I(E) = (I(C) + (F_0)) : I(D)$$

corresponds to an effective divisor E of degree $\deg E = de - \deg D$ by Bezout 9.4.1. Any $F \in I(E)_e \setminus I(C)_e$ cuts C in divisor $D' + E$ with $D' \geq 0$ and the rational function F/F_0 restricted to C lies in $L(D)$ because $(F/F_0)+D = D'+E-(D+E)+D = D' \geq 0$. Two rational functions F_1/F_0 and F_2/F_0 restrict to the same rational function on C if and only if $F_1 - F_2 \in I(C)_e$. Thus

$$I(E)_e / I(C)_e \to L(D), \overline{F} \mapsto F/F_0$$

is an inclusion. We estimate the dimension of this subspace. Let S denote the homogeneous coordinate ring of $\mathbb{P}^n$. Then $I(E)_e \subset S_e$ has codimension at most $\deg E$. Since $I(C)_e$ has codimension precisely $p_C(e) = ed + 1 - p_a(C)$ by condition 1) we obtain

$$\begin{aligned}\ell(D) &\geq \dim_K I(E)_e/I(C)_e \\ &\geq p_C(e) - \deg E = ed + 1 - p_a(C) - (de - \deg D) \\ &= \deg D + 1 - p_a(C).\end{aligned}$$

If $D = D_1 - D_2$ is the difference of two effective divisors then

$$\ell(D) \geq \ell(D_1) - \deg D_2 \geq \deg D_1 + 1 - p_a(C) - \deg D_2 = \deg D + 1 - p_a(C)$$

by Corollary 16.1.8 and the first part of the proof. □

Corollary 16.1.16. *Let C be an irreducible smooth projective curve of arithmetic genus $p_a(C) = 0$. Then C is isomorphic to $\mathbb{P}^1$.*

Proof. Consider a point $p_0 \in C$ and the divisor $D = p_0$. Then

$$2 = \deg D + 1 \geq \ell(D) \geq \deg D + 1$$

by Corollary 16.1.8 and Riemann's inequality 16.1.15. Thus equality holds. Since $\deg(D - p - q) = -1 < 0$ we have $0 = \ell(D - p - q) = \ell(D) - 2$. Thus $\varphi_D : C \to \mathbb{P}^1$ is an embedding by Theorem 16.1.14, which in this case is an isomorphism. □

Remark 16.1.17. Let g^* denote the smallest integer such that

$$\ell(D) \geq \deg D + 1 - g^*$$

holds for all divisors D on C. Thus $0 \leq g^* \leq p_a(C)$ by Theorem 16.1.15 and Corollary 16.1.8. In particular, $p_a(C) \geq 0$.

We will see in the course of the proof of the Riemann-Roch theorem that $g^* = p_a(C)$. In particular, the arithmetic genus of C does not depend on the embedding. Furthermore g^* also coincides with the geometric genus g of a plane model of C with only ordinary singular points. When $K = \mathbb{C}$ the integer g^* also coincides with the topological genus g_{top} of the underlying 2-dimensional real manifold of the Riemann surface of C, as introduced in Remark 14.2.3. Usually Riemann's inequality is written in the form

$$\ell(D) \geq \deg D + 1 - g.$$

16.2 Rational differentials and canonical divisors

Definition 16.2.1. We define the $K(C)$-vector space $\Omega(C)$ of **rational differential forms** on C as follows. Consider for each function $f \in K(C)$ a symbol Df and the

free $K(C)$-vector space
$$M = \bigoplus_{f \in K(C)} K(C)Df$$
with basis $\{Df \mid f \in K(C)\}$. Consider the subspace $N \subset M$ generated by the expressions

1) $\{D(f+g) - Df - Dg \mid f, g \in K(C)\}$
2) $\{D(\lambda f) - \lambda Df \mid \lambda \in K, f \in K(C)\}$
3) $\{D(fg) - fDg - gDf \mid f, g \in K(C)\}$.

Then we define $\Omega(C) = M/N$. Let $df \in \Omega(C)$ denote the class represented by DF. The the map
$$d \colon K(C) \to \Omega(C), f \mapsto df$$
is K-linear by 1) and 2) and $d\lambda = 0$ because $d(1) = d(1 \cdot 1) = d(1) + d(1)$ implies $d(1) = 0$. However, although $K(C)$ and $\Omega(C)$ are both $K(C)$-vector spaces, the map d is not $K(C)$-linear. Instead, we have the product rule
$$d(fg) = fdg + gdf.$$

The quotient rule
$$d(\frac{f}{g}) = \frac{gdf - fdg}{g^2}$$
holds as well because $0 = d(1) = d(g \cdot \frac{1}{g}) = \frac{1}{g}dg + gd(\frac{1}{g})$ implies $d(\frac{1}{g}) = -\frac{dg}{g^2}$.

Proposition 16.2.2. *$\Omega(C)$ is a 1-dimensional $K(C)$-vector space.*

Proof. Let $f \in K[x, y]$ be the equation of an affine plane model $V(f) \subset \mathbb{A}^2$ of C. Then $K(C) = Q(K[x, y]/(f))$ and the coordinate functions $\overline{x}, \overline{y}$ on C generate the function field $K(C) = K(\overline{x}, \overline{y})$. Thus any function $\overline{g} \in K(C)$ is a rational function in $\overline{x}$ and $\overline{y}$. Let $g \in K[x, y]$. Then
$$d(g(\overline{x}, \overline{y})) = \frac{\partial g}{\partial x}(\overline{x}, \overline{y})d\overline{x} + \frac{\partial g}{\partial y}(\overline{x}, \overline{y})d\overline{y}$$
since $d(\overline{x}^i \overline{y}^j) = i\overline{x}^{i-1}\overline{y}^j d\overline{x} + j\overline{x}^i \overline{y}^{j-1} d\overline{y}$ follows from the product rule by induction on i and j. In view of the quotient rule we conclude that $d\overline{x}$ and $d\overline{y}$ generate $\Omega(C)$ as a $K(C)$-vector space.
$$0 = d(f(\overline{x}, \overline{y})) = \frac{\partial f}{\partial x}(\overline{x}, \overline{y})d\overline{x} + \frac{\partial f}{\partial y}(\overline{x}, \overline{y})d\overline{y}$$
give a dependency relation between $d\overline{x}$ and $d\overline{y}$. This relation is nontrivial: Since $\deg \frac{\partial f}{\partial x} \in K[x, y]$ has smaller degree than f in x we have $\frac{\partial f}{\partial x}(\overline{x}, \overline{y}) = 0$ if and only if $\frac{\partial f}{\partial x} = 0 \in K[x, y]$. So the relation is trivial if and only if both $\frac{\partial f}{\partial x}$ and $\frac{\partial f}{\partial y}$ are zero in $K[x, y]$ and this holds if and only if $\text{char}(K) = p$ and $f \in K[x^p, y^p]$. Since K is

16.2 Rational differentials and canonical divisors

algebraically closed this implies that f is a p-th power, compare 10.5.9, contradicting the irreducibility of f. Hence $\Omega(C)$ is at most 1-dimensional.

It remains to prove $\Omega(C) \neq 0$. When $K = \mathbb{C}$ there is an easy argument using the underlying Riemann surface of C. Rational functions are meromorphic functions and the differential $d_\mathcal{M} h$ of a non-constant meromorphic function h, in the sense calculus of one complex variable, is a non trivial meromorphic 1-form. Let $\mathcal{M}(C)$ be the field of meromorphic functions on the Riemann surface C and $\mathcal{M}^1(C)$ the space of meromorphic 1-forms. So the composition

$$K(C) \hookrightarrow \mathcal{M}(C) \to \mathcal{M}^1(C), f \mapsto d_\mathcal{M} f$$

is non-trivial. On the other hand $d_\mathcal{M}$ satisfies the same rules as the rules for d induced by 1), 2), 3). So the map defined by

$$M \to \mathcal{M}^1(C), Df \mapsto d_\mathcal{M} f$$

has $N \subset M$ in the kernel. Here M and N denote the vector spaces in the definition of $\Omega(C)$ 16.2.1. Hence we get a well-defined map $\Omega(C) \to \mathcal{M}^1(C)$ and a commutative diagram

$$\begin{array}{ccc} K(C) & \longrightarrow & \mathcal{M}(C) \\ d \downarrow & & \downarrow d_\mathcal{M} \\ \Omega(C) & \longrightarrow & \mathcal{M}^1(C) \end{array}$$

and $\Omega(C)$ cannot be trivial.

For arbitrary fields we take the argument above as a clue. It suffices to find an $K(C)$-vector space $\mathcal{N}$ and a non-trivial K-linear map $d_\mathcal{N}: K(C) \to \mathcal{N}$ which satisfies the product rule $d_\mathcal{N}(\overline{g}\overline{h}) = \overline{g}d_\mathcal{N}(\overline{h}) + \overline{h}d_\mathcal{N}(\overline{g})$ because such $d_\mathcal{N}$ factors over $\Omega(C)$.

Suppose $\frac{\partial f}{\partial y} \neq 0 \in K[x, y]$. Then $\frac{\partial f}{\partial y}(\overline{x}, \overline{y}) \in K(C)$ is non-zero and $d\overline{y} = ud\overline{x}$, where

$$u = \frac{\partial f}{\partial x}(\overline{x}, \overline{y}) / \frac{\partial f}{\partial y}(\overline{x}, \overline{y}).$$

We take $\mathcal{N} = K(C)$ and define $d_\mathcal{N}$ on the subspace $K[x, y]/(f) \subset K(C)$ by

$$d_\mathcal{N}(\overline{g}) = \frac{\partial g}{\partial x}(\overline{x}, \overline{y}) - u\frac{\partial g}{\partial y}(\overline{x}, \overline{y})$$

for $\overline{g} = g(\overline{x}, \overline{y}) = g + (f) \in K[x, y]/(f)$. Then $d_\mathcal{N}$ satisfies the product rule since partial derivatives do this and it is well-defined by the definition of u. Moreover $d_\mathcal{N}(\overline{x}) = 1$, since $\frac{\partial x}{\partial x} = 1$ and $\frac{\partial x}{\partial y} = 0$. Finally we extend $d_\mathcal{N}$ to the quotient field $K(C) = Q(K[x, y]/(f))$ using the quotient rule

$$d_\mathcal{N}(\overline{g}/\overline{h}) := \frac{\overline{h}d_\mathcal{N}(\overline{g}) - \overline{g}d_\mathcal{N}(\overline{h})}{\overline{h}^2}.$$

Remark 16.2.3. If $\mathrm{char}(K) = p > 0$ then it is possible that a non-constant rational function $\bar{g}$ has trivial differential. For example $d(\bar{x}^p) = p\bar{x}^{p-1}d\bar{x} = 0$. It turns out that a non-constant function $\bar{g}$ has a non-trivial differential $d\bar{g} \neq 0$ if and only if $K(C)$ is a separable field extension of $K(\bar{g})$, see [18] Appendix B.

Proposition 16.2.4. *Let* $t \in \mathcal{O}_{C,p}$ *be a generator of the maximal ideal. Then the following holds:*

i) dt is a basis of $\Omega(C)$.
ii) If $s \in \mathfrak{m}_{C,p}$ is a further generator then $ds = u dt$ with $u \in \mathcal{O}_{C,p}$ a unit.
iii) If $g \in \mathcal{O}_{C,p}$ and $dg = g' dt$ then also $g' \in \mathcal{O}_{C,p}$.

Proof. We work with an affine plane model $V(f) \subset \mathbb{A}^2$ of C and assume in addition that p is a smooth point of the model and corresponds to the origin. This can be achieved by a Cremona transformation and coordinate changes. Thus

$$f = ax + by + \text{ higher degree terms.}$$

We may assume that $b \neq 0$. Then $\frac{\partial f}{\partial y} \neq 0 \in K[x,y]$, $d\bar{x} \in \Omega(C)$ is a basis and $\bar{x}$ generates $\mathfrak{m}_{C,p}$. We prove iii) first for $t = \bar{x}$. The elements $\bar{g} = g(\bar{x}, \bar{y}) \in K[x,y]/(f)$ lie in $\mathcal{O}_{C,p}$. We have

$$d\bar{g} = \frac{\partial g}{\partial x}(\bar{x}, \bar{y}) d\bar{x} + \frac{\partial g}{\partial y}(\bar{x}, \bar{y}) d\bar{y}$$

and

$$d\bar{y} = -\frac{\partial f}{\partial x}(\bar{x}, \bar{y}) / \frac{\partial f}{\partial y}(\bar{x}, \bar{y}) d\bar{x}.$$

$\frac{\partial f}{\partial y}(\bar{x}, \bar{y})$ is a unit in $\mathcal{O}_{C,p}$ since $\frac{\partial f}{\partial y}(0,0) = b \neq 0$. Thus $d\bar{g} = \bar{g}' dx$ with

$$\bar{g}' = \frac{\partial g}{\partial x}(\bar{x}, \bar{y}) - \frac{\partial g}{\partial y}(\bar{x}, \bar{y}) \frac{\partial f}{\partial x}(\bar{x}, \bar{y}) / \frac{\partial f}{\partial y}(\bar{x}, \bar{y})$$

satisfies $\bar{g}' \in \mathcal{O}_{C,p}$.

For an arbitrary element $\bar{g}/\bar{h} \in \mathcal{O}_{C,p}$ with $h(0,0) \neq 0$ we have

$$(\bar{g}/\bar{h})' = \frac{\bar{h}\bar{g}' - \bar{g}\bar{h}'}{\bar{h}^2} \in \mathcal{O}_{C,p}$$

because $\bar{h}$ is a unit in $\mathcal{O}_{C,p}$.

ii) If $s = u\bar{x}$ with u a unit in $\mathcal{O}_{C,p}$ then $ds = (u+u'\bar{x})d\bar{x}$. Observe that $v_p(u+u'\bar{x}) = v_p(u) = 0$ holds, because $v_p(u'\bar{x}) = v_p(u') + v_p(\bar{x}) \geq 0+1$ by iii) for $\bar{x}$. Thus $u + u'\bar{x}$ is a unit, establishing ii) for s and $t = \bar{x}$. In particular, ds is a basis for $\Omega(C)$, which proves i).

Finally writing $dt = \tilde{u} d\bar{x}$ with $\tilde{u}$ a unit, in $\mathcal{O}_{C,p}$ we have $d\bar{x} = \tilde{u}^{-1} dt$. Thus ii) and iii) holds also for t. □

16.2 Rational differentials and canonical divisors

We are now ready to define canonical divisors.

Definition 16.2.5. Let $\omega = g\,dh \in \Omega(C)$ be a non-zero rational differential form. Then we define

$$v_p(\omega) = v_p(f)$$

when $\omega = f\,dt$ for a generator t of $\mathfrak{m}_{C,p}$. This is well-defined because ds and dt, for two generators s, t, differ by a unit by ii) above. We call

$$W = \sum_{p \in C} v_p(\omega) p$$

the divisor of ω and any divisor of this form is called a **canonical divisor** on C. A rational differential with $v_p(\omega) \geq 0$ for all points $p \in C$ is called **regular**. Note that W is indeed a divisor, i.e., only finitely many $v_p(\omega)$ are non-zero: With the notation from the proof of Proposition 16.2.4 the function $\bar{x} - \bar{x}(p)$ is a generator of $\mathfrak{m}_{C,p}$ for all but finitely many $p \in C$ and $d\bar{x} = d(\bar{x} - \bar{x}(p))$ since $\bar{x}(p)$ is a constant. Thus we may take the same expression $\omega = f\,d\bar{x}$ for all but finitely many $p \in C$ and the sum is finite because f has only finitely many zeroes or poles.

Any two canonical divisors are linearly equivalent: For two non-zero rational differential forms we have $\omega_1 = g\omega_2$ for some $g \in K(C)$ because $\Omega(C)$ is a one-dimensional $K(C)$-vector space. Hence $W_1 = (g) + W_2$. What is really canonical is the linear equivalence class of a canonical divisor.

Our next goal is to compute the degree of a canonical divisor. Consider a plane model C' of C. Using Cremona transformations we may assume that $C' \subset \mathbb{P}^2$ has only ordinary singular points. If $q_1, \ldots, q_s$ are these points, then the strict transform of C' in the blow-up $X = \mathbb{P}^2(q_1, \ldots, q_s)$ of $\mathbb{P}^2$ in the points $q_1, \ldots, q_s$ is isomorphic to C. Let $\sigma : C \to C'$ be the corresponding morphism and let r_q denote the multiplicity of a point $q \in C'$. Then

$$E = \sum_{p \in C} (r_{\sigma(p)} - 1) p$$

is called the **divisor of multiple points** of C' on C.

$$\deg E = \sum_{i=1}^{s} r_i(r_i - 1),$$

where $r_i = r_{q_i}$ since each q_i has precisely r_i preimage points in C.

Proposition 16.2.6. *Let $C' \subset \mathbb{P}^2$ be an irreducible plane curve of degree d with only ordinary singular points. Let E be the divisor of multiple points of C'. Let H be the divisor cut out on C by a linear form of $\mathbb{P}^2$ and W a canonical divisor on C. Then*

$$W \sim (d-3)H - E.$$

Proof. We assume that our coordinates are chosen such that the line at infinity $L = V(z)$ intersects C' transversally in d distinct points and that the point $(0 : 1 : 0)$

is not among them. Moreover none of the r_q tangent lines to a multiple point q of C' should pass through $(0 : 1 : 0)$.

The idea of the proposition is to compare the divisor of $d\bar{x}$ with the divisor cut out by $\frac{\partial F}{\partial y}$ on C, where $F = f^h$ is homogeneous form defining C'.

The complement $\mathbb{A}^2$ of the line at infinity contains all singular points. Let $f \in K[x, y]$ be the affine equation of C' in this chart. Then the projection onto the x-axis corresponds in this chart to the projection of C' from $(0 : 1 : 0)$. We compute the divisor of $d\bar{x} \in K(C)$ building on the computation in the proof of Proposition 16.2.4.

For smooth points $p \in C'$ with $\frac{\partial f}{\partial y}(p) \neq 0$ we have that $\bar{x} - \bar{x}(p)$ is a generator of $\mathfrak{m}_{C,p}$. The same is true for p over a multiple point of C' since the corresponding tangent line does not pass through $(0 : 1 : 0)$. Thus $v_p(d\bar{x}) = 0$ at those points. However $v_p(\frac{\partial f}{\partial y}) = r_q - 1$ because for

$$f = \prod_{j=1}^{r_q} \ell_j + \text{ higher order terms}$$

the form $\frac{\partial \ell_i}{\partial y} \prod_{j \neq i} \ell_j$, where ℓ_i defines the tangent line corresponding to p, is the unique summand of $\frac{\partial f}{\partial y}$ which vanishes at p only to order $r_q - 1$.

At smooth points $p \in C' \cap \mathbb{A}^2$ where $\frac{\partial f}{\partial y}(p) = 0$ the partial derivative $\frac{\partial f}{\partial x}(p) \neq 0$ and $\bar{y} - \bar{y}(p)$ is a generator of $\mathfrak{m}_{C,p}$. Since

$$d\bar{x} = \frac{\partial f}{\partial y}(\bar{x}, \bar{y}) / \frac{\partial f}{\partial x}(\bar{x}, \bar{y}) d\bar{y},$$

we find that $v_p(d\bar{x}) = v_p(\frac{\partial f}{\partial y})$.

It remains to compute the vanishing order of $d\bar{x}$ at the points p at infinity. There $w = \frac{1}{x}$ is a generator of $\mathfrak{m}_{C,p}$ and $v_p(d\bar{x}) = -2$ because $d\bar{x} = -\frac{dw}{w^2}$. On the other hand $v_p(\frac{\partial F}{\partial y}) = 0$ because $F(x, y, 0)$ has no multiple roots.

In summary, we obtain

$$W = \sum_{p \in C} v_p(d\bar{x}) p = \text{div}(\frac{\partial F}{\partial y}) - E - 2C'.L.$$

Since $\deg \frac{\partial F}{\partial y} = d - 1$, and we have $\text{div}(\frac{\partial F}{\partial y}) \sim (d - 1)H$ and $C'.L \sim H$ we conclude

$$W \sim (d - 3)H - E.$$

□

Corollary 16.2.7. *Let g denote the geometric genus of a plane model C' with only ordinary points and let W be a canonical divisor on C. Then*

$$\deg W = 2g - 2.$$

16.2 Rational differentials and canonical divisors

In particular, the geometric genus does not depend on the choice of the plane model with only ordinary points.

From now on we speak of the geometric genus of an irreducible smooth projective curve.

Proof. Let $d = \deg C'$. Then

$$\deg W = (d-3)d - \sum_{q \in C'} r_q(r_q - 1)$$

$$= 2(\binom{d-1}{2} - \sum_{q \in C'} \binom{r_q}{2}) - 2 = 2g - 2.$$

$\square$

The formula for the degree of a canonical divisor allows us to generalize the Riemann-Hurwitz formula to arbitrary fields K.

Definition 16.2.8. Let $\varphi: C \to E$ be non-constant morphism between smooth irreducible projective curves of degree $d = [K(C) : K(E)]$. Let $q \in C$ be a point and let $\bar{q} = \varphi(q) \in E$ be the image point. Consider generators $t \in \mathfrak{m}_{E,\bar{q}}$ and $s \in \mathfrak{m}_{C,q}$. Then $\varphi^* t = u s^{e_q}$ with $u \in \mathcal{O}_{C,q}$ a unit, and e_q is called the **ramification index** of φ at q.

The pullback of divisors $\varphi^*: \text{Div}(E) \to \text{Div}(C)$ is defined by

$$D = \sum_{\bar{q} \in E} n_{\bar{q}} \bar{q} \mapsto \varphi^* D = \sum_{q \in C} (n_{\varphi(q)} e_q) q.$$

In particular, $\deg \varphi^* D = d \deg D$ holds, and for the principal divisor of a rational function $f \in K(E)$ we have $\varphi^*(f) = (\varphi^* f)$.

The morphism $\varphi: C \to E$ is called **separable** if the field extension $K(E) \subset K(C)$ is separable. For a non-zero rational differential form $\omega = g dh \in \Omega(E)$ the pullback is defined by $(\varphi^* g) d(\varphi^* h)$. Hence the diagram

$$\begin{array}{ccc} K(E) & \xrightarrow{\varphi^*} & K(C) \\ d \downarrow & & \downarrow d \\ \Omega(E) & \xrightarrow{\varphi^*} & \Omega(C) \end{array}$$

commutes. If φ is separable then the pullback of a non-zero form $\omega \in \Omega(E)$ is non-zero, i.e., $\varphi^* \omega \neq 0$ by Remark 16.2.3.

The **ramification divisor** R of a separable morphism $\varphi: C \to E$ is defined as follows. Consider $\varphi^*(dt) = (u e_q s^{e_q - 1} + u' s^{e_q}) ds = f ds$. Then set $r_q = v_q(f)$ and define $R = \sum_{q \in C} r_q q$.

If $p = \text{char } K$ does not divide e_q then $r_q = e_q - 1$ and we call the ramification of φ at q **tame**. If $p = \text{char } K$ divides e_q then $r_q > e_q - 1$ and we call the ramification at q **wild**.

Theorem 16.2.9 (Riemann-Hurwitz formula, second version). *Let $\varphi\colon C \to E$ be a separable morphism between irreducible smooth projective curves of degree $d = [K(C) : K(E)]$. Let $R = \sum_{q \in C} r_q q$ be the ramification divisor. Then*

$$2g_C - 2 = d(2g_E - 2) + \deg R,$$

where g_C and g_E denote the geometric genus of C and E respectively.

Proof. Consider a non-zero rational differential form ω on E. The divisor W_E of ω has degree $2g_E - 2$. The divisor W_C of $\varphi^*\omega$ has degree $2g_C - 2$. The computation above shows that

$$W_C = \varphi^* W_E + R$$

and the result follows, because $\deg \varphi^* W_E = d \deg W_E = d(2g_E - 2)$. □

Corollary 16.2.10. *Let C be a smooth projective curve defined over $\mathbb{C}$. Then the geometric genus of C and the topological genus of the underlying Riemann surface of C coincide.*

Proof. Combine versions 14.2.5 and 16.2.9 of the Riemann-Hurwitz formula. □

16.3 Proof of the Riemann-Roch theorem

Theorem 16.3.1 (Riemann-Roch). *Let C be an irreducible smooth projective curve of geometric genus g. Let D be a divisor on C and let W be a canonical divisor. Then*

$$\ell(D) - \ell(W - D) = \deg D + 1 - g.$$

Remark 16.3.2. If C is rational, i.e., $C \cong \mathbb{P}^1$, then the theorem holds because any two divisors of the same degree are linearly equivalent on $\mathbb{P}^1$ and $\deg W = -2$. Thus

$$\begin{array}{lll}
\ell(D) = \deg D + 1 & \text{and } \ell(W - D) = 0 & \text{if } \deg D \geq 0, \\
\ell(D) = 0 & \text{and } \ell(W - D) = 0 & \text{if } \deg D = -1, \\
\ell(D) = 0 & \text{and } \ell(W - D) = -\deg D - 1 & \text{if } \deg D \leq -2.
\end{array}$$

Recall that C is rational if it has arithmetic genus $p_a(C) = 0$ by Corollary 16.1.16 or if it has a plane model with only ordinary points and geometric genus $g = 0$ by Theorem 15.1.2. Thus we only have to prove the theorem in the case when C has geometric genus $g > 0$.

Let $C' \subset \mathbb{P}^2$ be an irreducible curve of degree d with only ordinary singular points $q_1, \ldots, q_s$ of multiplicities $r_1, \ldots, r_s$. Let $\sigma\colon C \to C'$ be a resolution of singularities, e.g. the strict transform C of C' in the blow-up $X = \mathbb{P}^2(q_1, \ldots, q_s)$ of $\mathbb{P}^2$ in the points $q_1, \ldots, q_s$. Let

16.3 Proof of the Riemann-Roch theorem

$$E = \sum_{j=1}^{s} \sum_{p \in \sigma^{-1}(q_j)} (r_j - 1)p$$

denote the **divisor of multiple points** of C' on C.

Any form $G \in L(e;(r_1-1)q_1,\ldots,(r_s-1)q_s) \subset K[x,y,z]_e$ which does not vanish on C' is called an **adjoint form** of C' and $V(G)$ an **adjoint curve** of degree e.

Proposition 16.3.3. *With the notation and assumptions as above, suppose G is an adjoint form of C' then $D = \mathrm{div}(G) - E$ is an effective divisor on C.*

Proof. For $p \in \sigma^{-1}(q)$ we have $v_p(G) \geq r_q - 1$ since G has multiplicity $r_q - 1$ in q. Thus the effective divisor $\mathrm{div}(G)$ satisfies $\mathrm{div}(G) \geq E$. □

Theorem 16.3.4 (Completeness of the adjoint systems). *Let $C' \subset \mathbb{P}^2$ be a plane model of an irreducible smooth projective curve C with only ordinary singular points. Let $H \in \mathrm{Div}(C)$ denote a divisor cut out on C by a linear form on $\mathbb{P}^2$ and let E be the divisor of multiple points. For every effective divisor D linearly equivalent to $eH - E$ there exists an adjoint form G of degree e such that $\mathrm{div}(G) - E = D$.*

A sketch of how this follows from general results on coherent sheaves and cohomology is given in Appendix A in Section A.4. A complete proof without sheaves can be found in Fulton's book on algebraic curves [26].

Corollary 16.3.5. *With the notation as above, suppose G_0 is an adjoint form with $\mathrm{div}(G_0) = D + E$. Then the map*

$$L(e;(r_1-1)q_1,\ldots,(r_s-1)q_s) \to L(D), G \mapsto G/G_0$$

induces an isomorphism $L(e;(r_1-1)q_1,\ldots,(r_s-1)q_s)/\mathrm{I}(C')_e \cong L(D)$.

Proof. Suppose $f \in L(D)$ and $D' = (f) + D \geq 0$. By the theorem there exists an adjoint form G such that $\mathrm{div}(G) - E = D'$. Regarding G/G_0 as a rational function on C we see that $(f) = D' - D = (G'/G_0)$, i.e., f and G'/G_0 have the same zeroes and poles. Thus the quotient $f/(G/G_0) = \lambda$ is a constant function on C and $\lambda G \in L(e;(r_1-1)q_1,\ldots,(r_s-1)q_s)$ is a form which maps to f. So the map is surjective. The result follows because the kernel of the map $L(e;(r_1-1)q_1,\ldots,(r_s-1)q_s) \to L(D)$ coincides with $\mathrm{I}(C')_e$. □

Corollary 16.3.6. *Let C be a smooth projective curve of geometric genus g and let W be a canonical divisor. Then $\ell(W) \geq g$ holds.*

Proof. Let C' be a plane model with only ordinary singularities. Then $W \sim (d-3)H - E$ holds by Proposition 16.2.6 and

$$\ell(W) = \dim L(d-3;(r_1-1)q_1,\ldots,(r_s-1)q_s) \geq \binom{d-1}{2} - \sum_{i=1}^{s}\binom{r_i}{2} = g$$

holds by Proposition 13.2.2. □

Algorithm 16.3.7 (Riemann-Roch space).
Input. An irreducible plane curve $C' = V(f)$ with only ordinary singularities given by an irreducible homogeneous form $f \in K[x_0, x_1, x_2]$.
A divisor $D = \sum n_i p_i$ with support disjoint from the singularities of C'.
Output. $\ell(D)$ and a basis of $L(D)$.

1. If the divisor is given by a list of pairs of multiplicities and points $\{(n_i, p_i)\}$ compute

$$I(D_1) = \bigcap_{n_i > 0} (I(p_i)^{n_i} + (f)) \text{ and } I(D_2) = \bigcap_{n_i < 0} (I(p_i)^{-n_i} + (f)).$$

2. Compute the adjoint ideal

$$J_{\text{adj}} = (f, \frac{\partial f}{\partial x_0}, \frac{\partial f}{\partial x_1}, \frac{\partial f}{\partial x_2}) : (x_0, x_1, x_2)^{\infty}.$$

3. Verify that C' has only ordinary singularities, if necessary.
4. Compute $I = I(D_1) \cap J_{\text{adj}}$.
5. Choose $e > 0$ such that $I_e \supsetneq (f)_e$ and a form $h \in I_e \setminus (f)_e$.
6. Compute the residual ideal $I' = (f, h) : I$.
7. If $V(I' + I(D_2)) = \emptyset$ then $J = I' \cap I(D_2)$, else $J = (I' \cdot I(D_2) + (f)) : (x_0, x_1, x_2)^{\infty}$.
8. Compute $\ell = \ell(D) = \dim J_e/(f)_e$ and if $\ell > 0$ forms $h_1, \ldots h_\ell$, which represent a basis of $J_e/(f)_e$.
9. Return $\ell(D)$ and if $\ell(D) > 0$ the rational functions $h_1/h, \ldots, h_\ell/h$.

Remarks 16.3.8. 1) If D is effective, i.e., $D = D_1$, then we may choose $h_1 = h$ so that the first fraction represents $1 \in L(D) \subset K(C)$.
2) If D and C' are defined over a subfield $k \subset K$, then $L(D)$ is defined over k as well. For D to be defined over k it is not necessary that all points p_i are defined over k. It suffices that $I(D_1)$ and $I(D_2)$ are defined over k. This is the case if and only if for each integer n the algebraic set $\Gamma_n = \{p_i \mid n_i = n\}$ is defined over k.

Example 16.3.9. Consider a smooth plane quartic C with 3 points. What is the dimension of $L(p_1 + p_2 + p_3)$? To give a concrete example, we take C to be the smooth quartic curve with affine equation

$$2x(x^2 - 1)(x - 2) + x^2 y + y(y^2 - 1)(y + 2) = 0$$

and the points $p = (0, -1)$, $q = (0, 0)$, $r = (0, -2)$ and $s = (2, 0)$ on C.

16.3 Proof of the Riemann-Roch theorem

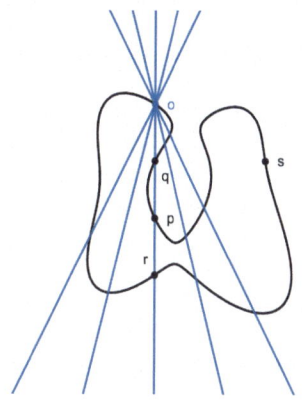
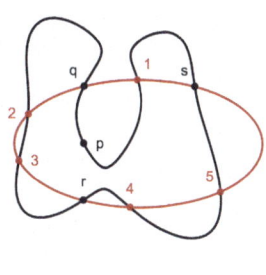

The space of rational functions $L(p+q+r)$ is two-dimensional, because the 3 points lie on a line. The pencil of lines through the fourth intersection point $o = (0,1)$ cut out the complete linear system and $L(p+q+r)$ is spanned by 1 and $\frac{y-1}{x} \in K(C)$.

On the other hand, $L(q+r+s)$ is only one-dimensional. The conic E defined by the affine equation

$$(x-1)^2 + 4(y+1)^2 = 5$$

intersects C in five further points, and any conic which passes through these points coincides with E by Bézout's theorem. Thus $L(q+r+s) = K \subset K(C)$.

To compute complete linear systems on smooth projective curves and to prove the equality of the arithmetic genus with the geometric genus, we need two results, whose proofs follow immediately from the Serre vanishing A.2.6 of coherent sheaves on $\mathbb{P}^n$.

Theorem 16.3.10. *Let $C \subset \mathbb{P}^n$ be a smooth irreducible projective curve not contained in the hyperplane $V(x_0)$ at infinity and let $H \in \text{Div}(C)$ be the divisor cut out by x_0. Then there exists an $e_0 \geq 1$ such that the map*

$$L(n,e) \to L(eH), g \mapsto (g/x_0^e)|_C$$

is surjective for all $e \geq e_0$.

Theorem 16.3.11. *Let $q_1, \ldots, q_s$ be distinct points in $\mathbb{P}^2$ and let $r_1, \ldots, r_s$ be positive integers. Then there exists an $e_0 \geq 0$ such that*

$$\dim L(e; (r_1-1)q_1, \ldots, (r_s-1)q_s) = \binom{e+2}{2} - \sum_{i=1}^{s} \binom{r_i}{2}$$

holds for all $e \geq e_0$.

Corollary 16.3.12. *Let $C \subset \mathbb{P}^n$ be an irreducible smooth projective curve and let $C' \subset \mathbb{P}^2$ be a plane model of C with only ordinary singularities. The smallest integer g^* such that $\ell(D) \geq \deg D + 1 - g^*$ holds for all divisor D coincides with the arithmetic genus of $p_a(C)$ and the geometric genus g of C'.*

In particular, the arithmetic genus of C does not depend on the embedding $C \hookrightarrow \mathbb{P}^n$ and the geometric genus g does not depend on the choice of a plane model C' with only ordinary points.

Proof. We already established $g^* \leq p_a(C)$ in Remark 16.1.17. To show equality, we consider the hyperplane class H cut out on $C \subset \mathbb{P}^n$ by a linear form. Then $\ell(eH) = p_C(e) = e \deg C + 1 - p_a(C)$ holds for e sufficiently large by Theorem 16.3.10. Since $\deg eH = e \deg C$ we obtain $g^* = p_a(C)$.

For g we argue similarly. Let f be a form of degree d with $I(C') = (f) \subset K[x_0, x_1, x_2]$. Choose e sufficiently large and an adjoint form $G_0 \notin I(C')$ of degree e. Then $\text{div}(G_0) = D + E$, where E denotes the divisor of multiple points of C'. By the completeness of the adjoint systems 16.3.4 and Theorem 16.3.11 we obtain the following.

$$\ell(D) = \binom{e+2}{2} - \sum_i \binom{r_i}{2} - \binom{e-d+2}{2}$$
$$= ed - \sum_i r_i(r_i - 1) + 1 - \left(\binom{d-1}{2} - \sum_i \binom{r_i}{2}\right)$$
$$= \deg D + 1 - g$$

since $\deg D = ed - \sum_i r_i(r_i - 1)$ and $g = \binom{d-1}{2} - \sum_i \binom{r_i}{2}$. Thus $g^* \leq g$. To show the inequality $\ell(D') \geq \deg D' + 1 - g$ for all divisors D' we argue as in the proof of Theorem 16.1.15. Choose e so large that in addition $\ell(D) > \deg D'$ holds. Then $D'' = D - D'$ is linearly equivalent to an effective divisor, and

$$\ell(D') \geq \ell(D) - \deg D'' = \deg D + 1 - g - (\deg D - \deg D')$$
$$= \deg D' + 1 - g$$

follows from Corollary 16.1.8. This proves Riemann's inequality with the geometric genus. □

Theorem 16.3.13 (Residue Theorem). *Let C be a smooth projective curve, let W be a canonical divisor and $p \in C$ be a point. Then $L(W + p) = L(W)$ holds.*

Proof. If C is rational, then both spaces are 0. Thus we may assume that C has geometric genus $g > 0$ and hence we may assume that W is effective by Corollary 16.3.6. Consider a plane model C' of degree d with only ordinary singularities such that the multiple point divisor E is disjoint from p. (Applying a suitable Cremona transformation we can achieve $p \notin \text{supp } E$.)

By Theorem 16.2.6 $W \sim (d-3)H - E$. Let $L \subset \mathbb{P}^2$ be a line through p transversal to C'. Then $C.L = p + p_2 + \cdots + p_d \sim H$ and

$$W + p \sim (d-2)H - E - (p_2 + \cdots + p_d).$$

By the completeness of the adjoint systems, Theorem 16.3.4, the linear system

16.3 Proof of the Riemann-Roch theorem

$$|(d-2)H - E - (p_2 + \cdots + p_d)|$$

is cut out by adjoint curves of degree $d-2$ which pass through the points $p_2, \ldots, p_d$. Since $d-2 < d-1$ the line L is a component of any of these adjoint curves by Bézout's theorem 9.3.1 and p is a base point of the linear system $|(d-2)H - E - (p_2 + \cdots + p_d)|$. Thus

$$\begin{aligned} L(W + p) &\cong L((d-2)H - E - (p_2 + \cdots + p_d)) \\ &= L((d-2)H - E - (p + p_2 + \cdots + p_d)) \cong L((d-3)H - E) \\ &\cong L(W) \end{aligned}$$

and the natural inclusion $L(W) \subset L(W + p)$ is an isomorphism. □

Remark 16.3.14. Let C be a smooth irreducible projective curve of genus g defined over $\mathbb{C}$. The C is also a compact connected Riemann surface. Let p be a point and ω a meromorphic differential form on C. If $t \in \mathfrak{m}_{C,p}$ is a local parameter and $\varphi: U \to V \subset \mathbb{C}$ the analytic chart defined by the function t in a euclidean neighborhood of p, then we can expand

$$\omega|_U = \sum_{n \geq N} a_n t^n dt$$

into a Laurent series. The coefficient a_{-1} of the term $t^{-1} dt$ is called the **residue** of ω at C, which we denote by $\mathrm{res}_p(\omega)$. This is well-defined, i.e., independent from the choice of t. Within the theory of one complex variable we have the following residue theorem, whose proof is a simple application of the Cauchy Integral formula.

Theorem (Residue Theorem [25]). *Let C be a compact connected Riemann surface and ω a meromorphic differential form on C. Then*

$$\sum_{p \in C} \mathrm{res}_p(\omega) = 0.$$

Note that the sum is finite because ω has a discrete, hence finite, set of poles.

Examples. 1) Consider $\mathbb{P}^1(\mathbb{C})$ and the differential form $\omega = dz$ on U_0. On U_1 with coordinate $w = 1/z$ we find $dz = d(\frac{1}{w}) = -w^{-2} dw$ and $\mathrm{res}_{(0:1)}(\omega) = 0$.

2) For the meromorphic form $\omega = z^{-1} dz$ on $U_0 \subset \mathbb{P}^1(\mathbb{C})$ we find $\omega|_{U_1} = -w^{-1} dw$. Thus

$$\mathrm{res}_{(1:0)}(\omega) + \mathrm{res}_{(0:1)}(\omega) = 1 + (-1) = 0.$$

3) A meromorphic differential form with at most one pole of order 1 on a compact Riemann surface is automatically holomorphic by the residue theorem. This explains the name of Theorem 16.3.13.

Corollary 16.3.15 (Noether's Reduction Lemma). *Let C be a smooth irreducible projective curve and let W denote a canonical divisor on C. If D is a divisor with $\ell(W - D - p) \neq \ell(W - D)$ then $\ell(D + p) = \ell(D)$.*

Proof. If C is rational, then one of the spaces $L(W - D)$ or $L(D + p)$ is zero by Remark 16.3.2. Thus we may assume that C has geometric genus $g > 0$. So there exists a regular differential form ω with $W = \sum_{q \in C} v_q(\omega) q \geq 0$.

Notice that both spaces $L(W - D - p) \subset L(W - D)$ and $L(D) \subset L(D + p)$ have codimension at most 1 by Proposition 16.1.7 2). We prove that these inclusions cannot both be strict. Assume the contrary. Consider a rational function $g \in L(W - D) \setminus L(W - D - p)$ and a rational function $f \in L(D + p) \setminus L(D)$. Then $fg \in L(W + p) \setminus L(W)$ (and $fg\omega$ is a rational differential form which has a simple pole at p and is regular otherwise). This contradicts the residue theorem 16.3.13. □

Proof of the Riemann-Roch theorem 16.3.1. For each divisor D we consider the equation

$$\ell(D) = \deg D + 1 - g + \ell(W - D). \qquad (*)_D$$

Case 1: $\ell(W - D) = 0$. We apply induction on $\ell(D)$. If $\ell(D) = 0$ then Riemann's inequality applied to D and $W - D$ gives $0 \geq \deg D + 1 - g$ and $0 \geq 2g - 2 - \deg D + 1 - g$, hence $\deg D = g - 1$ and $(*)_D$ holds. If $\ell(D) = 1$ then we may assume that $D \geq 0$. Choose a point $p \in C$ such that $\ell(D - p) = \ell(D) - 1$. Since D is effective we have $g \leq \ell(W) \leq \ell(W - D) + \deg D$ by Corollary 16.3.6 and Proposition 16.1.8. So $\deg D \geq g$ and Riemann's inequality implies $\ell(D) \geq \deg D + 1 - g \geq 1$ with equality if $\ell(D) = 1$. If $\ell(D) > 1$ then $\ell(D - p) = \ell(D) - 1$ implies $0 = \ell(W - D) = \ell(W - D + p)$ by Noether's reduction lemma 16.3.15 applied to $D - p$. Thus $(*)_{D-p}$ holds by the induction hypothesis and $(*)_D$ follows.

Case 2: $\ell(W - D) > 0$. This happens only if $\deg D \leq 2g - 2$ by Proposition 16.1.7. We argue by descending induction on the degree $\deg D$. By case 1, the assertion holds for all divisors of degree $\deg D > 2g - 2$. Suppose the assertion $(*)_{D'}$ for all divisors D' of degree larger than $\deg D$. Choose $p \in C$ such that $\ell(W - D - p) = \ell(W - D) - 1$. Then $\ell(D + p) = \ell(D)$ holds by Noether's reduction lemma 16.3.15 and $(*)_{D+p}$ implies $(*)_D$ because

$$\ell(D) = \ell(D + p) = \deg D + 1 + 1 - g + \ell(W - D - p)$$
$$= \deg D + 1 - g + \ell(W - D).$$

□

Corollary 16.3.16. *A canonical divisor W on a smooth projective curve of genus g satisfies $\ell(W) = g$ and $\deg W = 2g - 2$.*

Proof. The formula gives $1 = \ell(0) = 0 + 1 - g + \ell(W)$. The statement on the degree (which we already proved) follows once more from $g = \ell(W) = \deg W + 1 - g + \ell(W - W) = \deg W + 2 - g$. □

16.3 Proof of the Riemann-Roch theorem

Thus g has a further interpretation. It is the dimension of the K-vector space of regular differential forms in $\Omega(C)$, i.e., rational differential forms without poles.

Corollary 16.3.17. *Let D be a divisor of degree $D \geq 2g - 1$. Then*

$$\ell(D) = \deg D + 1 - g.$$

If $\deg D \geq 2g$ then $|D|$ is base point free and if $\deg D \geq 2g+1$ then $\varphi_D : C \to \mathbb{P}^{\ell(D)-1}$ is an embedding.

Proof. The vanishing $\ell(W - D) = 0$ for $\deg D > \deg W = 2g - 2$ gives the first assertion. The second assertion follows from Proposition 16.1.10 and Theorem 16.1.14 □

Corollary 16.3.18. *Let C be a smooth projective curve of genus $g \geq 2$ and let W be a canonical divisor on C. Then $|W|$ is base point free. The morphism $\varphi_W : C \to \mathbb{P}^{g-1}$ is an embedding unless the curve C is hyperelliptic.*

Definition 16.3.19. A smooth **canonical curve** C of genus g is a non-hyperelliptic smooth irreducible projective curve in its canonical embedding $\varphi_W : C \hookrightarrow \mathbb{P}^{g-1}$.

Proof. Let $p \in C$ be a point. Since C is not rational we have $\ell(p) = 1$. Hence

$$\ell(W - p) = 2g - 3 + 1 - g + \ell(p) = g - 1 = \ell(W) - 1,$$

and W is base point free by Proposition 16.1.10. The morphism $\varphi_W : C \to \mathbb{P}^{g-1}$ is not an embedding if and only if there exist two points $p, q \in C$ with $\ell(W - p - q) = g - 1$ by Theorem 16.1.14. But then $\ell(p + q) = 2 + 1 - g - \ell(W - p - q) = 2$ and

$$h = \varphi_{p+q} : C \to \mathbb{P}^1$$

defines a 2 to 1 morphism to $\mathbb{P}^1$. By definition, a smooth projective curve of genus $g \geq 2$ over an algebraically closed field K which has a 2 to 1 morphism to $\mathbb{P}^1$ is called **hyperelliptic**. □

Remark 16.3.20. For hyperelliptic curves the canonical map factors as follows.

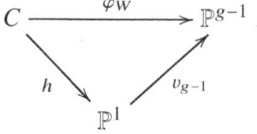

where v_{g-1} denotes the $g - 1$-uple embedding in suitable coordinates.

Thus the hyperelliptic map h is up to coordinate changes of $\mathbb{P}^1$ uniquely determined by C. Every curve of genus $g = 2$ is hyperelliptic. For $g \geq 3$ this is not the case. A smooth plane quartic $C \subset \mathbb{P}^2$ is a canonically embedded curve because $W \sim H$ by Proposition 16.2.6.

Exercise 16.3.21. 1) Let C be a hyperelliptic curve of genus g over a field of $\text{char}(K) \neq 2$. Prove: The $2:1$ map $h: C \to \mathbb{P}^1$ has $2g+2$ branch points.

2) Prove that there exist smooth hyperelliptic curves of any genus $g \geq 2$. Hint: Consider a curve of bi-degree $(2, g+1)$ on $\mathbb{P}^1 \times \mathbb{P}^1$.

Exercise 16.3.22. Let $C \subset \mathbb{P}^n$ be a smooth projective curve of genus g and degree d, which is not contained in the hyperplane $V(x_0)$ at infinity. Suppose e_0 is an integer such that

a) the Hilbert function h_C and the Hilbert polynomial p_C have the same value $h_C(e) = p_C(e)$ for $e \geq e_0$, and

b) $e_0 \cdot d > 2g - 2$

holds. Then the map

$$L(e, n) \to L(eH), G \mapsto (G/x_0^e)|_C$$

is surjective for all $e \geq e_0$, where H denotes the divisor cut out by x_0 on C.

Exercise 16.3.23. Describe an algorithm and prove its correctness with the following input and output.

Algorithm 16.3.24 (Riemann-Roch space on a smooth projective curve).
Input. An irreducible smooth projective curve $C \subset \mathbb{P}^n$ specified by its homogeneous ideal $I(C)$ and a divisor $D = \sum n_i p_i$ on C.
Output. $\ell(D)$ and a basis of $L(D)$.

Exercise 16.3.25 (Elliptic curves). Let C be an irreducible smooth projective curve of genus $g = 1$ and let $p \in C$ be a fixed point. Prove:

1) C is isomorphic to a smooth plane cubic E. Hint: $\varphi_{3p}: C \to \mathbb{P}^2$ is an embedding.
2) In suitable coordinates on $\mathbb{P}^2$ the cubic has an equation

$$y^2 + a_1 xy = x^3 + b_1 x^2 + b_2 x + b_3,$$

where the image of p is the unique point $(0:1:0)$ on the line $V(z)$ at infinity.

3) Let $D \in \text{Div}(C)$ be a divisor of degree 0. Then $D \sim q - p$ for a unique point q on C.

4) The map

$$\varphi: C \to \text{Div}^0(C)/\sim, \quad q \mapsto q - p$$

is a bijection and addition of divisors on C induces on C the structure of an abelian group

$$+: C \times C \to C, (q_1, q_2) \mapsto q_3,$$

where $q_3 \in C$ is the unique point such that $q_1 - p + q_2 - p \sim q_3 - p$. Note that p corresponds to the zero element of this group structure.

5) The group law has the following geometric description on E, where we assume that the coordinates are chosen as in 2). If q_4 denotes the third intersection point of the line $\overline{q_1 q_2}$ with the cubic, then the line $\overline{q_4 q_3}$ is a vertical line whose

16.3 Proof of the Riemann-Roch theorem

third intersection point is the point p at infinity. In other words, q_3 is the third intersection point of $\overline{pq_4}$ with E. If $q_1 = q_2$ one has to replace the secant by the projective tangent line $T_{q_1}E \subset \mathbb{P}^2$.

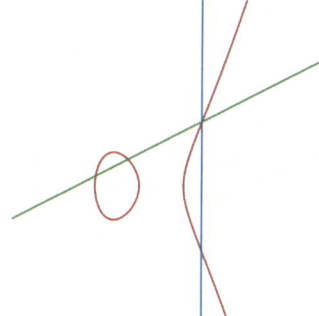

6) If the cubic E from 2) is defined over $\mathbb{Q}$ then $E(\mathbb{Q})$ is a subgroup of $E(\mathbb{C})$.

The subgroup $E(\mathbb{Q})$ is finitely generated by a famous theorem of Mordell [63].

Exercise 16.3.26. (Weierstrass points) Let C be a smooth curve of genus g and let $p \in C$ be a point over a field K of characteristic zero. Prove:

1) The subset of integers $\Gamma = \{n \in \mathbb{N} \mid \ell(np) > \ell((n-1)p)\} \subset \mathbb{N}$ is a semi-group.
2) The complement $\mathbb{N} \setminus \Gamma$ consists of g elements $\{\gamma_1 < \gamma_2 < \ldots < \gamma_g\}$.

The γ_i are called the Weierstrass **gaps** at p. The sum $wt(p) = \sum_{i=1}^{g}(\gamma_i - i)$ is called the Weierstrass weight of p and a point $p \in C$ with $wt(p) > 0$ is called a **Weierstrass point**. Prove:

3) A ramification point p of a hyperelliptic curve $C \xrightarrow{2:1} \mathbb{P}^1$ has the gap sequence $\{1, 3, 5, \ldots, 2g - 1\}$ and weight $wt(p) = \frac{1}{2}g(g+1)$.
4) The largest gap is bounded by $\gamma_g \leq 2g - 1$.

Let $\omega_1 \ldots, \omega_g$ be rational differentials without poles which correspond to a basis of $L(W)$. Let $t \in \mathfrak{m}_p \in \mathcal{O}_{C,p}$ be a generator of $\mathfrak{m}_p$ and let $\omega_i = f_i dt$ be the local description. Then

$$W(f_1, \ldots, f_g) = \det \begin{pmatrix} f_1 & f_2 & \cdots & f_g \\ f_1' & f_2' & \cdots & f_g' \\ f_1'' & f_2'' & \cdots & f_g'' \\ \vdots & & & \vdots \\ f_1^{(g-1)} & f_2^{(g-1)} & \cdots & f_g^{(g-1)} \end{pmatrix}$$

is called the **Wronski determinant**. Prove:

5) $W(f_1, \ldots, f_g)$ vanishes of order $wt(p)$ at p.
6) Counted with weights there are precisely $g^3 - g$ Weierstrass points:

$$\sum_{p \in C} wt(p) = g^3 - g.$$

7) The Weierstrass points on a hyperelliptic curve are precisely the ramification points of $C \xrightarrow{2:1} \mathbb{P}^1$.

Hint for 5): Change basis such that $v_p(f_1) < v_p(f_2) < \ldots < v_p(f_g)$ and consider the power series expansion of the $f_i \in K[[t]]$.

Hint for 6): Since $f_i' = \frac{df_i}{dt}$ the zero loci of $W(f_1, \ldots, f_g)(dt)^{(g+1)g/2}$ is a divisor in $L((g+1)g/2W)$.

Exercise 16.3.27. It was a long-standing open question [47] whether every semi-group $\Gamma \subset \mathbb{N}$ with g gaps occurs as a Weierstrass semi-group. This is not the case, as the following example due to Buchweitz [14] shows. Prove: The complement of $\{1, 2, \ldots, 12, 19, 21, 24, 25\}$ is a semi-group but not a Weierstrass semi-group.

Hint: Suppose ω_i are rational differential forms without poles on a curve of genus $g = 16$ with vanishing order $v_p(\omega_i) = \gamma_i - 1$. Consider the vanishing orders of the products $\omega_i \cdot \omega_j \in L(2W)$.

See [50] for further investigations in this direction.

Exercise 16.3.28. Prove computationally that a general smooth quartic curve $C \subset \mathbb{P}^2$ has 24 ordinary flexes and that these points are all Weierstrass points on this curve of genus $g = 3$. They have gap sequence $\{1, 2, 4\}$.

What kind of singular points does the dual curve $\check{C}$ of a smooth quartic with a non-ordinary flex have? These flexes correspond to Weierstrass points with gap sequence $\{1, 2, 5\}$.

It is known [19] that a general curve of genus g has only ordinary Weierstrass points, i.e., points with gap sequence $\{1, 2, \ldots, g-1, g+1\}$.

16.4 Riemann's count

A famous result of Riemann says that irreducible smooth projective curves of genus g depend on $3g - 3$ moduli, i.e., parameters. One ingredient of this count is Riemann's existence theorem.

Theorem 16.4.1 (Riemann's Existence Theorem). *Let X be a compact connected Riemann surface. Then there exists a smooth projective curve over $K = \mathbb{C}$, such that the underlying Riemann surface of C is biholomorphic to X. Moreover, the meromorphic function field $\mathcal{M}(X)$ of X is isomorphic to the rational function field $K(C)$ of C.*

Example 16.4.2. Every meromorphic function on $\mathbb{P}^1(\mathbb{C})$ is actually a rational function [25]. Thus $\mathcal{M}(\mathbb{P}^1(\mathbb{C})) = \mathbb{C}(z)$.

Example 16.4.3. Consider a lattice $\Lambda \subset \mathbb{C}$ and the Riemann surface $E = \mathbb{C}/\Lambda$. The Weierstrass $\wp$-function

$$\wp(z) = \frac{1}{z^2} + \sum_{\ell \in \Lambda \setminus \{0\}} \left(\frac{1}{(z-\ell)^2} - \frac{1}{\ell^2} \right)$$

16.4 Riemann's count

is double periodic with respect to Λ, thus descends to a meromorphic function on E. Actually,
$$\mathcal{M}(E) = \mathbb{C}(\wp, \wp')$$
and $z \mapsto (1 : \wp(z) : \wp'(z)) \in \mathbb{P}^2$ identifies E with underlying Riemann surface of the elliptic curve $V(y^2 - 4x^3 + g_2 x + g_3)$, where g_2, g_3 are the constants in Weierstrass's equation
$$(\wp')^2 = 4\wp^3 - g_2\wp - g_3,$$
see e.g. [48]. The group structure on the elliptic curve according to Exercise 16.3.25 coincides with the group structure on $\mathbb{C}/\Lambda$ induces by the group $(\mathbb{C}, +)$.

Remark 16.4.4. The hardest part of the proof of Theorem 16.4.1 is to show the existence of a non-constant meromorphic function on X. Once this is done, the proof builds upon a version of the Riemann-Roch theorem for compact Riemann surfaces, see [25], [52] or [60].

Riemann mainly considered the case of branched coverings of $\mathbb{P}^1(\mathbb{C})$. Let $b_1, \ldots, b_r \in \mathbb{P}^1(\mathbb{C})$ be points, and consider $B = \mathbb{P}^1(\mathbb{C}) \setminus \{b_1, \ldots, b_r\}$ with its euclidean topology. Fix a base point $b_0 \in B$. The fundamental group $\pi_1(B, b_0)$ is freely generated by the r paths γ_i from b_0 which loop around the b_i clockwise

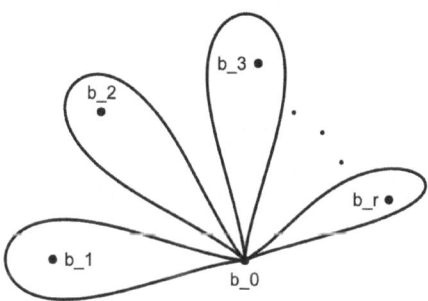

with one relation: $\gamma_1 \cdot \gamma_2 \cdot \ldots \cdot \gamma_r$ is null homotopic. Given an integer d and permutations $\sigma_i \in S_d$ there exists a d-sheeted branched cover $X \to \mathbb{P}^1(\mathbb{C})$ with branching behavior σ_i above b_i and unramified otherwise if and only if $\sigma_1 \cdot \sigma_2 \cdot \ldots \cdot \sigma_r = \text{id} \in S_d$: Consider the the group homomorphism
$$\rho: \pi_1(B, b_0) \to S_d, \gamma_i \mapsto \sigma_i.$$

Let $\widetilde{B}$ be the universal covering space of B. The fundamental group $\pi_1(B, b_0)$ acts on $\widetilde{B}$ via path lifting once we fix a point $\widetilde{b}_0$ over b_0. Consider $\widetilde{B} \times \{1, \ldots, d\}$, where the action of $\pi_1(B, b_0)$ on $\{1, \ldots, d\}$ is induced by ρ. The one-dimensional complex manifold

$$X_0 = (\widetilde{B} \times \{1,\ldots,d\})/\pi_1(B,b_0)$$

is connected if and only if $\sigma_1,\ldots,\sigma_r$ generate a transitive subgroup of S_d. Since $\widetilde{B}/\pi_1(B,b_0) \cong B$ the space X_0 is a d-sheeted unramified covering space of B. The restriction of this covering $X_0 \to B$ to a small punctured disk $\{z \mid |z-b_i| < \epsilon\} \setminus \{b_i\}$ is determined by the action of σ_i on $\{1,\ldots,d\}$.

We can compactify $X_0 \to B$ to a compact Riemann surface X together with a morphism $X \to \mathbb{P}^1$ by gluing in points p_{ij} over b_i with analytic chart $(z - b_i)^{1/s_j}$ for each orbit O_j of the action σ_i on $\{1,\ldots,d\}$ of length s_j. See [25] for more details.

With this structure $X \to \mathbb{P}^1(\mathbb{C})$ becomes a holomorphic map with the pre-described ramification behaviour. The Riemann surface X depends on r parameters given by $b_1,\ldots,b_r \in \mathbb{P}^1(\mathbb{C})$ and discrete data given by the tuple $\sigma_1,\ldots,\sigma_n$ up to conjugation in S_d.

Riemann's count follows by calculating the dimension of the space of the degree d maps $C \to \mathbb{P}^1$ for large d.

Let C be a smooth projective curve of genus g and $d \geq 2g + 1$ an integer. Then any effective divisor D of degree d defines an embedding $\varphi_D: C \to \mathbb{P}^{d-g}$ by Riemann-Roch. Moreover the map

$$C^{(d)} = \underbrace{C \times \ldots \times C}_{d \text{ copies}}/S_d \to \mathrm{Div}^d(C)/\sim$$

from the d-th symmetric product of C to the set of divisor classes of degree d has fibers the linear space $\mathbb{P}(L(D)) \cong \mathbb{P}^{d-g}$, from which we conclude that $\mathrm{Div}^d(C)/\sim$ should carry the structure of a g-dimensional variety. This is indeed the case. It is isomorphic to the Jacobian of C, see [3]. Linear projections of $C \subset \mathbb{P}^{d-g}$ to $\mathbb{P}^1$ of degree d up to automorphisms of $\mathbb{P}^1$ correspond to the choice of a point in an open subset of the Grassmannian $\mathbb{G}(2, d+1-g) = \mathbb{G}(2, L(D))$.

Thus for a given curve C of genus g we have a $g + 2(d-1-g)$-dimensional family of degree d morphisms to $\varphi: C \to \mathbb{P}^1$. On the other hand such a morphism has $2d + 2g - 2$ branch points counted with multiplicity. For a general pair (C, φ) the branch points are simple. Thus up to automorphism of $\mathbb{P}^1$ they depend on $2d + 2g - 5$ parameters. Thus

$$3g - 3 = 2d + 2g - 5 - (g + 2(d-1-g))$$

is the number of moduli on which an abstract (non-embedded) smooth irreducible algebraic curve depends. This proves the following.

Theorem 16.4.5 (Riemann). *A general smooth irreducible projective curve of genus $g \geq 2$ depends on $3g - 3$ moduli.*

Remarks 16.4.6. 1) The statement is true for curves of genus $g \geq 2$ over arbitrary algebraically closed fields.

2) Mumford [66] showed that there exists a quasi-projective variety $\mathfrak{M}_g$ of dimension $3g - 3$ whose points correspond to isomorphism classes of smooth irreducible projective curves of genus g.

16.4 Riemann's count

3) For $K = \mathbb{C}$ Lüroth [57] showed that this moduli space is irreducible. For arbitrary fields this was proved by Deligne and Mumford [17] by considering a projective closure of $\mathfrak{M}_g$ whose points correspond to so-called stable curves.

4) For $g = 0, 1$ the formula $\dim \mathfrak{M}_g = 3g - 3$ does not hold due to automorphism: $\dim \mathfrak{M}_1 = 1$ since a curve C of genus $g = 1$ has a one-dimensional group of automorphisms by Exercise 16.3.25. By Corollary 16.1.16 the space $\mathfrak{M}_0$ consists of a single point and $\dim \mathrm{PGL}(2, K) = 3$.

Corollary 16.4.7 (Max Noether)**.** *Suppose $g \geq 2$ and $d \geq 2g + 1$. The Hilbert scheme $\mathrm{Hilb}_{dt+1-g}(\mathbb{P}^3)$ has a component of dimension $4d$, whose points of a dense open part correspond to irreducible smooth projective curves of degree d and genus g.*

Proof. To specify a smooth curve $C \subset \mathbb{P}^3$ we have to choose an abstract curve C, which depends on $3g - 3$ parameters, a divisor class D depending on g parameters and 4 sections $f_0, \ldots, f_3 \in L(D)$ up to a common factor $\lambda \in K^*$. Thus altogether we have
$$3g - 3 + g + 4(d + 1 - g) - 1 = 4d$$
parameters. □

The dimension computation is valid in a wider range. We already know the following:

1) The Grassmannians $\mathbb{G}(2, 4)$ of lines in $\mathbb{P}^3$ is 4-dimensional.
2) The Hilbert scheme $\mathrm{Hilb}_{3t+1}(\mathbb{P}^3)$ has a 12-dimensional component whose general points correspond to twisted cubics curves.

Concerning the case $g = 1$ and $d = 4$ we have the following.

Example 16.4.8. *A smooth irreducible projective curve $C \subset \mathbb{P}^3$ of genus $g = 1$ and degree $d = 4$ is the complete intersection of two quadrics.*

Let $S = K[x_0, \ldots, x_3]$ denote the homogeneous coordinate ring of $\mathbb{P}^3$, and let D denote the divisor cut out by the hyperplane $V(x_0)$ at infinity. Note that C is not contained in a hyperplane, because plane curves of degree 4 and geometric genus $g = 1$ have singularities. The map
$$S_2 \to L(2D), \ q \mapsto q/x_0^2$$
has an at least 2-dimensional kernel, because
$$\dim S_2 - \dim L(2D) = \binom{3+2}{2} - (2 \deg D + 1 - 1) = 10 - 8 = 2.$$

Thus C is contained in two quadrics q_1, q_2 and these quadrics are irreducible, because C is not contained in a hyperplane. Thus $V(q_1, q_2)$ is a curve of degree 4 by Bézout 9.4.1, which contains C. Thus $C = V(q_1, q_2)$ for degree reasons.

We conclude that $\mathrm{Hilb}_{4t}(\mathbb{P}^3)$ has a component of dimension $\dim \mathbb{G}(2, 10) = 2 \cdot 8 = 4 \cdot 4$ corresponding to smooth curves of genus 1 and degree 4.

Exercise 16.4.9. Consider smooth irreducible curves $C \subset \mathbb{P}^3$ of degree $d = 6$ and genus $g = 3$. Curves of bidegree $(2, 4)$ on quadrics form a family of such curves of dimension $9 + (3 \cdot 5 - 1) = 23 < 24$. Since $10 = 12 + 1 - 3$ a general curve $C \subset \mathbb{P}^3$ of degree 6 and genus 3 does not lie on a quadric. Prove:

A general curve $C \subset \mathbb{P}^3$ of this kind lies on $\binom{3+3}{3} - (3 \cdot 6 + 1 - 3) = 4$ cubics which have 3 linear syzygies. C lies on a quadric if and only if C is hyperelliptic.

16.5 The Clifford index and syzygies of canonical curves

Let $C \subset \mathbb{P}(L(W)^*) = \mathbb{P}^{g-1}$ be a canonically embedded irreducible smooth projective curve and D an effective divisor. Then $\mathbb{P}(L(W - D)) \subset \mathbb{P}(L(W))$ can be interpreted as the space of hyperplanes in $\mathbb{P}^{g-1}$ which intersect C in divisor $W' \geq D$. Thus

$$\overline{D} = \bigcap_{H \in \mathbb{P}(L(W-D))} H \subset \mathbb{P}^{g-1}$$

is the linear span of D, i.e., the smallest linear subspace which contains the subscheme $D \subset C$.

Theorem 16.5.1 (Geometric version of Riemann-Roch)**.** *Let D be an effective divisor on C. Then with the above notation we have*

$$\dim |D| = \deg D - 1 - \dim \overline{D}.$$

Proof. By definition $\operatorname{codim} \overline{D} = \ell(W - D)$. Hence $\dim \overline{D} = g - 1 - \ell(W - D)$ and the equality is equivalent to the Riemann-Roch formula. □

Remarks 16.5.2. 1) The formula is also true for hyperelliptic curves if we interpret $\overline{D}$ as the span of the image of D on the rational normal curve

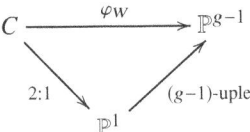

2) Let $D = p_1 + \cdots + p_d$ be a divisor which is the sum of d general points on C. Then

$$\dim |D| = \begin{cases} 0 & \text{if } d \leq g \\ d - g & \text{if } d \geq g \end{cases}$$

since d general points on C span a $\mathbb{P}^{d-1} \subset \mathbb{P}^{g-1}$ if $d \leq g$ and the whole space for $d \geq g$.

Definition 16.5.3. An effective divisor D on a smooth irreducible projective curve C with $\ell(W - D) > 0$ is called **special**. Here W denotes a canonical divisor on C.

16.5 The Clifford index and syzygies of canonical curves

Theorem 16.5.4 (Clifford). *Let D be a special divisor on a smooth projective curve C. Then*
$$\dim |D| \leq \frac{1}{2} \deg D$$
holds. Equality holds if $D = 0$ or $D = W$ or if C is hyperelliptic and D is linearly equivalent to a multiple of the g_2^1.

Proof. By assumption $|W - D|$ is non-empty. The map
$$|D| \times |W - D| \to |W|, (D', E') \mapsto D' + E'$$
has finite fibers, because an effective divisor is a sum of two effective divisors only in finitely many ways. Identifying $|D| = \mathbb{P}(L(D))$, $|W - D| = \mathbb{P}(L(W - D))$ and $|W| = \mathbb{P}(L(W)) \cong \mathbb{P}^{g-1}$, we see that $\dim(\mathbb{P}(L(D)) \times \mathbb{P}(L(W - D))) \leq g - 1$. Thus
$$\ell(D) + \ell(W - D) \leq g + 1$$
and combined with the Riemann-Roch formula
$$\ell(D) - \ell(W - D) = \deg D + 1 - g$$
gives the desired inequality
$$2\ell(D) - 2 = 2 \dim |D| \leq \deg D.$$

The second statement is clear if $D = 0$ or $D \sim W$. It remains to consider the case of a divisor D with $0 < \deg D < 2g - 2$ of even degree. If $\deg D = 2$ then $|D| = g_2^1$ is the hyperelliptic series. Suppose $\deg D \geq 4$. Consider an effective divisor $E \in |W - D|$ and two points p, q on C such that $E - p$ is effective and $E - q$ is not. Since $\dim |D| \geq 2$ we can choose $D_2 \in |D - p - q|$ and can replace D by $D_2 + p + q$. Then $D - p - q$ is effective. Consider now $D' = \min(D, E)$, which is effective as the minimum of effective divisors. Moreover, $\deg D' < \deg D$ since $q \notin \operatorname{supp} D'$.

There is an exact sequence
$$0 \to L(D') \to L(D) \oplus L(E) \to L(D + E - D')$$
where the first map is the diagonal inclusion and the second map is induced by subtraction of rational functions. Exactness follows from $L(D') = L(D) \cap L(E) \subset K(C)$. Thus
$$\dim |D| + \dim |E| \leq \dim |D'| + \dim |D + E - D'|.$$

Since $E = W - D$ we get $\dim |D| + \dim |E| = g - 1$ from the first part of the proof. Since $\ell(D + E - D') = \ell(W - D')$ the inequality $\dim |D'| + \dim |W - D'| \leq g - 1$ is also an equality for D'. Thus $\dim |D'| = \frac{1}{2} \deg D'$. Since $1 \leq \deg D' < \deg D$ we obtain that C is hyperelliptic by induction.

Finally, by the geometric version of Riemann-Roch $\frac{1}{2}\deg D = \dim |D| = \deg D - 1 - \dim \overline{D}$ holds if and only if the image of D on the rational normal curve spans a linear space of dimension $\frac{1}{2}\deg D - 1$. Since any effective divisor of degree $e \leq g$ on a rational normal curve $\mathbb{P}^1 \hookrightarrow \mathbb{P}^{g-1}$ spans a $\mathbb{P}^{e-1}$ this is only possible if D is the sum of e divisors of the g_2^1. $\square$

Definition 16.5.5. Let C be a smooth irreducible projective curve of genus $g \geq 2$ and let D be a special divisor. Then the difference

$$\mathrm{Cliff}(D) = \deg D - 2\dim |D|$$

is called the **Clifford index** of D. Note that $\mathrm{Cliff}(D) = \mathrm{Cliff}(W - D)$. The **Clifford index** of a curve C of genus $g \geq 3$ is

$$\mathrm{Cliff}(C) = \min\{\mathrm{Cliff}(D) \mid \ell(D) \geq 2 \text{ and } \ell(W - D) \geq 2\}.$$

For a curve of genus $g = 2$ one defines $\mathrm{Cliff}(C) = 0$. The **gonality** $\mathrm{gon}(C)$ of C is the minimal degree of a morphism $C \to \mathbb{P}^1$. Thus 2-gonal curves are hyperelliptic.

Remarks 16.5.6. 1) The Clifford index is non-negative, $\mathrm{Cliff}(C) \geq 0$, and equality holds if and only if C is hyperelliptic.
2) $\mathrm{Cliff}(C) = 1$ if and only if C is trigonal or C is isomorphic to a smooth plane quintic.
3) In nearly all cases the Clifford index is computed by a pencil, in which case $\mathrm{Cliff}(C) = \mathrm{gon}(C) - 2$ holds. There are two series of exceptions:

- smooth plane curves of degree $d \geq 5$ and hence genus $g = \binom{d-1}{2}$.
- half-canonical curves $C \subset \mathbb{P}^r$ of degree $d = 4r - 3$ and genus $g = 4r - 2$ which are not contained in a rank ≤ 4 quadric.

In [21] it is conjectured that these are the only exceptions. It is known [16] that

$$\mathrm{gon}(C) - 3 \leq \mathrm{Cliff}(C) \leq \mathrm{gon}(C) - 2$$

always holds.

The study of special divisors was initiated by Brill and Noether [12]. A major role is played by the Petri map

$$\mu: L(D) \otimes L(W - D) \to L(W), f \otimes g \mapsto fg.$$

For example,

Theorem 16.5.7 (Brill-Noether, Griffiths-Harris [37]). *Let C be a general curve of genus g. There exists a g_d^r on C if and only if the Brill-Noether number*

$$\rho(g,r,d) = g - (r+1)(g - d + r)$$

is non-negative.

16.5 The Clifford index and syzygies of canonical curves

For a proof and the beautiful geometry of curves, see the book [3].

Corollary 16.5.8. *A general smooth irreducible projective curve C of genus g has gonality $gon(C) = d$ for $d = \lceil \frac{g+2}{2} \rceil$.*

Proof. $g - 2(g - d + 1) \geq 0 \Leftrightarrow 2d \geq g + 2$. □

Note that if $|D|$ is a g_d^r then $\ell(D) = r + 1$ and $\ell(W - D) = g - d + r$ by Riemann-Roch. So for a special divisor D on a general curve the map μ has a source which never has a larger dimension than the target. A famous conjecture attributed to Petri [70] and proved in [29] and also in [55] is that for a general curve and any special divisor the map μ is always injective.

Another place where μ plays a role is for the canonical embedding of $C \hookrightarrow \mathbb{P}^{g-1}$. Identifying $L(W)$ with linear forms on $\mathbb{P}^{g-1}$ the map μ defines a $(r + 1) \times (g - d + r)$ matrix of linear forms, whose 2×2-minors vanish on C. Syzygies among these quadrics influence the numerical type of the minimal free resolution of a canonical curve.

To display the numerical invariants of a free resolution we introduce Betti tables.

Definition 16.5.9. Let

$$0 \longleftarrow M \longleftarrow F_0 \xleftarrow{\varphi_1} F_1 \xleftarrow{\varphi_2} \cdots \xleftarrow{\varphi_{c-1}} F_{c-1} \xleftarrow{\varphi_c} F_c \longleftarrow 0$$

be a finite free resolution of a graded $S = k[x_0, \ldots, x_n]$-module M with free modules $F_i = \bigoplus_j S(-j)^{\beta_{ij}}$. The table with entries $b_{ij} = \beta_{i,i+j}$ is called the **Betti table** of F. If F is the minimal free resolution of M then we speak of the Betti table of M. The Betti table of M is a numerical invariant which refines the Hilbert function and the Hilbert polynomial of M by the formula given in the proof of Theorem 8.4.2.

Example 16.5.10. Consider the homogeneous coordinate ring of the Veronese surface $V = V_{2,2} \subset \mathbb{P}^5$. By Example 11.4.6 the homogeneous ideal of I_V has the ideal of leading terms $\text{Lt}(I_V) = (x_1, x_3, x_4)^2$. Thus the resolution computed by Algorithm 8.3.7 has shape

$$0 \longleftarrow S_V \longleftarrow S \xleftarrow{\varphi_1} S^6(-2) \xleftarrow{\varphi_2} S^8(-3) \xleftarrow{\varphi_3} S^3(-4) \longleftarrow 0$$

and Betti table

	0	1	2	3
0	1	.	.	.
1	.	6	8	3

In this table a dot represents a zero Betti number. The information of the matrices φ_i is lost in this notation except that we still can read off the degrees of the entries. The first matrix is given by a 1×6 matrix of quadrics. The remaining matrices are all linear since the corresponding Betti numbers are displayed on the same row (line). In particular, we see that there are no degree 0 entries. So F is the minimal free resolution of S_V.

Example 16.5.11. The resolution of Example 8.3.2 has the Betti table

	0	1	2	3
0	1	.	.	.
1	.	5	5	1
2	.	.	1	1

It is not a minimal resolution because the third matrix contains a non-zero degree 0 entry. The minimal resolution has Betti table

	0	1	2	3
0	1	.	.	.
1	.	5	5	.
2	.	.	.	1

i.e., the corresponding entries cancel. Two of the matrices in the minimal resolution have quadric entries, the middle matrix is a linear 5×5 matrix. By a general structure theorem for codimension 3 Gorenstein rings of Buchsbaum and Eisenbud [13] this matrix can be chosen to be skew-symmetric, see also [18] Corollary 22.16.

Example 16.5.12 (Syzygies of canonical curves of low genus). Let $C \subset \mathbb{P}^{g-1}$ be a smooth canonical embedded curve. So C is not hyperelliptic. By a result of Max Noether [69] the homogeneous coordinate of a canonical canonical curve is Cohen-Macaulay and hence Gorenstein in the sense of [18, Chapter 18 and 19]. In particular the length of the minimal free resolution of C is $c = g - 2$ by the formula of Auslander-Buchsbaum and Serre [18, Theorem 19.25] and symmetric by [18, Corollary 22.16]. This explains the symmetry of the Betti tables in 16.1.

Theorem 16.5.13. *A canonically embedded smooth projective curve C of genus $g = 5, 6$ and 7 over a field k of characteristic $\neq 2$ has one the following Betti tables listed in 16.1. Which table occurs depends on the existence of the indicated special linear series on C over the algebraic closure of K of k.*

A proof of this theorem can be found in [76]. For fields k of characteristic char$(k) = 2$ the list is nearly the same with the exception of the Betti table for general curves of genus 7, where the Betti table is

	0	1	2	3	4	5
0	1	.	.	.	.	.
1	.	10	16	1	.	.
2	.	.	1	16	10	.
3	.	.	.	.	.	1

instead.

Remarks 16.5.14. 1) Currently the possible Betti table for canonical curves over a field of characteristic 0 are known up to genus 9 by Mukai's structure theorems [64] and [74]. For some experimental results see [78].

16.5 The Clifford index and syzygies of canonical curves

Genus 5:

general:

	0	1	2	3
0	1	.	.	.
1	.	3	.	.
2	.	.	3	.
3	.	.	.	1

$\exists g_3^1$:

	0	1	2	3
0	1	.	.	.
1	.	3	2	.
2	.	2	3	.
3	.	.	.	1

Genus 6:

general:

	0	1	2	3	4
0	1	.	.	.	.
1	.	6	5	.	.
2	.	.	5	6	.
3	.	.	.	.	1

$\exists g_3^1$ or g_5^2:

	0	1	2	3	4
0	1	.	.	.	.
1	.	6	8	3	.
2	.	3	8	6	.
3	.	.	.	.	1

Genus 7:

general:

	0	1	2	3	4	5
0	1	.	.	.	.	.
1	.	10	16	.	.	.
2	.	.	.	16	10	.
3	.	.	.	.	.	1

$\exists! g_4^1$:

	0	1	2	3	4	5
0	1	.	.	.	.	.
1	.	10	16	3	.	.
2	.	.	3	16	10	.
3	.	.	.	.	.	1

$\exists g_6^2$:

	0	1	2	3	4	5
0	1	.	.	.	.	.
1	.	10	16	9	.	.
2	.	.	9	16	10	.
3	.	.	.	.	.	1

$\exists g_3^1$:

	0	1	2	3	4	5
0	1	.	.	.	.	.
1	.	10	20	15	4	.
2	.	4	15	20	10	.
3	.	.	.	.	.	1

Table 16.1 All possible Betti tables of smooth canonical curves of genus 5, 6 and 7 over fields of characteristic $\neq 2$.

2) The linear systems might not be defined over field of definition of C. For example, a curve of genus $g = 7$ which is a degree 4 cover of a smooth conic has the second Betti table. But for the g_4^1 to be defined over k we need that the conic contains a k-rational point. The same is true for triple covers of conics in case of genus $g = 6$.

On the other hand, if a curve of genus $g = 6$ has a g_5^2 over K then the degree 5 plane model of C is defined over the field of definition of C.

3) Betti numbers are semi continuous in families of modules M_t with constant Hilbert function, see e.g., [9]. Moreover, for each j the alternating sums $\sum_{i=0}^{c}(-1)^i \beta_{i,j}$ depend only on the Hilbert function of M. This explains why the alternating sums along the diagonals in the Betti tables of 16.1 are constant.

4) The following Betti table decomposes

$$\begin{array}{|cccccc|} \hline 1 & . & . & . & . & . \\ . & 10 & 16 & 9 & . & . \\ . & . & 9 & 16 & 10 & . \\ . & . & . & . & . & 1 \\ \hline \end{array} = \begin{array}{|ccccc|} \hline 1 & . & . & . & . \\ . & 9 & 16 & 9 & . \\ . & . & . & . & 1 \\ \hline \end{array} \otimes \begin{array}{|cc|} \hline 1 & . \\ . & 1 \\ \hline \end{array}$$

This indicates that a curve with this Betti table is the complete intersection of a surface S of degree 6 with a quadric hypersurface. The surface S could be a del Pezzo surface, i.e., isomorphic to the blow-up of $\mathbb{P}^2$ in three points embedded by the rational map induced by the linear system $L(3; p_1, p_2, p_3)$. In that case C has a plane model of degree 6 with three double points, and each of the projections from a double point gives a g_4^1 on C. However, there are other possibilities. The surface

$S \subset \mathbb{P}^6$ could be a cone over an elliptic curve E of degree 6 in $\mathbb{P}^5$. In that case the projection from the vertex of the cone gives a double cover $C \to E$ and any g_6^2 on C defines a 2 to 1 morphism onto a plane cubic model of E.

Conjecture 16.5.15 (Green 1984, [35])**.** Let $C \subset \mathbb{P}^{g-1}$ be a smooth canonical curve of genus g defined over a field of characteristic 0 and let p be an integer. Then the Betti table of C satisfies $b_{p,2} = 0 \iff \mathrm{Cliff}(C) > p$.

$$\overbrace{}^{g-1}$$

1							
	b_{11}	…					
					…	$b_{g-3,2}$	
							1

$$\underbrace{}_{p}$$

Remarks 16.5.16. 1) Green and Lazarsfeld proved the direction from geometry to syzygies, i.e., $\mathrm{Cliff}(C) = p \Rightarrow b_{p,2} \neq 0$ in the appendix to [35]. For example, if $|D|$ is a complete base point free pencil of divisors of degree d, then the map $\mu : L(D) \times L(W-D) \to L(W)$ defines a $2 \times (g-d+1)$-matrix φ of linear forms on $\mathbb{P}^{g-1}$. The Eagon-Northcott complex of φ, see [18, Appendix A2], which resolves the ideal generated by the 2×2 minors of φ, has Betti table

	0	1	…	i	…	$g-d$
0	1	.	.	.	.	.
1	.	$\binom{g-d+1}{2}$	…	$i\binom{g-d+1}{i+1}$	…	$(g-d)\binom{g-d+1}{g-d+1}$

This complex contributes to the linear strand of C. Since $b_{g-d,1} = b_{d-2,2}$ this implies $b_{p,2} \neq 0$ for $p = \mathrm{Cliff}(D)$.

2) For a general curve of genus g the Conjecture was proved by Claire Voisin in the landmark papers [84], [85] for even and odd genus. For a general curve Green's conjecture says that for each j only one of the Betti numbers $b_{j,1}$ and $b_{j-1,2}$ on a diagonal of the Betti table can be non-zero. In particular, the Betti table depends only on the Hilbert function of the coordinate ring, hence it depends only on g.

3) A different proof of the generic Green's conjecture was given in [2], which establish the generic Green's conjecture also over fields K of finite characteristic provided $\mathrm{char}(K) \geq \frac{g+2}{2}$. By the example above, Green's conjecture does not hold curves of genus $g = 7$ and $\mathrm{char}(K) = 2$.

4) As a corollary of Voisin's theorem for odd genus $g = 2k+1$ one obtains that the condition $b_{k-1,2} \neq 0$ is a codimension 1 condition on the moduli space $\mathfrak{M}_g$ because $b_{k-1,2} = b_{k,1}$ holds by the symmetry of the Betti tables.

16.5 The Clifford index and syzygies of canonical curves

Theorem 16.5.17 (Hirschowtz-Ramanan-Voisin [45, 85]). *Let K be a field of characteristic 0 and let $g = 2k + 1$ be an odd genus. Let*

$$\mathfrak{D}_k = \{C \in \mathfrak{M}_g \mid \exists\, g^1_{k+1} \text{ on } C\}$$

denote the codimension 1 subset of $\mathfrak{M}_g$ of curves with non-maximal gonality. A general curve $C \in \mathfrak{D}_k$ has Betti numbers $b_{k-1,2} = b_{k,1} = k + 1$. Conversely every curve of odd genus $g = 2k + 1$ with $b_{2,k-1} \neq 0$ is contained in the closure of $\mathfrak{D}_k$.

The proof is based on a divisor class computation on the moduli space of curves. The so-called Koszul divisor of curves with extra syzygies coincides with $(k+1)\mathfrak{D}_k$, where the factor $k + 1$ reflects the fact that a general curve in $\mathfrak{D}_k$ has a $b_{k,1} = (k+1)$-dimensional space of extra syzygies.

Green's conjecture was inspired by a classical result of Petri.

Theorem 16.5.18 (Petri 1923, [70]). *The homogeneous ideal of a smooth canonical curve $C \subset \mathbb{P}^{g-1}$ is generated by quadrics unless C is trigonal or isomorphic to a smooth plane quintic. In the exceptional cases the quadrics generate the homogeneous ideal of a surface $S \subset \mathbb{P}^{g-1}$ of minimal degree $\deg S = g - 2$.*

Remarks 16.5.19. 1) In the exceptional cases of Petri's theorem the geometric version of Riemann-Roch 16.5.1 implies that the homogeneous ideal cannot be generated by quadrics alone. If $D = p_1 + p_2 + p_3$ is a divisor of degree 3 which moves in a pencil, then the linear span $\overline{D}$ is a line, and any quadric containing C contains that line as well by Bezout's theorem 9.4.1. The intersection of the quadrics turns out to be a ruled surface as in Exercise 11.2.7. Similarly, if C is isomorphic to a smooth plane quintic, then the five points of a divisor $D \in g^2_5$ span only a $\mathbb{P}^2$, and the conic through these five points in this $\mathbb{P}^2$ is contained in any quadric which contains C by Bézout again. In this case the intersection of the quadrics is the Veronese surface $V_{2,2} \subset \mathbb{P}^5$.

2) Petri proves, using a carefully chosen coordinate system, that

- either the quadrics form a Gröbner basis of a surface of minimal degree, which by the classification of Bertini, see [20], is a rational normal surface scroll or the Veronese surface and the homogeneous ideal needs $g - 3$ cubic generators, or
- the ideal is generated by quadrics alone.

3) Petri's argument has been given in several treatments, e.g. [68, 3]. An argument in the spirit of Gröbner bases is given in [7], which makes it very clear where the irreducibility of the curve is used.

4) I find it surprising that for a smooth canonical curve one needs either no or $g - 3$ cubic generators of the ideal. This is not true for reducible canonically embedded curves. For example, consider a reducible curve consisting of two smooth components $C = C_1 \cup C_2$ of genus g_1 and g_2 glued to each other by identifying three points on C_1 with three points on C_2. Such a curve C has arithmetic genus $g = g_1 + g_2 + 2$. Then the theory of canonical sheaves on stable curves and the residue

theorem implies that the three double points of C lie on a line. For g_1, g_2 not too small and C_1 and C_2 sufficiently general one can show that the homogeneous ideal of $C \subset \mathbb{P}^{g-1}$ needs precisely one cubic generator, see [77].

5) The case $p = 2$ of Green's conjecture was established in [83, 77], i.e., $b_{2,2} \neq 0 \Rightarrow \text{Cliff}(C) \leq 2$ holds. The case $p = 3$ of Green's conjecture is open.

6) Loose [56] establishes Green's conjecture for smooth plane curves.

7) By the work of Aprodu and Farkas we know that the conjecture is true in many other cases. For example in [1] they prove that Green's conjecture is true for all smooth curves which lie on a K3-surface.

8) Recently, Kemeny proved that a curve C which has precisely m pencils of degree $d = \text{gon}(C) \leq \lfloor \frac{g+1}{2} \rfloor$ satisfies $b_{d-2,2} = b_{g-d,1} = m(g - d)$ under some additional mild transversality conditions [51]. So each of the g_d^1's contributes $(g - d)$ independent syzygies.

9) We do not have a conjectural complete list of Betti tables for a given genus $g \geq 11$. One reason is that there will be gaps for the possibility for curves to have m pencils g_d^1. For example, in genus $g = 7$ one might wonder why there are no curves with $b_{2,2} = b_{1,3} = 6$, indicating that such a curve has two g_4^1. The reason is that if a curve C of genus $g = 7$ has two base point free pencils $|D_1|$ and $|D_2|$ of degree 4 then the image of

$$C \xrightarrow{\varphi_{D_1} \times \varphi_{D_2}} \mathbb{P}^1 \times \mathbb{P}^1$$

is a curve of bidegree $(4, 4)$ unless it is an elliptic curve E of bidegree $(2, 2)$ and $C \to E$ is $2 : 1$. Since a curve of bi-degree $(4, 4)$ on $\mathbb{P}^1 \times \mathbb{P}^1$ has arithmetic genus $9 = 7 + 2$ by Exercise 11.1.5 the image has a double point and the projection from the double point yields a g_6^2 on C. The plane model of degree 6 has three double points, counted with multiplicities. Hence there are three g_4^1 (counted with multiplicities), which fits with $b_{1,3} = 9$.

For a smooth canonical curve C of genus $g = 9$ and Clifford index $\text{Cliff}(C) = 3$ we know by work of Mukai [64] and Sagraloff [74] that C has one, two or three g_5^1 counted with multiplicities or a g_7^2, in which case the plane model of degree 7 has 6 double points counted with multiplicities, and there are six g_5^1 counted with multiplicities. The curve C has the Betti number $b_{3,2} \in \{4, 8, 12, 24\}$ accordingly. Green's conjecture fails in genus $g = 9$ for fields of characteristic $\text{char}(K) = 3$.

10) Base point free pencils $|D|$ of non-prime degree $d = d_1 \cdot d_2$ are possibly composed, i.e., the morphism might factor over some curve E

$$C \xrightarrow{d_1:1} E \xrightarrow{d_2:1} \mathbb{P}^1 .$$

16.5 The Clifford index and syzygies of canonical curves

For example, a curve of genus $g = 11$ can be a triple cover of an elliptic curve or a double cover of a plane quartic. In both cases we get infinitely many g_6^1. (Actually there are smooth curves of genus $g = 11$ which are simultaneously a triple cover of an elliptic curve and a double cover of a plane quartic.) If C is a cover of a curve E with a special Brill-Noether configuration, then this might be reflected in the Betti table of C.

11) One might ask whether all syzygies in the linear strand are spanned by geometric syzygies, i.e., those constructed by Green and Lazarsfeld. This is certainly not true if one does not take into account non-reduced structures coming from multiplicities in the count of g_d^r's. For example a curve of genus 6 might have only a single g_4^1 with high multiplicity. In that case the corresponding del Pezzo surface of degree 5 obtained from the decomposition

$$\begin{bmatrix} 1 & . & . & . & . \\ . & 6 & 5 & . & . \\ . & . & 5 & 6 & . \\ . & . & . & . & 1 \end{bmatrix} = \begin{bmatrix} 1 & . & . & . \\ . & 5 & 5 & . \\ . & . & . & 1 \end{bmatrix} \otimes \begin{bmatrix} 1 & . \\ . & 1 \end{bmatrix}$$

will be highly singular.

12) Even if we take multiplicity into account, the hope that all syzygies are spanned by geometric syzygies is most likely wrong. I conjectured that the syzygies of a general 5-gonal curve of genus $g = 13$ can be deduced from the Eagon-Northcott complex of the g_5^1 and symmetry, i.e., I expected the following Betti table

	0	1	2	3	4	5	6	7	8	9	10	11
0	1	.	.	.	.	.	.	.	.	.	.	.
1	.	55	320	891	1416	1218	216	63	8	.	.	.
2	.	.	.	8	63	216	1218	1416	891	320	55	.
3	.	.	.	.	.	.	.	.	.	.	.	1

Note that $216 = 6\binom{9}{7}$ is the contribution of the Eagon-Northcott complex by Remark 16.5.16 1). Christian Bopp [8] proved that this expectation is not true. A general 5-gonal curve of genus 13 has Betti table

	0	1	2	3	4	5	6	7	8	9	10	11
0	1	.	.	.	.	.	.	.	.	.	.	.
1	.	55	320	891	1416	1218	222	63	8	.	.	.
2	.	.	.	8	63	222	1218	1416	891	320	55	.
3	.	.	.	.	.	.	.	.	.	.	.	1

So there are 6 additional syzygies for which we have no Brill-Noether explanation.

Exercise 16.5.20. Prove that curves odd genus $g = 2k + 1$ which have a g_{k+1}^1 depend on $3g - 4$ moduli.

Exercise 16.5.21. Consider the family of curves

$$C_t = V(f_t) \subset \mathbb{P}^2$$

defined by the affine equations

$$\begin{aligned}
f_t &= xy(x+y) - 3(x^5 + y^5) - 2(x^6 + y^6) \\
&+ t[-2xy + 12(x^4 + y^4) + x^3y + xy^3 - 20x^2y^2 + 8(x^5 + y^5) - 12(x^4y + xy^4) \\
&\quad + 6(x^3y^2 + x^2y^3) - 5(x^5y + xy^5) - 2(x^4y^2 + x^2y^4) + 14x^3y^3] \\
&+ t^2[-12(x^3 + y^3) - 2(x^2y + xy^2) - 8(x^4 + y^4) + 24(x^3y + xy^3) \\
&\quad - 44x^2y^2 + 10(x^4y + xy^4) + 20(x^3y^2 + x^2y^3) + 10(x^4y^2 + x^2y^4) + 24x^3y^3].
\end{aligned}$$

The curve C_t has four ordinary double points at $p_0 = (0,0), p_1 = (2t, 0), p_2 = (0, 2t)$ and $p_3 = (-1, -1)$, and no further singularities for general values for t.

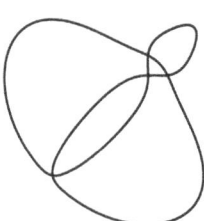

The curve C_0 has an ordinary triple point at p_0 and a double point at p_3. Hence, this is a family of curves of geometric genus $g = 6$. Projection from the triple points yields a g_3^1 on C_0. Hence C_0 is trigonal.

Compute the canonical images of C_0 and C_t for $t = \frac{1}{10}$. Conclude that C_0 is trigonal while $C_{0.1}$ (plotted above) is 4-gonal.

The curve $C_{0.1}$ has five g_4^1's. Four of them from come the projections from the double points. Where is the fifth one?

Exercise 16.5.22 (General curves of genus $g = 10$). Since $\rho = 10 - 3 \cdot 3 \geq 0$ we know by Theorem 16.5.7 that a general curve C of genus $g = 10$ has a g_9^2. Moreover by the Gieseker-Petri theorem [29] there is a ρ-dimensional family of g_9^2's on C. In this exercise we establish that for a general pair (C, g_9^2) the corresponding plane curve of degree $d = 9$ has $\delta = \binom{8}{2} - 10 = 18$ ordinary double points as its only singularities computationally: Choose a (not so small) prime field $k = \mathbb{F}_p$ and δ distinct k-rational points $p_1, \ldots, p_\delta \in \mathbb{P}^2(k)$. Consider the ideals

$$I_\Gamma = \bigcap_{j=1}^{\delta} I_{p_j} \text{ and } J = \bigcap_{j=1}^{\delta} I_{p_j}^2 \subset k[x_0, x_1, x_2].$$

16.5 The Clifford index and syzygies of canonical curves

The degree 9 part of J is the linear system

$$J_9 \cong L(9; 2p_1, \ldots, 2p_\delta).$$

Pick a random element f in J_9. Prove that f is absolutely irreducible. Verify computationally that f defines a degree 9 curve with only ordinary nodes $p_1, \ldots, p_\delta$. Let C be the resolution of singularities of $V(f)$.

Compute the canonical image of $C \subset \mathbb{P}^{g-1}$, and verify that C satisfies the generic Green conjecture.

Let W denote a canonical divisor on C and let $D \in \mathrm{Div}(C)$ be the divisor cut out on C by a general line $L \subset \mathbb{P}^2$. Check that the map $\mu \colon L(W-D) \otimes L(D) \to L(W)$ is injective. Use semi-continuity and Theorem 11.6.5 to prove that all this is also true for general curves of genus $g = 10$ defined over $\mathbb{Q}$ or $\mathbb{C}$.

Since $\dim J_9$ has the expected dimension $\binom{11}{2} - 18 \cdot 3 = 1$ we can conclude that $\mathrm{Hilb}_\delta(\mathbb{P}^2)$ contains a component which maps rationally onto a component of $\mathcal{G}^r_{g,d} = \{(C, g^r_d)\}$ which dominates $\mathfrak{M}_g$ with generically ρ-dimensional fibers for $g = 10, d = 9$ and $r = 2$: The injectivity of μ says that the fiber of

$$\mathcal{G}^2_{10,9} \to \mathfrak{M}_{10}$$

over the given curve C is smooth of dimension ρ at the point $g^2_9 = |D|$. See [5] for a construction of the moduli spaces $\mathfrak{M}_g$ and $\mathcal{G}^r_{g,d}$ and their basic theory. The paper [4] contains their application in the context above.

This gives a computational proof of Severi's theorem [81] that $\mathfrak{M}_{10}$ is unirational. Notice that $2 \cdot 18 - \dim \mathrm{PGL}(3) = 36 - 8 = 3 \cdot 9 + 1 = \dim \mathcal{G}^2_{10,9}$ confirms Riemann's count.

Exercise 16.5.23. The approach of Exercise 16.5.22 works to prove the unirationality for all $\mathfrak{M}_g$ with $g \leq 10$. Why does it not work for $\mathfrak{M}_{11}$?

Appendix A
A glimpse of sheaves and cohomology

This appendix outlines how Theorems 16.3.4, 16.3.10 and 16.3.11 follow from general results on coherent sheaves and their cohomology. As a main reference we use the book [42], where all definitions and nearly all results of this appendix can be found.

Theorems 16.3.10 and 16.3.11 are immediate consequences of Serre vanishing. The proof of the completeness of the adjoint systems 16.3.4 is more involved. One has to generalize the concept of divisors and differentials to arbitrary dimensions. The desired surjectivity then follows from the H^1 vanishing of a certain coherent sheaf in the adjunction sequence on the blow-up of $\pi \colon X \to \mathbb{P}^2$ of the singular points of the plane model of C with only ordinary singularities. Using the intersection theory of curves on smooth surfaces one sees that the exceptional curve $E_i \cong \mathbb{P}^1$ has self-intersection number $E_i^2 = -1$. The final step in the proof uses that $\pi_* \mathcal{O}_X \cong \mathcal{O}_{\mathbb{P}^2}$ and $R^1 \pi_* \mathcal{O}_X = 0$, a result which Hartshorne proves within his treatment of Castelnuovo's Contraction criterion in the final chapter of [42, Proposition V.3.4]. Fulton's treatment explains nicely how certain aspects in the work of Brill-Noether lead to cohomology theory. He also needs Corollary A.4.2, which of course is interesting on its own.

I hope very much that this appendix will serve as a good motivation for further studies.

A.1 Sheaves

The standard references for sheaf theory are Godement [30] and Iversen [49].

Definition A.1.1. Let X be a topological space. A **presheaf** $\mathcal{F}$ of abelian groups on X associates to every open subset $U \subset X$ an abelian group $\mathcal{F}(U)$, where $\mathcal{F}(\emptyset) = 0$, and to every inclusion $V \subset U$ of open sets a **restriction**

$$\rho_{U,V} \colon \mathcal{F}(U) \to \mathcal{F}(V),$$

which is a group homomorphism, such that $\rho_{U,U} = \mathrm{id}_{\mathcal{F}(U)}$ for any open set U and $\rho_{V,W} \circ \rho_{U,V} = \rho_{U,W}$ for any three open subsets $W \subset V \subset U$.

In other words, a pre-sheaf is a contravariant functor from the category of open sets of X with inclusions as the only morphisms into the category of abelian groups. An element $s \in \mathcal{F}(U)$ is called a **section** of $\mathcal{F}$ on U.

Definition A.1.2. A presheaf $\mathcal{F}$ on X is a **sheaf** if the following two axioms hold for every open covering $\mathfrak{U} = \{U_i\}$ of every open set $U = \bigcup_i U_i \subset X$.

1. A section $s \in \mathcal{F}(U)$ with $\rho_{U,U_i}(s) = 0$ for all i is the zero section, i.e., $s = 0 \in \mathcal{F}(U)$.
2. For any collection of sections $s_i \in \mathcal{F}(U_i)$ which satisfy

$$\rho_{U_i, U_i \cap U_j}(s_i) = \rho_{U_j, U_i \cap U_j}(s_j)$$

for all intersection $U_i \cap U_j$ of two open sets $U_i, U_j \in \mathfrak{U}$, there exists a section $s \in \mathcal{F}(U)$ whose restrictions $\rho_{U,U_i}(s)$ coincide with s_i for all i.

In other words, sections in a sheaf are determined by local data, and if local data are compatible, then they glue.

Examples A.1.3. 1) Let N be a differentiable manifold. The presheaf $\mathcal{C}_N^\infty$ of differentiable functions on N with $\mathcal{C}_N^\infty(U) = C^\infty(U, \mathbb{R})$ is a sheaf because being a C^∞-function is a local condition.
2) Likewise, if M is a complex manifold then the presheaf $\mathcal{O}_M$ of holomorphic functions is a sheaf.
3) A quasi-projective algebraic set $X \subset \mathbb{P}^n$ equipped with the Zariski-topology carries the sheaf of regular functions $\mathcal{O}_X$. Regular functions form a sheaf by the local nature of Definition 11.2.2:

$\mathcal{O}_X(U) = \{s \colon U \to K \mid \forall p \in U\ \exists$ an open neighborhood $V \subset U$ and homogeneous
$$\text{polynomials } g, h \text{ of the same degree such that } s(q) = \frac{g(q)}{h(q)}\ \forall q \in V\}.$$

In all three examples the sheaves are sheaves of commutative rings with 1. The restrictions are ring homomorphisms.

We call $\mathcal{C}_N^\infty$, $\mathcal{O}_M$ and $\mathcal{O}_X$ the **structure sheaf** of the differentiable manifold N, the complex manifold M and the quasi-projective algebraic set X, respectively.

Definition A.1.4. Let X be topological space, let $\mathcal{F}$ be a presheaf on X and let $p \in X$ be a point. The **stalk of** $\mathcal{F}$ **at** p is defined by

$$\mathcal{F}_p = \Big(\bigcup_{U \text{ containing } p} \mathcal{F}(U) \Big) / \sim,$$

where $s \in \mathcal{F}(U)$ and $t \in \mathcal{F}(V)$ are equivalent if there exists an open neighborhood $W \subset U \cap V$ of p such that $\rho_{U,W}(s) = \rho_{V,W}(t)$. The equivalence class $s_p \in \mathcal{F}_p$ represented by $s \in \mathcal{F}(U)$ is called the **germ** of s at p and

A.1 Sheaves

$$\rho_{U,p} \colon \mathcal{F}(U) \to \mathcal{F}_p, \ s \mapsto s_p$$

denotes the map to the equivalence classes.

Note that $\rho_{U,p} = \rho_{V,p} \circ \rho_{U,V}$ holds for open neighborhoods $V \subset U$ of p.

Examples A.1.5. 1) Let be M a complex manifold of dimension n, and let $p \in M$ be a point. The the stalk of $\mathcal{O}_M$ at p is isomorphic to a ring of convergent power series,

$$\mathcal{O}_{M,p} \cong \mathbb{C}\{x_1,\ldots,x_n\},$$

where $x_1,\ldots,x_n$ are the coordinate functions of a holomorphic chart $\varphi \colon U \to V \subset \mathbb{C}^n$, with $\varphi(p) = 0 \in \mathbb{C}^n$. This follows from the identity theorem for holomorphic functions. The map $\rho_{U,p}$ maps a holomorphic function $f \in \mathcal{O}_M(U)$ to its power series expansion at p.

2) For a C^∞-manifold N and $\mathcal{C}_N^\infty$ the sheaf of C^∞-functions, the stalk $\mathcal{C}_{N,p}^\infty$ at a point p is the ring of germs of infinitely often differentiable functions. It is a local ring whose maximal ideal $\mathfrak{m}$ is the ideal of germs of functions which vanish at p. The $\mathfrak{m}$-adic completion of $\mathcal{C}_{N,p}^\infty$ is isomorphic to $\mathbb{R}[[x_1,\ldots,x_n]]$ and the map

$$T \colon \mathcal{C}_{N,p}^\infty \to \mathbb{R}[[x_1,\ldots,x_n]]$$

maps a germ f to its Taylor series expansion. The ring $\mathcal{C}_{N,p}^\infty$ is not Noetherian by Krull's intersection theorem 10.1.5, because, for example, the germ of the function

$$g_1(x) = \begin{cases} \exp(x^{-2}) & \text{if } x > 0 \\ 0 & \text{if } x \leq 0 \end{cases}$$

in one variable lies in the kernel of the completion map. Note that this kind of function plays an important role in the theory of C^∞-manifolds. The function $g(x_1,\ldots,x_n) = g_1(1 - \|x\|^2)$ is a non-negative function with support the unit ball. These functions are used to define partitions of unity on manifolds.

Proposition A.1.6. *Let $A \subset \mathbb{A}^n$ be an affine algebraic set. For $f \in K[A]$ let $U_f = A \setminus V(f)$ denote the Zariski-open subset of A where f does not vanish. Then the sheaf $\mathcal{O}_A$ of regular function on A satisfies $\mathcal{O}_A(U_f) = K[A]_f$. The stalk of $\mathcal{O}_A$ at a point $p \in A$ is $\mathcal{O}_{A,p} = K[A]_{\mathfrak{m}_p}$.*

Thus for an affine algebraic set A, the structure sheaf $\mathcal{O}_A$ systematically organizes the localizations of $K[A]$ which we have considered so far.

Proof. The set U_f is an affine algebraic set as well, and $K[x_1,\ldots,x_n,y]/(I(A),yf-1)$ is its coordinate ring. If A is irreducible, then U_f is irreducible as well. Thus, if A is a variety, then the first assertion follows from Theorem 5.1.6. The statement about the stalk holds by Definition 5.1.5. For the general case, see Proposition II.2.2 of [42], where Hartshorne gives a proof which works more generally for affine schemes. □

Definition A.1.7. A morphism $\varphi \colon \mathcal{F} \to \mathcal{G}$ between presheaves on a topological space X consists of a collection of group homomorphisms $\varphi(U) \colon \mathcal{F}(U) \to \mathcal{G}(U)$

for every open subset $U \subset X$ which are compatible with the restrictions, i.e., the diagrams

$$\begin{array}{ccc} \mathcal{F}(U) & \xrightarrow{\varphi(U)} & \mathcal{G}(U) \\ \rho_{U,V} \downarrow & & \downarrow \rho_{U,V} \\ \mathcal{F}(V) & \xrightarrow{\varphi(V)} & \mathcal{G}(V) \end{array}$$

commute for all inclusions $V \subset U$.

Given a presheaf $\mathcal{F}$ there exists a sheaf $\mathcal{F}^+$ and a morphism $\iota \colon \mathcal{F} \to \mathcal{F}^+$ with following universal property: Any morphism $\varphi \colon \mathcal{F} \to \mathcal{G}$ to a sheaf factors uniquely over $\mathcal{F}^+$:

$$\begin{array}{ccc} \mathcal{F} & \xrightarrow{\iota} & \mathcal{F}^+ \\ & \varphi \searrow & \downarrow \exists! \varphi^+ \\ & & \mathcal{G} \end{array}$$

The construction is similar to the construction of regular functions on an algebraic set. We define

$$\mathcal{F}^+(U) = \{s \colon U \to \bigcup_{p \in U} \mathcal{F}_p \mid s(p) \in \mathcal{F}_p \text{ and } \forall\, p \in U\, \exists \text{ an open neighborhood } V$$
$$\text{and a section } \tilde{s} \in \mathcal{F}(V) \text{ such that } s(q) = \tilde{s}_q \forall\, q \in V\}.$$

Example A.1.8. For the constant presheaf $\mathcal{F}$ on X with $\mathcal{F}(U) = \mathbb{Z}$ for all nonempty open subsets U, the sheaf $\mathcal{F}^+$ is the sheaf of locally constant functions, i.e.,

$$\mathcal{F}^+(U) = \{f \colon U \to \mathbb{Z} \mid f \text{ is locally constant}\}.$$

Usually one denotes this sheaf by $\mathbb{Z}_X$.

Proposition A.1.9. *A sequence*

$$0 \to \mathcal{E} \to \mathcal{F} \to \mathcal{G} \to 0$$

of sheaves on X is a short exact sequence of sheaves if and only if the sequences on stalks

$$0 \to \mathcal{E}_p \to \mathcal{F}_p \to \mathcal{G}_p \to 0$$

is exact for all $p \in X$. □

Definition A.1.10. Let X be a quasi-projective algebraic set. A sheaf $\mathcal{F}$ of $\mathcal{O}_X$-modules is **coherent** if for each point $p \in X$ there exists an open neighborhood U with $p \in U \subset X$ and a presentation

$$\mathcal{O}_U^b \to \mathcal{O}_U^a \to \mathcal{F}_U \to 0,$$

A.1 Sheaves

where $\mathcal{F}_U$ denotes the restriction of $\mathcal{F}$ to U, i.e., $\mathcal{F}_U(V) = \mathcal{F}(V)$ for open subsets $V \subset U$.

We frequently use the notation $\Gamma(U, \mathcal{F}) = \mathcal{F}(U)$ for the space sections. In particular, $\Gamma(X, \mathcal{F})$ denotes the space of global sections of $\mathcal{F}$.

Remark A.1.11. A similar conditions defines coherent sheaves of $\mathcal{O}_M$-modules on a complex manifold M. Unlike the situation of quasi-projective algebraic sets X the fact that the kernel of a morphism $\mathcal{F} \to \mathcal{G}$ of coherent sheaves of $\mathcal{O}_M$-modules is coherent again is a non-trivial fact which relies on Oka's coherence theorem [33] for the structure sheaf of a complex manifold M.

Example A.1.12. Let A be an affine algebraic set and let M be a finitely generated $K[A]$-module. Then $\widetilde{M}$ defined by

$$\widetilde{M}(U) = \{s: U \to \bigcup_{p \in U} M_{\mathfrak{m}_p} \mid s(p) \in M_{\mathfrak{m}_p} \text{ and } \forall p \in U \exists \text{ an open neighborhood } V \text{ and}$$

elements $m \in M$ and $f \in K[A]$ such that $s(q) = \frac{m}{f}$ for all $q \in V\}$

is a coherent sheaf of $\mathcal{O}_A$-modules.

Theorem A.1.13 (Serre [79]). *Let A be an affine algebraic set. Then the functors*

$$coh(\mathcal{O}_A) \longleftrightarrow mod(K[A])$$

$$\mathcal{F} \mapsto \Gamma(A, \mathcal{F}), \quad \widetilde{M} \mapsfrom M$$

define equivalences between the category of coherent sheaves of $\mathcal{O}_A$-modules and finitely generated $K[A]$-modules.

Definition A.1.14. Let $X \subset \mathbb{P}^n$ be a projective algebraic set, $S_X = S/I(X)$ its homogeneous coordinate ring and let M be a finitely generated graded S_X-module. The **associated sheaf** $\widetilde{M}$ of M is the sheaf defined by the presheaf

$$\widetilde{M}^-(U) = \{\frac{m}{f} \mid m \in M \text{ is homogeneous of some degree } d$$

and $f \in (S_X)_d$ is nowhere vanishing on $U\}$.

Notice that if M is a module which is annihilated by a power of the homogeneous ideal $(x_0, \ldots, x_n)$ then $\widetilde{M} = 0$ because $\frac{m}{f} = \frac{x_i^N m}{x_i^N f} = 0 \in \widetilde{M}_p$ for $N \gg 0$ sufficiently large and x_i such that $p \notin V(x_i)$. In particular, M and any truncation $M_{\geq r} = \oplus_{d \geq r} M_d$ define the same coherent sheaf.

Examples A.1.15. 1. For $\mathbb{P}^n$ and $S = K[x_0, \ldots, x_n]$ the graded S-module $S(d)$ gives rise to the sheaf

$$\mathcal{O}(d) = \widetilde{S(d)}$$

of $\mathcal{O} = \mathcal{O}_{\mathbb{P}^n}$-modules. Note that $\mathcal{O}(d) \otimes_{\mathcal{O}} \mathcal{O}(e) \cong \mathcal{O}(d + e)$ holds. In particular, $\mathcal{O}(d)$ is invertible in the sense that $\mathcal{O}(d) \otimes \mathcal{O}(-d) \cong \mathcal{O}$.

2. For a coherent sheaf $\mathcal{F}$ on a projective variety $X \subset \mathbb{P}^n$ we denote $\mathcal{F} \otimes \mathcal{O}(d)$ by $\mathcal{F}(d)$. If $\mathcal{F} = \widetilde{M}$ is the sheafification of a graded S_X-module M, then $\mathcal{F}(d) \cong \widetilde{M(d)}$.
3. S_X gives the structure sheaf $\mathcal{O}_X$. Regarding S_X as an S-module, then we can regard $\mathcal{O}_X$ also as a sheaf of $\mathcal{O}_{\mathbb{P}^n}$-modules.
4. The homogeneous ideal $I(X) \subset S$ sheafifies to a sheaf of ideals $\mathcal{I}_X \subset \mathcal{O}_{\mathbb{P}^n}$ and we have a short exact sequence

$$0 \to \mathcal{I}_X \to \mathcal{O}_{\mathbb{P}^n} \to \mathcal{O}_X \to 0.$$

Lemma A.1.16. *Let $\mathcal{F}$ be a coherent sheaf of $\mathcal{O}_X$-modules on a projective variety $X \subset \mathbb{P}^n$. Let $f \in S_X$ be a homogeneous element of degree d and let $X_f = X \setminus V(f)$ be the corresponding open subset. For a section $s \in \Gamma(X_f, \mathcal{F})$ there exists an integer $e \geq 0$ such that $s f^e \in \Gamma(X_f, \mathcal{F} \otimes \mathcal{O}(de))$ is the restriction of a section $t \in \Gamma(X, \mathcal{F} \otimes \mathcal{O}(de))$.*

Theorem A.1.17 (Theorem A, [79]). *Let $\mathcal{F}$ be a coherent sheaf on a projective algebraic set $X \subset \mathbb{P}^n$. There exists an integer a such that $\mathcal{F}(a)$ is generated by finitely many global sections, i.e., there exists a finite-dimensional subvector space $V \subset \Gamma(X, \mathcal{F}(a))$ such that the natural morphism*

$$V \otimes \mathcal{O}_X \to \mathcal{F}(a)$$

is surjective.

For the proofs of Theorem A.1.13, Lemma A.1.16 and Theorem A.1.17 see [42, Section II.5], where these results are proven in the more general setting of schemes.

Theorem A.1.18. *Let S denote the homogeneous coordinate ring of $\mathbb{P}^n$. Sheafification defines an exact functor*

$$gr\text{-}mod(S) \to coh(\mathbb{P}^n), \quad M \mapsto \widetilde{M}$$

between the category of finitely generated graded S-modules and the category of coherent sheaves on $\mathbb{P}^n$. Conversely,

$$coh(\mathbb{P}^n) \to gr\text{-}mod(S), \quad \mathcal{F} \mapsto \Gamma_{\geq 0}(\mathcal{F}) = \oplus_{d \geq 0} \Gamma(\mathbb{P}^n, \mathcal{F}(d))$$

gives for every coherent sheaf $\mathcal{F}$ a finitely generated graded S-module $M = \Gamma_{\geq 0}(\mathcal{F})$ with $\widetilde{M} = \mathcal{F}$.

The first part of the theorem follows from the exactness of localization. The converse uses, apart from Theorem A.1.17, the explicit knowledge of the cohomology of the invertible sheaves $\mathcal{O}_{\mathbb{P}^n}(d)$ summarized in Theorem A.2.4 and Hilbert's syzygy theorem 8.3.5. We sketch the essential steps in the proof of Theorem A.2.6 below.

Exercise A.1.19. Deduce Theorem A.1.17 from Lemma A.1.16 by extending germs $s_1, \ldots, s_r \in \mathcal{F}_p$ which generate $\mathcal{F}_p$ as an $\mathcal{O}_{X,p}$-module and by using an induction on the dimension of the support of cokernels of evaluation maps of type

A.2 Cohomology

$$V \otimes \mathcal{O}_X(-a) \to \mathcal{F}$$

for subspaces $V \subset \Gamma(X, \mathcal{F}(a))$.

A.2 Cohomology

In general, for a short exact sequence of sheaves

$$0 \to \mathcal{E} \to \mathcal{F} \to \mathcal{G} \to 0$$

on a topological space X and an open subset $U \subset X$ the sequence of sections

$$0 \to \mathcal{E}(U) \to \mathcal{F}(U) \to \mathcal{G}(U)$$

might not be surjective on the left. To measure the failure of exactness on the right one defines higher cohomology groups $H^i(X, -)$ as the right derived functors of

$$\Gamma(X, -) \colon \mathcal{F} \mapsto \Gamma(X, \mathcal{F})$$

as follows: Since the category of sheaves on X has enough injectives one can find an exact complex of injective sheaves

$$0 \to \mathcal{F} \to I^0 \to I^1 \to I^2 \to \ldots$$

and defines

$$H^0(X, \mathcal{F}) = \ker\left(I^0(X) \to I^1(X)\right) = \Gamma(X, \mathcal{F})$$

and

$$H^i(X, \mathcal{F}) = \ker\left(I^i(X) \to I^{i+1}(X)\right) / \operatorname{im}\left(I^{i-1}(X) \to I^i(X)\right)$$

for $i \geq 1$.

Proposition A.2.1. *A short exact sequence*

$$0 \to \mathcal{E} \to \mathcal{F} \to \mathcal{G} \to 0$$

of sheaves on X gives rise to a long exact sequence

$$
\begin{aligned}
0 &\to H^0(X, \mathcal{E}) \to H^0(X, \mathcal{F}) \to H^0(X, \mathcal{G}) \\
&\to H^1(X, \mathcal{E}) \to H^1(X, \mathcal{F}) \to H^1(X, \mathcal{G}) \\
&\to H^2(X, \mathcal{E}) \to \ldots
\end{aligned}
$$

of cohomology groups.

There are various different ways to define cohomology groups. The derived functor definition is due to Grothendieck. The first appearance of cohomology groups in algebraic geometry was Čech cohomology introduced by Serre [79].

Let $\mathfrak{U} = \{U_0, \ldots U_N\}$ be a finite open covering of an algebraic set X and $\mathcal{F}$ a sheaf on X. The Čech complex

$$0 \to C^0(\mathfrak{U}, \mathcal{F}) \to C^1(\mathfrak{U}, \mathcal{F}) \to C^2(\mathfrak{U}, \mathcal{F}) \to \cdots$$

is defined as follows. For $p \geq 0$ and $0 \leq i_0 < i_1 < \ldots < i_p \leq N$ we denote $U_{i_0} \cap \ldots \cap U_{i_p}$ by $U_{i_0 i_1 \ldots i_p}$ and the space of p-Čech chains by

$$C^p(\mathfrak{U}, \mathcal{F}) = \bigoplus_{0 \leq i_0 < i_1 < \ldots < i_p \leq N} \mathcal{F}(U_{i_0 i_1 \ldots i_p}).$$

The differential δs of a Čech chain $s = (s_{i_0 i_1 \ldots i_p})$ is defined by

$$(\delta s)_{i_0 i_1 \ldots i_{p+1}} = \sum_{v=0}^{p+1} (-1)^v s_{i_0 \ldots \widehat{i_v} \ldots i_{p+1}} \in \mathcal{F}(U_{i_0 i_1 \ldots i_{p+1}}),$$

where $\widehat{i_v}$ indicates that we drop the corresponding index and the alternating sum is taken after restricting these sections to $U_{i_0 i_1 \ldots i_{p+1}}$. The Čech differentials $\delta \colon C^p(\mathfrak{U}, \mathcal{F}) \to C^{p+1}(\mathfrak{U}, \mathcal{F})$ satisfy $\delta \circ \delta = 0$. Thus we get a complex.

Definition A.2.2. The group

$$\check{H}^p(\mathfrak{U}, \mathcal{F}) = \frac{\ker(C^p(\mathfrak{U}, \mathcal{F}) \to C^{p+1}(\mathfrak{U}, \mathcal{F}))}{\operatorname{im}(C^{p-1}(\mathfrak{U}, \mathcal{F}) \to C^p(\mathfrak{U}, \mathcal{F}))}$$

is called the p-th **Čech cohomology group** of $\mathcal{F}$ with respect to $\mathfrak{U}$.

Notice that $\check{H}^0(\mathfrak{U}, \mathcal{F}) = \Gamma(X, \mathcal{F})$ holds due to the sheaf axioms. In general, $H^p(X, \mathcal{F})$ and $\check{H}^p(\mathfrak{U}, \mathcal{F})$ for $p > 0$ do not coincide. However for an affine covering we have the following theorem.

Theorem A.2.3. *Let $\mathfrak{U} = \{U_0, \ldots U_N\}$ be a covering such that all U_i are affine algebraic subsets of X, and let $\mathcal{F}$ be a coherent sheaf on X. Then*

$$\check{H}^p(\mathfrak{U}, \mathcal{F}) \cong H^p(X, \mathcal{F}) \text{ for all } p \geq 0.$$

For a proof, see [42, Theorem III.4.5].

Theorem A.2.4. *The sheaves $\mathcal{O}(d)$ on $\mathbb{P}^n$ have the following cohomology groups*

1. $H^0(\mathbb{P}^n, \mathcal{O}(d)) = \begin{cases} S_d & \text{if } d \geq 0 \\ 0 & \text{if } d < 0 \end{cases}$

2. $H^i(\mathbb{P}^n, \mathcal{O}(d)) = 0$ *for $1 \leq i \leq n-1$ or $i > n$.*

A.2 Cohomology 257

3. $H^n(\mathbb{P}^n, \mathcal{O}(d)) = \begin{cases} 0 & \text{if } d \geq -n \\ (S_{-n-1-d})^* & \text{if } d \leq -n-1. \end{cases}$

This is usually proved with Čech cohomology. For a proof see, for example, [42, Theorem III.5.1]. Notice that

$$S = \oplus_{d \geq 0} H^0(\mathbb{P}^n, \mathcal{O}(d))$$

recovers the homogeneous coordinate ring in a somewhat more conceptual way.

An immediate consequence of the vanishing of the groups $H^1(\mathbb{P}^2, \mathcal{O}(a))$ is Max Noether's famous AF+BG Theorem.

Theorem A.2.5 (Noether's AF+BG Theorem). *Let F and G be homogeneous polynomials in $K[x_0, x_1, x_2]$ of degree d and e without common factor. Let H be a further homogeneous form of degree h. Suppose that for each point $p \in V(F,G)$ the form H satisfies the local condition $H_p \in (F_p, G_p) \subset \mathcal{O}_{\mathbb{P}^2, p}$. Then there exist forms A and B of degree $h - d$ and $h - b$ such that*

$$H = AF + BG.$$

Proof. Consider the ideal sheaf $\mathcal{I}$ which is the image of the morphism $\mathcal{O}_{\mathbb{P}^2}(-d) \oplus \mathcal{O}_{\mathbb{P}^2}(-e) \to \mathcal{O}_{\mathbb{P}^2}$ defined by the matrix (F, G). Then

$$0 \longrightarrow \mathcal{O}_{\mathbb{P}^2}(-d-e) \xrightarrow{\begin{pmatrix} G \\ -F \end{pmatrix}} \mathcal{O}_{\mathbb{P}^2}(-e) \oplus \mathcal{O}_{\mathbb{P}^2}(-d) \xrightarrow{(F,G)} \mathcal{I} \longrightarrow 0$$

is a short exact sequence of coherent sheaves, since F and G have no common factor. The form H is a section of $H^0(\mathbb{P}^2, \mathcal{I}(h))$, since the local conditions are satisfied. The result follows from the long exact cohomology sequence, because $H^1(\mathbb{P}^2, \mathcal{O}_{\mathbb{P}^2}(h - d - e)) = 0$ holds by Theorem A.2.4. □

We now sketch how Theorem A.2.4 combined with Theorem A.1.17 and Hilbert's syzygy theorem 8.3.5 implies the following fundamental finiteness result.

Theorem A.2.6 (Theorem B and Serre Vanishing). *Let $\mathcal{F}$ be a coherent sheaf on $\mathbb{P}^n$. There exists a finitely generated graded S-module M such that $\mathcal{F} = \widetilde{M}$. The cohomology groups $H^i(\mathbb{P}^n, \mathcal{F})$ are finite-dimensional K-vector spaces which vanish for $i > n$. Moreover, there exists an integer d_0 such that*

$$H^i(\mathbb{P}^n, \mathcal{F}(d)) = 0 \text{ for all } i \geq 1 \text{ and } H^0(\mathbb{P}^n, \mathcal{F}(d)) = M_d$$

holds for all $d \geq d_0$.

Proof. Let $\mathcal{F}$ be a coherent sheaf on $\mathbb{P}^n$. Choose a and V as in Theorem A.1.17 and consider the kernel

$$\mathcal{G} = \ker(V \otimes \mathcal{O}_{\mathbb{P}^n} \to \mathcal{F}(a)).$$

$\mathcal{G}$ is coherent again, and applying the theorem to $\mathcal{G}$ yields a presentation

$$W \otimes \mathcal{O}(-a-b) \to V \otimes \mathcal{O}(-a) \to \mathcal{F} \to 0.$$

The corresponding map

$$W \otimes S(-a-b) \to V \otimes S(-a)$$

is given by a matrix of homogeneous forms of degree b. The cokernel of this map is a finitely generated graded S-module M with $\widetilde{M} \cong \mathcal{F}$ by the exactness of the sheafification. By Hilbert's syzygy theorem the presentation extends to a finite free resolution

$$0 \to F_{n+1} \to F_n \to \ldots \to F_2 \to F_1 \to F_0 \to M \to 0.$$

Since each F_i is a finite direct sum of $S(-j)$'s there exists an integer d_0 such that the sheaves $\widetilde{F}_i(d)$ have vanishing higher cohomology $H^p(\mathbb{P}^n, \widetilde{F}_i(d)) = 0$ for all $d \geq d_0$ and $p \geq 1$ by Theorem A.2.4. Breaking the resolution into short exact sequences

$$0 \to M_{i+1} \to F_i \to M_i \to 0$$

we obtain from the long exact sequences connecting homomorphisms which are isomorphisms

$$H^i(\mathbb{P}^n, \mathcal{F}(d)) \cong H^{i+1}(\mathbb{P}^n, \widetilde{M}_1(d)) \cong \ldots \cong H^{i+n+1}(\mathbb{P}^n, \widetilde{M}_{n+1}(d))$$

for $i \geq 1$. Since $M_{n+1} = F_{n+1}$ the last cohomology group is zero as well. The same argument shows that $H^1(\mathbb{P}^n, \widetilde{M}_i(d)) = 0$ so

$$0 \to H^0(\mathbb{P}^n, \widetilde{M}_1(d)) \to H^0(\mathbb{P}^n, \widetilde{F}_0(d)) \to H^0(\mathbb{P}^n, \mathcal{F}(d)) \to 0$$

is exact and $H^0(\mathbb{P}^n, \widetilde{F}_0(d)) \to H^0(\mathbb{P}^n, \widetilde{M}(d))$ is surjective. Thus $(F_1)_d \to (F_0)_d \to M_d$ induces the isomorphism $H^0(\mathbb{P}^n, \mathcal{F}(d)) = M_d$.

Running the long exact sequences of the short exact sequences for $d = 0$ implies $H^i(\mathbb{P}^n, \mathcal{F}) = 0$ for $i > n$ and that all groups $H^i(\mathbb{P}^n, \mathcal{F})$ are finite-dimensional. The cohomology groups $H^i(\mathbb{P}^n, \mathcal{F}), H^{i+1}(\mathbb{P}^n, \widetilde{M}_1) \ldots, H^{i+n+1}(\mathbb{P}^n, \widetilde{M}_{n+1})$ differ by finite-dimensional vector spaces due to the finiteness in Theorem A.2.4. □

Proof of Theorem 16.3.10. Consider the short exact sequence

$$0 \to \mathcal{I}_C \to \mathcal{O}_{\mathbb{P}^n} \to \mathcal{O}_C \to 0.$$

The desired surjectivity $S_d \to L(dH)$ follows from the Serre vanishing applied to the ideal sheaf $\mathcal{I}_C$ and the identification $H^0(\mathbb{P}^n, \mathcal{O}_C(d)) = H^0(C, \mathcal{O}_C(d)) = L(dH)$. □

Proof of Theorem 16.3.11. Consider the ideal $I_\Gamma = \bigcap_{j=1}^s I(q_j)^{r_j-1}$, the corresponding ideal sheaf $\mathcal{I}_\Gamma = \widetilde{I}_\Gamma$ and the sequence

A.2 Cohomology

$$0 \to \mathcal{I}_\Gamma \to \mathcal{O}_{\mathbb{P}^2} \to \mathcal{O}_\Gamma \to 0.$$

The sheaf $\mathcal{O}_\Gamma$ is supported on the collection of points and

$$\dim H^0(\mathbb{P}^2, \mathcal{O}_\Gamma(d)) = \sum_{j=1}^{s} \binom{r_j}{2}$$

holds for any $d \in \mathbb{Z}$. The result follows from the Serre vanishing of $\mathcal{I}_\Gamma$ and the identification $H^0(\mathbb{P}^2, \mathcal{I}_\Gamma(e)) = L(e; (r_1 - 1)q_1, \ldots, (r_s - 1)q_s)$. □

Definition A.2.7. Let $\pi \colon X \to Y$ be a morphism between algebraic sets, and let $\mathcal{F}$ be a sheaf on X. We define the **direct image sheaf** $\pi_*\mathcal{F}$ by $(\pi_*\mathcal{F})(U) = \mathcal{F}(\pi^{-1}(U))$. This is a sheaf. If $\mathcal{F}$ is a sheaf of $\mathcal{O}_X$-modules, then $\pi_*\mathcal{F}$ is a sheaf of $\mathcal{O}_Y$-modules because $\mathcal{F}(\pi^{-1}(U))$ is a $\mathcal{O}_Y(U)$-module via the ring homomorphism $\mathcal{O}_Y(U) \to \mathcal{O}_X(\pi^{-1}(U))$.

Corollary A.2.8. *The cohomology groups $H^i(X, \mathcal{F})$ of a coherent sheaf $\mathcal{F}$ on projective variety $X \subset \mathbb{P}^n$ are finite-dimensional vector spaces, and vanish for $i > \dim X$.*

Proof. Let $\iota \colon X \to \mathbb{P}^n$ denote the inclusion. Then $\iota_*(\mathcal{F})$ is a sheaf of $\mathcal{O}_{\mathbb{P}^n}$-modules which is coherent since $\mathcal{O}_X = \iota_*\mathcal{O}_X$ is coherent. The cohomology groups

$$H^i(X, \mathcal{F}(d)) = H^i(\mathbb{P}^n, \iota_*\mathcal{F}(d))$$

do not change since the Čech complex does not change:

$$\iota_*\mathcal{F}(U_{i_0..i_p}) = \mathcal{F}(\iota^{-1}(U_{i_0..i_p})).$$

Thus the first statement follows from Theorem A.2.3 and Theorem A.2.6. For the vanishing we note that for $d = \dim X$ we can find homogeneous coordinates such that the standard charts $U_0, \ldots, U_d$ cover X, i.e., $X \subset U_0 \cup \ldots \cup U_d$. Thus for $\mathfrak{U} = \{U_0 \cap X, \ldots U_d \cap X\}$ we have $C^p(\mathfrak{U}, \mathcal{F}) = 0$ for $p > d$ and $H^p(X, \mathcal{F}) = \check{H}^p(\mathfrak{U}, \mathcal{F}) = 0$ for $p > d$ follows from Leray's theorem A.2.3. □

In particular, the **Euler characteristic**

$$\chi(X, \mathcal{F}) = \sum_{i=0}^{d} (-1)^i h^i(X, \mathcal{F}),$$

where $d = \dim X$ and $h^i(X, \mathcal{F}) = \dim H^i(X, \mathcal{F})$, is well-defined.

Let M be a finitely generated graded $S = K[x_0, \ldots, x_n]$-module. The Euler characteristic of the sheaf $\mathcal{F} = \widetilde{M}$ gives an interpretation of the Hilbert polynomial $p_M(t)$ for all values of t.

Proposition A.2.9. $p_M(t) = \chi(\mathbb{P}^n, \mathcal{F}(t))$ *for all* $t \in \mathbb{Z}$.

Proof. The assertion is true for the $M = S$, since

$$\chi(\mathbb{P}^n, \mathcal{O}(t)) = \frac{1}{n!} \prod_{i=1}^{n}(t+i)$$

holds by Theorem A.2.4: If $t < -n$ then all factors are negative and we get the sign $(-1)^n$ corresponding to the contribution of $h^n(\mathbb{P}^n, \mathcal{O}(t)) = \binom{-n-1-t+n}{n} = \binom{-t-1}{n}$. Consider a graded free resolution

$$0 \to F_{n+1} \to F_n \to \ldots F_i \to F_0 \to M \to 0$$

with $F_i = \sum_j S(-j)^{\beta_{ij}}$. Sheafification and the additivity of $\chi(\mathbb{P}^n, -)$ in short exact sequences, see Exercise A.2.10, yields that

$$\chi(\mathbb{P}^n, \mathcal{F}(t)) = \sum_{i=0}^{n+1}(-1)^i \sum_j \beta_{ij} \chi(\mathbb{P}^n, \mathcal{O}(t-j))$$

is a sum of polynomial functions. Since $\chi(\mathbb{P}^n, \mathcal{F}(t)) = p_M(t)$ holds for all $t \gg 0$ by Theorem A.2.6 we conclude that these polynomials coincide. □

Exercise A.2.10. Let
$$0 \to \mathcal{E} \to \mathcal{F} \to \mathcal{G} \to 0$$
be a short exact sequence of coherent sheaves on a projective variety X. Prove:
$$\chi(X, \mathcal{E}) + \chi(X, \mathcal{G}) = \chi(X, \mathcal{F}).$$

A.3 Differentials and the adjunction sequence

The proof of Theorem 16.3.4 uses the adjunction sequence. We need to generalize the concept of a divisor to higher-dimensional varieties.

Definition A.3.1. Let X be a variety. A **prime divisor** $P \subset X$ is a codimension 1 subvariety. We define the local ring $\mathcal{O}_{X,P}$ as the localization of $K[U]$ in the prime ideal $I(P \cap U) \subset K[U]$, where $U \subset X$ is an affine open subvariety which intersects P. The local ring $\mathcal{O}_{X,P}$ is independent from the choice of U.

Proposition A.3.2. *If P intersects the smooth part of X, then $\mathcal{O}_{X,P}$ is a discrete valuation ring.*

Proof. Let $q \in P \cap U$ be a smooth point of X. Then since the regular local rings are factorial [18, Theorem 19.19] the ideal $I(P \cap U)\mathcal{O}_{X,q} \subset \mathcal{O}_{X,q}$ is a principal ideal generated by a prime element $f_P \in \mathfrak{m}_{X,q} \subset \mathcal{O}_{X,p}$. We can define the valuation v_P for a non-zero $f \in \mathcal{O}_{X,q} \subset K(X)$ by $v_P(f) = n$ if $f = f_P^n \prod f_i^{n_i}$ is the factorization of f into prime factors in $\mathcal{O}_{X,q}$. Then v_P extends to a valuation $K(X)^* \to \mathbb{Z}$, whose valuation ring coincides with $\mathcal{O}_{X,P} \cong (\mathcal{O}_{X,q})_{(f_P)}$. □

A.3 Differentials and the adjunction sequence

Definition A.3.3. A **Weil divisor** $D = \sum_i n_i D_i$ is a finite formal sum of codimension 1 subvarieties $D_i \subset X$. The divisor D is effective if all $n_i \geq 0$. If the singular locus $\text{sing}(X)$ has codimension ≥ 2, then we can define the **principal divisor** of a rational function $f \in K(X)$ as

$$(f) = \sum_P v_P(f) P$$

and the Riemann-Roch space by

$$L(D) = \{f \in K(X)^* \mid (f) + D \geq 0\} \cup \{0\}.$$

The definition of the Riemann-Roch spaces sheafifies. We define for a Weil divisor $D = \sum_P n_P P$ the sheaf $\mathcal{O}_X(D)$ as the sheaf

$$\mathcal{O}_X(D): U \mapsto \{f \in K(X)^* \mid v_P(f) + n_P \geq 0 \text{ for all } P \text{ with } P \cap U \neq \emptyset\} \cup \{0\}.$$

This is a sheaf of $\mathcal{O}_X$-modules.

Proposition A.3.4. *If X is smooth, then the germ of $\mathcal{O}_X(D)$ at a point q of X is the $\mathcal{O}_{X,q}$-submodule of $K(X)$ generated by $f_{D,q} = \prod_{P \ni q} f_P^{-n_P}$, where $f_P \in \mathcal{O}_{X,q}$ is the generator of $I(P)\mathcal{O}_{X,q}$ and n_P is the coefficient of P in $D = \sum n_P P$. In particular, $\mathcal{O}_X(D)$ is an invertible sheaf if X is smooth.* □

Examples A.3.5. 1. Let $H = V(x_0) \subset \mathbb{P}^n$ be a hyperplane. Then

$$\mathcal{O}_{\mathbb{P}^n}(H) \cong \mathcal{O}_{\mathbb{P}^n}(1) \text{ via } f \mapsto f x_0.$$

2. For Weil divisors D_1 and D_2 on a smooth variety X one has

$$\mathcal{O}_X(D_1) \otimes_{\mathcal{O}_X} \mathcal{O}_X(D_2) \cong \mathcal{O}_X(D_1 + D_2).$$

3. If D is an effective divisor then $\mathcal{O}_X(-D)$ is a sheaf of ideals in $\mathcal{O}_X$. In particular, for a prime divisor P on X we have

$$\mathcal{I}_P \cong \mathcal{O}_X(-P).$$

Definition A.3.6. Let X be a quasi-projective variety. Similar to the techniques of Section 16.2 one defines the sheaf of Kähler differentials Ω^1_X together with a K-linear map

$$d: \mathcal{O}_X \to \Omega^1_X$$

satisfying the product rule $d(fg) = g df + f dg$. If $x_1, \ldots, x_n$ generate $\mathfrak{m}_{X,p} \subset \mathcal{O}_{X,p}$ then $dx_1, \ldots dx_n$ are generators of $\Omega^1_{X,p}$ as an $\mathcal{O}_{X,p}$-module. Moreover, if X is non-singular then Ω^1_X is locally free of rank $n = \dim X$. If X is non-singular of dimension n then one defines the **dualizing sheaf**

$$\omega_X = \Lambda^n \Omega^1_X$$

as the top exterior power of Ω_X^1. See [18, Section 16] and [42, Section II.8] for more details.

Proposition A.3.7. $\omega_{\mathbb{P}^n} \cong \mathcal{O}_{\mathbb{P}^n}(-n-1)$.

Proof. We compute how the section $dy_1 \wedge \ldots \wedge dy_n \in \omega_{\mathbb{P}^n}(U_0)$ extends to the other charts U_i. Since $y_j = \frac{x_j}{x_0}$ for $j \neq i$ and $y_i = \frac{1}{x_0}$ we obtain

$$dy_1 \wedge \ldots \wedge dy_i \wedge \ldots \wedge dy_n$$
$$= (\frac{dx_1}{x_0} - x_1 \frac{dx_0}{x_0^2}) \wedge \ldots \wedge \frac{-dx_0}{x_0^2} \wedge \ldots \wedge (\frac{dx_n}{x_0} - x_n \frac{dx_0}{x_0^2})$$
$$= -\frac{dx_1 \wedge \ldots \wedge dx_0 \wedge \ldots \wedge dx_n}{x_0^{n+1}}$$
$$= (-1)^i \frac{dx_0 \wedge \ldots \wedge dx_{i-1} \wedge dx_{i+1} \wedge \ldots \wedge dx_n}{x_0^{n+1}}.$$

Thus the extended section has a pole of order $n+1$ along the hyperplane H_0 at infinity, which implies $\omega_{\mathbb{P}^n} \cong \mathcal{O}_{\mathbb{P}^n}(-(n+1)H_0) \cong \mathcal{O}_{\mathbb{P}^n}(-n-1)$. □

If $C \subset X$ is a smooth subvariety of a smooth variety X and $\mathcal{I}_C \subset \mathcal{O}_X$ is the ideal sheaf of C in X then

$$0 \to \mathcal{I}_C/\mathcal{I}_C^2 \to \Omega_X^1 \otimes \mathcal{O}_C \to \Omega_C^1 \to 0$$

is a short exact sequence of locally free sheaves on C and $\mathcal{N}_{C/X} = (\mathcal{I}_C/\mathcal{I}_C^2)^*$ is the normal bundle of C in X, [42, Theorem II.8.17]. Taking exterior powers yields

$$\omega_C \cong \omega_X \otimes_{\mathcal{O}_X} \wedge^c \mathcal{N}_{C/X},$$

where $c = \mathrm{codim}_X C$ denotes the codimension of C in X.

In the special case when $C \subset X$ is a smooth codimension 1 subvariety of a smooth variety X we have

$$\mathcal{I}_C \cong \mathcal{O}_X(-C)$$

and

$$(\mathcal{I}_C/\mathcal{I}_C^2)^* \cong \mathcal{O}_X(C) \otimes \mathcal{O}_C \cong \mathcal{O}_C(C).$$

This implies

Theorem A.3.8 (Adjunction sequence). *Let $C \subset X$ be a smooth codimension 1 subvariety of a smooth variety X. Then*

$$0 \to \omega_X \to \omega_X(C) \to \omega_C \to 0$$

is a short exact sequence of coherent sheaves on X. □

A.4 Intersection theory of curves on smooth projective surfaces

Lemma A.4.1. *Let $C \subsetneq X$ be a curve in a smooth projective variety X of dimension $\dim X \geq 2$ and let $H = \sum n_i P_i$ be an effective divisor on X, whose support $\operatorname{supp} H = \bigcup_i P_i$ contains no component of C. Let $p \in C \cap \operatorname{supp} H$ be a point. Let $\pi \colon \tilde{C} \to C$ be a resolution of singularities of C. There exists an affine neighborhood U of $p \in X$ such that*

1. *H is defined by a single equation $g \in \mathcal{O}_X(U)$,*
2. *p is the only point of $C \cap \operatorname{supp} H \cap U$,*
3. *$\mathcal{O}_{\tilde{C}}(\tilde{U})$ is a finite $\mathcal{O}_C(U)$-module, where $\tilde{U} = \pi^{-1}(U)$, and*
4. *the quotient module $\Delta = \mathcal{O}_{\tilde{C}}(\tilde{U})/\mathcal{O}_C(U)$ has support in p and is a finite-dimensional K-vector space.* □

Corollary A.4.2. *Let $C \subsetneq X$ be a curve in a smooth projective variety X and let $H = \sum n_i P_i$ be an effective divisor on X, whose support $\operatorname{supp} H = \bigcup_i P_i$ contains no component of C. Let $p \in C \cap \operatorname{supp} H$ be a point. Let $\pi \colon \tilde{C} \to C$ be a resolution of singularities of C. Let $g \in \mathcal{O}_X(U)$ be a local equation of H. Then*

$$i(C, H; p) = \sum_{q \in \pi^{-1}(p)} v_q(g),$$

where v_q denotes the discrete valuation at q of the function field of the component of $\tilde{C}$ which contains q

Proof. Consider the commutative diagram with exact columns induced by multiplication by g

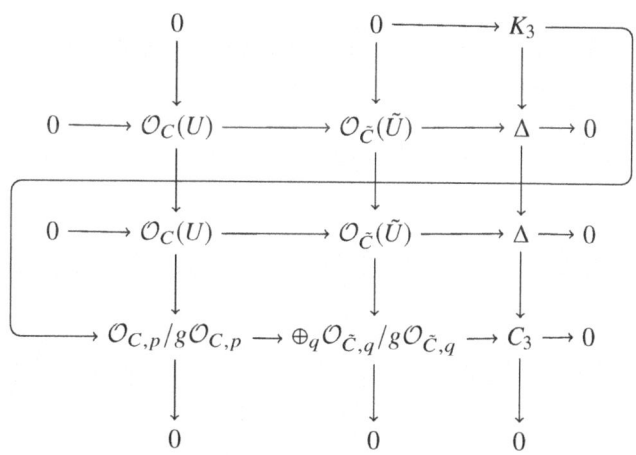

The Snake lemma A.4.18 says that the induced sequence

$$0 \to K_3 \to \mathcal{O}_{C,p}/g\mathcal{O}_{C,p} \to \oplus_q \mathcal{O}_{\tilde{C},q}/g\mathcal{O}_{\tilde{C},q} \to C_3 \to 0$$

is exact. Since the K-vector space Δ is finite-dimensional, the exactness of
$$0 \to K_3 \to \Delta \to \Delta \to C_3 \to 0$$
implies $\dim_K K_3 = \dim_K C_3$ and hence
$$i(C, H; p) = \dim_K \mathcal{O}_{C,p}/g\mathcal{O}_{C,p} = \dim_K \sum_q \mathcal{O}_{\tilde{C},q}/g\mathcal{O}_{\tilde{C},q} = \sum_{q \in \pi^{-1}(p)} v_q(g).$$

□

Example A.4.3. Consider the divisor $H = V(x^2 - 4y^3) \subset \mathbb{A}^2$ and the two curves

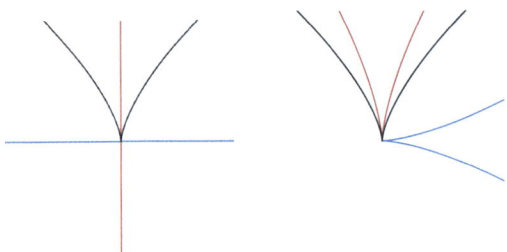

defined by $C = V(xy)$ and $C' = V((x^2 - y^3)(y^2 - x^3))$.

We compute the intersection multiplicity at the origin $o \in \mathbb{A}^2$. In the first case H intersects the two branches with multiplicity 2 and 3 respectively. Since
$$\mathcal{O}_{\mathbb{A}^2, o}/(xy, x^2 - 4y^3) \cong K[[x, y]]/x^2 - 4y^3, xy, y^4)$$
we obtain $i(C, H; o) = 5$, which coincides with $2 + 3$ by Corollary A.4.2.

In the second case we use the parametrization $\mathbb{A}^1 \to V(y^2 - x^3), t \mapsto (t^2, t^3)$ and $\mathbb{A}^1 \to V(x^2 - y^3), t \mapsto (t^3, t^2)$ for the two components. Substituting this into the equation of H we obtain $g_1 = t^4 - 4t^6$ and $g_2 = t^6 - 4t^6 = -3t^6$. Thus the intersection multiplicity is $10 = 4 + 6$. Since $\mathcal{O}_{\mathbb{A}^2, o}/(x^2y^2 - x^2y^3 - x^3y^2 + x^3y^3, x^2 - 4y^3) \cong K[[x, y]]/(x^2 - 4y^3, 4y^5 - 4y^6 - 4xy^5 + 4xy^6)$ we obtain
$$i(C', H; o) = \dim_K K[[x, y]]/(x^2, y^5) = 2 \cdot 5 = 10,$$
as predicted.

Corollary A.4.4. *Let $C \subsetneq X$ be a curve in a smooth projective variety X and let H_1 and H_2 be two effective divisors on X, whose supports contain no component of C. If $H_1 \sim H_2$ then*
$$\sum_{p \in C} i(C, H_1; p) = \sum_{p \in C} i(C, H_2; p).$$

Proof. Let $f \in K(X)$ be a rational function whose divisor is $(f) = H_1 - H_2$. Then f restricts to a rational function $f_i \in K(C_i)$ on each irreducible component C_i of C. On the desingularization $\tilde{C}_i \subset \tilde{C}$ the rational function f_i has the same number of poles and zeroes by 14.2.2. Hence the result follows from Corollary A.4.2. □

A.4 Intersection theory of curves on smooth projective surfaces

Remark A.4.5. The corollary allows us to define the **intersection number** between an arbitrary curve C and an arbitrary divisor D on a smooth projective variety X of dimension $\dim X \geq 2$.

$$C.D = \sum_{p \in C} i(C, D_1; p) - \sum_{p \in C} i(C, D_2; p),$$

where D_1, D_2 are two effective divisors whose supports contain no component of C such that their difference is linearly equivalent to D, i.e., $D \sim D_1 - D_2$. By Corollary A.4.4 this does not depend on the choice of D_1 and D_2.

If X is smooth projective surface and $D = \sum_i n_i C_i$ a divisor, then the prime components C_i are also irreducible curves. In particular, for $C \subset X$ a curve in a smooth projective surface X the self-intersection number $C.C$ is well-defined.

Proposition A.4.6. *Let $\sigma: X \to Y$ be the blow up of a smooth projective surface at a point $p \in Y$, and let $E = \sigma^{-1}(p)$ be the exceptional curve. Then*

$$E^2 = E.E = -1.$$

Proof. Let H be a smooth hyperplane section of X through p. The equation of H defines on X the divisor $\sigma^* H = E + H'$, where H' denotes the strict transform of H and E is the exceptional curve. Let H_1 be another hyperplane section of Y which does not pass through p. Then $H_1' \sim H'$ holds. Hence

$$0 = E.H_1' = E.(E + H') = E^2 + E.H' = E^2 + 1 \text{ and } E^2 = -1$$

holds, since E and H' intersect transversally. □

Remark A.4.7. With notation as in Proposition A.4.6 let $C \subset Y$ be a curve with multiplicity m at p. Then the equation of C defines on X the divisor

$$\sigma^* C = mE + C',$$

where C' denotes the strict transform of C. The self-intersection number of C' satisfies

$$C'.C' = C^2 - m^2.$$

Indeed, $C^2 = (\sigma^* C)^2 = (mE + C')^2 = -m^2 + 2m^2 + (C')^2$ since $C'.E = m$. Thus for $m = 1$ the self-intersection number drops by 1.

Definition A.4.8. Let X be a smooth projective variety. Choose a multiple of a hyperplane H such that $H^0(X, \omega_X(aH)) \neq 0$ and let W be the divisor of zeroes of a section of $s \in H^0(X, \omega_X(aH))$. Then a **canonical divisor** on X is defined by

$$K_X = W - aH.$$

The divisor class of K_X is well-defined.

Proposition A.4.9. *Let $\sigma \colon X \to Y$ be the blow-up of a smooth projective surface at a smooth point with exceptional divisor E. Then $K_X = \sigma^* K_Y + E$.*

Proof. Let $s \in H^0(Y, \omega_Y(aH))$ be a section which does not vanish at p and let x_1, x_2 be generators of the maximal ideal $\mathfrak{m}_{Y,p}$. Then locally at p the section s is of the form $s = f \, dx_1 \wedge dx_2$, where $f \in \mathcal{O}_{Y,p}$ does not vanish at p. Substituting $(x_1, x_2) = (u, uw)$ the equations for the blow-up we obtain

$$\sigma^* s = (f \circ \sigma) \cdot u \, du \wedge dw,$$

which vanishes additionally with multiplicity 1 along the exceptional divisor $E = V(u)$. $\square$

Proposition A.4.10. *Let $C \subset \mathbb{P}^2$ be an irreducible curve of degree d with ordinary singularities at $p_1, \ldots, p_s$ of multiplicities r_i. Let $\pi \colon X \to \mathbb{P}^2$ be the blow-up of $\mathbb{P}^2$ at these points and let $E_1, \ldots, E_s$ denote the exceptional curves. The strict transform C' is a divisor of class*

$$C' \sim dL - (r_1 E_1 + \ldots + r_s E_s),$$

where $L \subset X$ denotes the preimage of a general line in $\mathbb{P}^2$ and

$$\omega_{C'} \cong \mathcal{O}_{C'}((d-3)L - (r_1 - 1)E_1 + \ldots + (r_s - 1)E_s).$$

Proof. The canonical sheaf $\omega_{\mathbb{P}^2}$ on $\mathbb{P}^2$ is $\mathcal{O}(-3)$ by A.3.7. Hence $K_X \sim (-3)L + E_1 + \ldots + E_s$. Since the equation of C has multiplicity r_i at p_i, the total transform of C is $\pi^* C = C' + \sum_i r_i E_i \sim dL$, which proves the first formula. The second formula follows from the adjunction sequence

$$0 \to \omega_X \to \omega_X(C') \to \omega_{C'} \to 0$$

since $\omega_X(C') \cong \mathcal{O}_X(K_X + C')$.

Note that the adjunction formula

$$2g - 2 = \deg \omega_{C'} = C'.(C' + K_X) = d(d-3) - \sum_i r_i(r_i - 1)$$

follows once more from the intersection theory on X. $\square$

We are now ready to understand the crucial steps in the proof of Theorem 16.3.4. Let X be the blow-up of $\mathbb{P}^2$ in the singularities of our plane model of degree d with only ordinary singular points of multiplicities $r_1, \ldots, r_s$. Let C' denote the strict transform of the plane curve. Then C' is smooth and $C' \sim dL - (r_1 E_1 + \ldots + r_s E_s)$. Using the identification $H^0(X, aL - \sum_i (r_i - 1)E_i) = L(a; (r_1 - 1)p_1, \ldots, (r_s - 1)p_s)$ we see that the assertion follows from the vanishing $H^1(X, \omega_X((a - d + 3)L)) = 0$ and the long exact cohomology sequence of the adjunction sequence twisted:

$$0 \to \omega_X((a-d+3)L)) \to \mathcal{O}_X(aL - \sum_i (r_i - 1)E_i) \to \omega_{C'}((a-d+3)L) \to 0.$$

Since $E_i \cong \mathbb{P}^1$ we have $\omega_{E_i} \cong \mathcal{O}_{\mathbb{P}^1}(-2)$ by Proposition A.3.7 and $\omega_{E_i}(-E_i) \cong \mathcal{O}_{\mathbb{P}^1}(-1)$ by Proposition A.4.6. Using the adjunction sequence for the E_i

$$0 \to \omega_X((a-d+3)L - E_i)) \to \omega_X((a-d+3)L)) \to \omega_{E_i}(-E_i) \to 0,$$

we obtain $H^1(X, \omega_X((a-d+3)L) - E_i)) \cong H^1(\omega_X((a-d+3)L))$ from the vanishing $H^0(\mathbb{P}^1, \mathcal{O}_{\mathbb{P}^1}(-1)) = H^1(\mathbb{P}^1, \mathcal{O}_{\mathbb{P}^1}(-1)) = 0$. Repeating this argument for all exceptional divisors we obtain

$$H^1(\omega_X((a-d+3)L)) \cong H^1(X, \omega_X((a-d+3)L - \sum_{i=1}^s E_i)) \cong H^1(X, \mathcal{O}_X((a-d)L)).$$

The final step is $H^1(X, \mathcal{O}_X((a-d)L)) \cong H^1(\mathbb{P}^2, \mathcal{O}_{\mathbb{P}^2}(a-d)) = 0$, where the first equality follows from the facts $\pi_* \mathcal{O}_X = \mathcal{O}_{\mathbb{P}^2}$ and $R^i \pi_* \mathcal{O}_X = 0$ for $i \geq 1$ [42, Proposition V.3.4] and the Leray spectral sequence [42, Exercise III.8.1] or [30, Theorem II,4.17.1]. Here π_* is the direct image functor and $R^i \pi_*$ are its derived functors [42, Section III.1]. Note that $R^i \pi_* \mathcal{O}_X = 0 \Rightarrow R^i \pi_* \mathcal{O}_X((a-d)L) = 0$ for $i \geq 1$, because π is an isomorphism in a neighborhood of L. Hence we have

$$\begin{aligned}
H^1(X, \mathcal{O}_X((a-d)L)) &\cong H^1(\mathbb{P}^2, \pi_*(\mathcal{O}_X(L(a-d)))) & \text{by Leray's spectral sequence} \\
&\cong H^1(\mathbb{P}^2, \pi_* \mathcal{O}_X \otimes \mathcal{O}_{\mathbb{P}^2}(a-d)) & L \text{ is disjoint from the } E_i\text{'s} \\
&\cong H^1(\mathbb{P}^2, \mathcal{O}_{\mathbb{P}^2}(a-d)) & \text{since } \pi_* \mathcal{O}_X = \mathcal{O}_{\mathbb{P}^2} \\
&= 0 & \text{by Theorem A.2.4.}
\end{aligned}$$

□

In the language of coherent sheaves, the Riemann Roch theorem is usually broken into two parts. The first part is a formula for the Euler characteristic of $\mathcal{O}_X(D)$, which in the case of a smooth irreducible projective curve reads

$$\chi(C, \mathcal{O}_C(D)) = h^0(C, \mathcal{O}_C(D)) - h^1(C, \mathcal{O}_C(D)) = \deg D + 1 - g.$$

The second part is Serre duality, which in the case of a curve implies our formula 16.3.1 from the formula above: We have $h^1(C, \mathcal{O}_C(D)) = h^0(C, \omega_C(-D)) = \ell(W-D)$ and $h^0(C, \mathcal{O}_C(D)) = \ell(D)$ because $H^0(C, \mathcal{O}_C(D)) = L(D)$ and $H^0(C, \omega_C(-D)) \cong L(W-D)$ hold, where $W = K_C$ is a canonical divisor on C.

Theorem A.4.11 (Serre duality). *Let X be a smooth projective variety of dimension n and D a divisor on X. There are perfect pairings*

$$H^i(X, \mathcal{O}_X(D)) \times H^{n-i}(X, \omega_X(-D)) \to H^n(X, \omega_X) \cong K.$$

Proof. See [42, Theorem III.7.6] for a more general version. □

Theorem A.4.12 (Riemann-Roch on surfaces, [42, Theorem V.1.6]). *Let X be a smooth projective surface and D a divisor on X. Then*

$$\chi(X, \mathcal{O}_X(D)) = \frac{1}{2} D.(D - K_X) + \chi(X, \mathcal{O}_X).$$

Remark A.4.13. In the case when D is a smooth irreducible curve of genus g this formula follows from the long exact cohomology sequence of

$$0 \to \mathcal{O}_X \to \mathcal{O}_X(D) \to \mathcal{O}_D(D) \to 0,$$

Riemann-Roch for curves and the adjunction formula $2g - 2 = D.(D + K_X)$. We have

$$\chi(X, \mathcal{O}_X(D)) = \chi(D, \mathcal{O}_D(D)) + \chi(X, \mathcal{O}_X)$$
$$= D^2 + 1 - g + \chi(\mathcal{O}_X)$$
$$= \frac{1}{2} D.(D - K_X) + \chi(X, \mathcal{O}_X).$$

Riemann-Roch formulas on surfaces were discovered by the Italian school of algebraic geometry in the 19th century. They could understand $h^2(X, \mathcal{O}_X(D)) = \ell(K_X - D)$ in terms of divisors, but $h^1(X, \mathcal{O}_X(D))$, which they called the index of speciality of D, remained mysterious before the introduction of sheaf cohomology. They called $p_g = h^2(X, \mathcal{O}_X) = h^0(X, \omega_X)$ and $q = h^1(X, \mathcal{O}_X) = h^1(X, \omega_X)$ the **geometric genus** and the **irregularity** of X respectively. These turned out to be birational invariants by Theorem 12.2.2 and their invariance under blow-ups, see Exercise A.4.20.

Theorem A.4.14 (Castelnuovo's contraction criterion). *Let X be a smooth projective surface and $E \subset X$ an irreducible curve with $E^2 = E.K_X = -1$. Then there exists a smooth projective algebraic surface Y and a morphism $\pi \colon X \to Y$ with $\pi(E) = p$ a point such that π coincides with the blow-up of Y at p.*

We call an irreducible curve E with $E.K_X = E^2 = -1$ a (-1)-**curve** on X. Note that E has genus 0 by the adjunction formula: $\deg \omega_E = E.(E + K_X) = -2$ and hence $E \cong \mathbb{P}^1$.

Proof. See [42, Theorem V.5.7]. □

Remark A.4.15. Let X be a smooth projective surface, and let $\pi \colon X \to Y$ be a birational morphism which induces an isomorphism $X \setminus \pi^{-1}(p) \cong Y \setminus \{p\}$, where p is a not necessarily smooth point of Y such that the completion $\widehat{\mathcal{O}}_{Y,p}$ is a domain. Assume that π is not an isomorphism. Then $E = \bigcup_{i=1}^r E_i = \pi^{-1}(p)$ is a connected collection of curves on X such that the intersection matrix

$$(E_i.E_j)_{i,j=1,\ldots,r}$$

is negative definite, [42, Corollary III.11.4 and Theorem V.1.9] and [65, Section 1, page 6].

A.4 Intersection theory of curves on smooth projective surfaces

When $K = \mathbb{C}$ we have the following converse.

Theorem A.4.16 (Grauert, [31]). *Let X be a smooth complex 2-dimensional manifold and let $E = \bigcup_{i=1}^{r} E_i$ be a compact connected collection of 1-dimensional complex submanifolds E_i of X. If the intersection matrix $(E_i.E_j)$ is negative definite, then there exists a 2-dimensional complex space Y with a point p and a holomorphic map $\pi: X \to Y$ which contracts E to a point $p \in Y$ and restricts to a biholomorphic map $X \setminus E \to Y \setminus \{p\}$.*

Note however, that even in the case when X is the underlying holomorphic manifold of a projective algebraic surface, the resulting surface Y might not be the underlying analytic space of a singular projective algebraic surface. For an example, due to Hironaka, see [42, Example V.5.7.3].

Exercise A.4.17. Let $C \subsetneq X$ be a curve in a smooth projective variety X and let P be a prime divisor on X. Prove that there exist effective divisors D_1, D_2 whose support does not contain any component of C such that $P \sim D_1 - D_2$.

Exercise A.4.18 (Snake Lemma). Given a commutative diagram

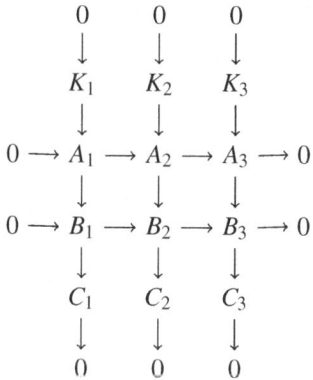

of abelian groups with exact rows and columns. Prove that there is an induced six-term exact sequence

$$0 \to K_1 \to K_2 \to K_3 \to C_1 \to C_2 \to C_3 \to 0$$

between the kernels and cokernels.

Exercise A.4.19. Let $\pi: X \to \mathbb{P}^2$ be the blow-up in s points, and let $L, E_1, \ldots, E_s$ denote the total transform of a line in $\mathbb{P}^2$ and the exceptional divisors. Prove: Every divisor D on X is linearly equivalent to a divisor of type $aL - \sum_{i=1}^{s} b_i E_i$. The integers $a, b_1, \ldots, b_s$ are uniquely determined by D.

Exercise A.4.20. Let $\sigma: X \to Y$ be the blow-up of a smooth projective surface Y at a point p. Prove: $H^i(X, \omega_X) \cong H^i(Y, \omega_Y)$ for all i.

Appendix B
Code for Macaulay2 computations

In this appendix we provide Macaulay2 code for the solutions of those exercises which require computer algebra. The code can also be downloaded from

https://www.math.uni-sb.de/ag-schreyer/index.php/computeralgebra.

In emacs, the code has syntax highlightening. Of course this code gives no complete solutions. The explanation, why the execution of this code proves whatever it is supposed to prove is not given. If the exercise involves complicated equations, it is convenient for the students and the teachers to use cut and paste to avoid cumbersome typing.

The computation of Section 13.5 needs the package HilbertSchemeStrata.m2, which can also be found at

https://www.math.uni-sb.de/ag-schreyer/index.php/computeralgebra.

B.1 Solutions to Exercises in Chapter 1, 2, 3, 4 and 6

Exercise 1.3.28

```
kk=QQ
R=kk[x,y]
f1=28*x^3+63*x^2*y+14*x*y^2+7*y^3+63*x*y+28*y^2+
    28*x+49*y+58;
f2=109*x^3+99*x^2*y+41*x*y^2+16*y^3+27*x^2+108*x*y+
    46*y^2+28*x+103*y+88;
f3=35*x^3+30*x^2*y+13*x*y^2+5*y^3+9*x^2+33*x*y+
    14*y^2+8*x+32*y+32
gens gb ideal(f1,f2,f3)
-- => no solution
f1=5*x^3+8*x^2*y+7*x*y^2+7*y^3-7*x^2+x*y-
    6*y^2+x-2*y-14
```

```
f2=3*x^3+15*x^2*y+4*x*y^2+y^3-3*x^2-10*x*y+
   3*y^2-4*x-3*y-6;
f3=8*x^3+14*x^2*y+8*x*y^2+4*y^3-9*x^2-6*x-4*y-15
gens gb ideal(f1,f2,f3)
sub(ideal(f1,f2,f3),{x=>1,y=>1})
-- => p=(1,1) in AA2(CC) is the only solution
```

Exercise 1.4.11

```
R1=QQ[x,y,z]
f1=x^2-y*z+z
f2 = y^3-2*y^2+y-z^2
I=ideal(f1,f2)
gens gb I
R2=QQ[y,z,x,MonomialOrder=>Lex]
J=sub(I,R2)
L=flatten entries gens gb J
reverse L
```

Exercise 2.2.8

```
R=QQ[t]
S=QQ[x,y]
param=matrix{{t^2+1,t^3+t}}
phi=map(R,S,param)
I=ker phi
sub(I,{x=>0,y=>0})
-- the point p=(0,0) lies in V(I) but
-- is not an image point because the x-coordinate
-- of a point in the parametrization is always >=1.
```

Exercise 3.2.14

```
R=QQ[x,y,z]
I=ideal(x*y,x*z,y*z)
primaryDecomposition I
I2=I^2
cI2=primaryDecomposition(I2)
apply(cI2,q->(radical q,q))
q012=intersect apply(3,i->cI2_i)
J=trim ideal (gens q012 %I2)
I2:J
ass(I2)
ass(R^1/I2)
```

Exercise 4.4.7

```
p=nextPrime random 10^3
d=8
```

B.2 Solutions to Exercises in Chapter 8, 9, 10, 11 and 12

```
Fp=ZZ/p
R=Fp[x,y]
f=random(d,R)
cf=decompose ideal f
apply(cf,c->degree c)
elapsedTime tally apply(1000,i->(f=random(d,R);
cf=decompose ideal f;
L=apply(cf,c->degree c);
     member(1,L)
))
d=20
p=nextPrime 10^4
Fp=ZZ/p
R=Fp[x,y]
elapsedTime tally apply(1000,i->(f=random(d,R);
cf=decompose ideal f;
L=apply(cf,c->degree c);
     member(1,L)
))
1-exp(-1)
```

Exercise 6.4.9

```
R=QQ[e_2,e_3]
T=QQ[t_1,t_2]
m=matrix{{t_1*t_2-(t_1+t_2)^2,t_1*t_2*(t_1+t_2)}}
RT=QQ[gens R|gens T]
I=ideal(sub(vars R,RT)-sub(m,RT))
eliminate(I,t_2)
J=I+ideal(t_1-t_2)
disc=eliminate(J,{t_1,t_2})
ramifi=(eliminate(disc+I,{e_2,e_3}))
factor ramifi_0
factor (radical ramifi)_0
```

B.2 Solutions to Exercises in Chapter 8, 9, 10, 11 and 12

Exercise 8.1.5

```
R=QQ[x,y,z]
f=y^2*z+x^3-x^2*z
(cos(pi/6),sin(pi/6))
(sqrt 3)/2
p0=matrix{{-866/1000,-1/2,1}}
p1=matrix{{866/1000,-1/2,1}}
```

```
  p2=matrix{{0,1,1}}
  p3=1/3*sum({p0,p1,p2})
  normp0=sqrt(1/2^2+(866/1000)^2)
  -- fairly good approximation of 3 points
  -- on the unit circle
  A=transpose(1/3*(p0||p1||p2))
  phi=map(R,R,vars R*A)
  g=phi f
  sub(g,{z=>1})
```

Exercise 8.1.7

```
  R=QQ[x,y,z]
  S2=ideal(x^2+y^2+z^2-1)
  m=matrix{{x*y,x*z,y*z}}
  phi=map(R/S2,R,m)
  J=ker phi
  cTermsJ=primaryDecomposition ideal terms  J_0
  apply(cTermsJ,c-> radical c)
```

Exercise 8.2.6

```
  R=QQ[x_0..x_3]
  Iaff= ideal(x_2-x_1^3,x_3-x_1^4)
  Iproj=homogenize(Iaff,x_0)
  I0=ideal(homogenize(Iaff_0,x_0),homogenize(Iaff_1,x_0))
  Irest=I0:Iproj
  Ired=radical Irest
  degree I0 ,degree Iproj, degree Irest
  cI0=primaryDecomposition I0
  apply(cI0,c-> radical c)
  ass(trim (Iproj+ideal x_0))
```

Exercise 8.3.6

```
  kk=ZZ/nextPrime 10^2
  N=6
  netList apply(toList(2..N),n->(
          x=symbol x;
          R=kk[x_1..x_n];
          I=ideal gens R;
          minimalBetti I))
  _*
       +----------------------+
       |         0 1 2        |
  o3 = |total: 1 2 1          |
       |    0: 1 2 1          |
       +----------------------+
```

B.2 Solutions to Exercises in Chapter 8, 9, 10, 11 and 12   275

```
       |         0 1 2 3           |
       |total: 1 3 3 1             |
       |    0: 1 3 3 1             |
       +---------------------------+
       |         0 1 2 3 4         |
       |total: 1 4 6 4 1           |
       |    0: 1 4 6 4 1           |
       +---------------------------+
       |         0 1  2  3 4 5     |
       |total: 1 5 10 10 5 1       |
       |    0: 1 5 10 10 5 1       |
       +---------------------------+
       |         0 1  2  3  4 5 6|
       |total: 1 6 15 20 15 6 1|
       |    0: 1 6 15 20 15 6 1|
       +---------------------------+
*-
netList apply(toList(2..N),n->(
        x=symbol x;
        R=kk[x_1..x_n];
        I=ideal gens R;
        minimalBetti I^2))
-*
       +---------------------------+
       |         0 1 2             |
o4 =   |total: 1 3 2               |
       |    0: 1 . .               |
       |    1: . 3 2               |
       +---------------------------+
       |         0 1 2 3           |
       |total: 1 6 8 3             |
       |    0: 1 . . .             |
       |    1: . 6 8 3             |
       +---------------------------+
       |         0 1  2  3 4       |
       |total: 1 10 20 15 4        |
       |    0: 1 .  .  . .         |
       |    1: . 10 20 15 4        |
       +---------------------------+
       |         0 1  2  3  4 5    |
       |total: 1 15 40 45 24 5     |
       |    0: 1 .  .  .  . .      |
       |    1: . 15 40 45 24 5     |
       +---------------------------+
       |         0 1  2  3  4 5 6|
```

```
    |total: 1 21 70 105 84 35 6|
    |    0: 1  .  .   .  .  . .|
    |    1: . 21 70 105 84 35 6|
    +--------------------------+
*_
```

Exercise 9.3.6

```
-- 1)
R=QQ[x,y,t]
f=-3*x^5-2*x^4*y-3*x^3*y^2+x*y^4+3*y^5+6*x^4+7*x^3*y+
  3*x^2*y^2-2*x*y^3-6*y^4-3*x^3-5*x^2*y+x*y^2+3*y^3
singf=ideal f+ ideal jacobian ideal f
singfrad=ideal gens gb trim radical singf
-- the pencil of conics
Dt=t*(x^2-x)+(y^2-y)
movingPt1=(gens saturate (ideal(f,Dt),singfrad))
--select the equation which are linear in x and y
m3x5=contract( transpose gens (ideal(x,y))^2,movingPt1)
s=syz(m3x5,DegreeLimit=>0)
linearEqs=transpose (movingPt1*s)
A=contract( gens ideal(x,y),linearEqs)
constantTerms=linearEqs-A*matrix{{x},{y}}
T=QQ[t]
cramersRule1=exteriorPower(2,sub(A|constantTerms,T))
xt=(cramersRule1)_(0,2)/(cramersRule1)_(0,0)
yt=-(cramersRule1)_(0,1)/(cramersRule1)_(0,0)
sub(f,matrix{{xt,yt,t}})

--2)
use R
f=-2*x^4-2*x^3*y+x^2*y^2+3*x*y^3+4*y^4+4*x^3+x^2*y-
  4*x*y^2-8*y^3-2*x^2+x*y+4*y^2
singf=ideal f+ ideal jacobian ideal f
singfrad=ideal gens gb trim radical singf
basePts=ideal(x^2-x,y^2-y)
movingPt1=(gens saturate (ideal(f,Dt),basePts))
--select the equation which are linear in x and y
m3x5=contract( transpose gens (ideal(x,y))^2,movingPt1)
s=syz(m3x5,DegreeLimit=>0)
linearEqs=transpose (movingPt1*s)
A=contract( gens ideal(x,y),linearEqs)
constantTerms=linearEqs-A*matrix{{x},{y}}
cramersRule1=exteriorPower(2,sub(A|constantTerms,T))
use T
xt=(cramersRule1)_(0,2)/(cramersRule1)_(0,0)
```

B.2 Solutions to Exercises in Chapter 8, 9, 10, 11 and 12   277

```
yt=-(cramersRule1)_(0,1)/(cramersRule1)_(0,0)
sub(f,matrix{{xt,yt,t}})
```

Exercise 9.3.7

```
R=QQ[y,z,x,MonomialOrder=>Lex]
f1=x^2-y*z+z
f2 = y^3-2*y^2+y-z^2
I=ideal(f1,f2)
J=ideal gens gb I
J_0
T=QQ[t]
RT=R**T
m1=matrix{{y,x*t,x}}
p1=factor sub(J_0,m2)
first (toList p1)_1
m2=matrix{{y,t*(t^5-t),t^5-t}}
J2=trim sub(J,m2)
factor J2_0
use T
m3=matrix{{t^4,t*(t^5-t),t^5-t}}
sub(I,m3)
ker map(T,R,m3)==I
```

Exercise 9.3.8

```
P1=QQ[t_0,t_1]
P2=QQ[x_0..x_2]
m=matrix{{t_0^4,t_0^3*t_1-t_0*t_1^3,t_1^4}}
phi=map(P1,P2,m)
I=ker phi
singI=ideal jacobian I
csingI=decompose singI
apply(csingI,c->saturate trim phi c)
-- => there is an ordinary node at (1:0:1)
--      with two real tangents
--      and two isolated nodes (1:+-sqrt 2:-1)
z=exp(sqrt(-1)*pi/4)
z-z^3,z^7-z^21
z^3-z^9,z^5-z^15
-- the points (1:z) and (1:z^7) map to the same point
-- (1:z-z^3:-1)
-- (1:z^3) and (1:z^5) map both to (1:z^3-z:-1)
```

Exercise 9.4.7

```
QQ[x_0..x_4]
```

```
m3x3=matrix apply(3,i->apply(3,j->x_(i+j)))
P3=QQ[x_0..x_3]
m3x3=sub(m3x3,P3)
f3=det m3x3
f2=det m3x3^{0,1}_{0,1}
I=ideal(f2,f3)
ideal gens gb I
p1=radical I
M=P3^1/I
M1=P3^1/p1
hom1=Hom(M1,M)
betti hom1
phi1=homomorphism hom1_{0}
prune ker phi1
Q1=coker phi1
p2=ideal(x_1,x_2,x_3)
N=P3^1/p2
hom2=Hom(N,Q1)
betti hom2
phi2=homomorphism hom2_{0}
prune ker phi2
Q2= coker phi2
prune Q2==M1
L={image phi1,image phi2,Q2};
sum(L,c->hilbertPolynomial c)==hilbertPolynomial M
-- second filtration
hom2b=Hom(M1,Q1)
betti hom2b
phi2b=homomorphism hom2b_{0}
prune ker phi2b
Q2b=prune coker phi2b
hilbertPolynomial Q2b
ass(Q2b)
degree Q2b, dim Q2b
ass(M)
apply(L,c->ass(c))
```

Exercise 10.1.9

```
R=ZZ[x_0..x_5]
I=ideal (x_0*x_1*x_2, x_1*x_2*x_3, x_0*x_1*x_4,
    x_0*x_3*x_4, x_2*x_3*x_4, x_0*x_2*x_5, x_0*x_3*x_5,
    x_1*x_3*x_5, x_1*x_4*x_5, x_2*x_4*x_5)
F2=ZZ/2
R2=F2[x_0..x_5]
minimalBetti sub(I,R2)
```

B.2 Solutions to Exercises in Chapter 8, 9, 10, 11 and 12

```
fI=res I
betti fI
fI.dd_4
fI.dd_3
cI=decompose I
use R
coordPts=c->select(toList(0..5),i-> not x_i%c==0)
apply(cI,c->coordPts c)
_*
o14 = {{3, 4, 5}, {0, 4, 5}, {2, 3, 5}, {1, 2, 5},
       -----------------------------------------
       {0, 1, 5}, {1, 3, 4}, {1, 2, 4}, {0, 2, 4},
       -----------------------------------------
       {0, 2, 3}, {0, 1, 3}}
*_
```

Exercise 10.5.16 and Exercise 10.5.17

```
P4=QQ[x_0..x_4]
eqs=matrix{{x_2^3-x_1^2*x_3+x_1*x_2*x_4-x_1*x_3*x_4-
x_2*x_4^2-x_1*x_2,
x_1*x_2^2-x_1*x_3^2+2*x_2*x_3*x_4-x_3^2*x_4-x_2*x_3,
x_1^3-x_1*x_2*x_3+x_2^2*x_4+x_1*x_4^2-x_4^3-x_1*x_4,
x_1^2*x_3-x_2*x_3^2+x_1*x_2*x_4+2*x_3*x_4^2-x_3*x_4}}
eqsh=homogenize(eqs,x_0)
IX=ideal eqsh
fX=res IX
p0=matrix{{1,0,0,0,0}}
affChart=matrix{{1,x_1,x_2,x_3,x_4}}
apply(5,i->rank sub(fX.dd_(i+1),p0))
betti fX
m4x9=sub(fX.dd_2,affChart)
eqs'=sub(fX.dd_1,affChart)
ideal eqs== ideal eqs'
m4x4=m4x9*syz transpose sub(fX.dd_3,p0)
m4x4%(ideal(x_1..x_4))^2
-- => the 4 syzygies between the lead terms
eqs%(ideal(x_1..x_4))^3
--   with respect to >_ldrlex extend
eqs'*m4x4==0
-- => the tangent cone is
Tp0X=ideal (eqs%(ideal(x_1..x_4))^3)
decompose Tp0X

degree IX==5
J0=ideal(x_1+x_2,x_3+x_4)
```

```
J1=ideal(x_1+x_2-x_0,x_3+x_4-x_0)
c0=decompose(IX+J0)
apply(c0,c->(degree c, dim c))
sum(c0,c-> degree c)==4
c1=decompose(IX+J1)
apply(c1,c->(degree c, dim c))
sum(c1,c-> degree c)==5
degree(IX+J0)==6
degree(IX+J1)==5
threePts=c0_1
localContr=(IX+J0):threePts
Ip0=radical localContr
degree localContr
```

Exercise 11.4.6

```
P5=QQ[w_0..w_5]
Delta'=genericSymmetricMatrix(P5,w_0,3)
Delta=sub(Delta',{w_2=>w_3,w_3=>w_2})
I=minors(2,Delta)
p=matrix{{0,1_P5,0,1,1,0}}
ISecant=ideal det Delta
sub(ISecant,p)==ideal 1_P5
Ip=ideal(vars P5 *syz p)
vars P5%Ip
w_4|gens Ip
A=diff(transpose vars P5,w_4|gens Ip)
coordChange=vars P5*inverse A
I'=sub(I,coordChange)
IprojectedV=eliminate(I',w_0)
P4=QQ[support IprojectedV]
J=sub(IprojectedV,P4)
betti res J
codim J
singJ=J+minors(codim J,jacobian J);
radical singJ == ideal gens P4
-- => J is smooth
eqs=gens trim I'
eqsLinInw0=eqs_{0..4}
linPart=contract(w_0,eqsLinInw0)
constantPart=eqsLinInw0-w_0*linPart
m2x5=constantPart||linPart
matrix{{1,w_0}}*m2x5==eqsLinInw0
W0=apply(5,i->m2x5_(0,i)/m2x5_(1,i))
radical ideal m2x5^{1}
-- => W0 is every where defined on V(J)
```

B.2 Solutions to Exercises in Chapter 8, 9, 10, 11 and 12   281

```
--    the projection is an isomorphism
gens minors(2,m2x5)%IprojectedV
minors(2,m2x5)==IprojectedV
m2x5'=sub(m2x5,P4)
m2x5''=sub(m2x5',P4/J)
i=0
m0=matrix{{-m2x5''_(0,i)/m2x5''_(1,i)}|gens P4}
phi=map(frac (P4/J),P5,m0)
I''=ker phi
I''==I'
```

Exercise 11.6.8

```
p=nextPrime random(10^3)
kk=ZZ/p
P1=kk[x_0,x_1]
P2=kk[y_0..y_2]
d=6
phi=map(P1,P2,random(P1^1,P1^{3:-d}))
C=ker phi;
genus C
singC=ideal jacobian C;
degree singC, degree radical singC
singPts=primaryDecomposition singC
apply(singPts,c->degree c)
kkrationalPts=select(singPts,c->degree c==1)
if #kkrationalPts ==0 then print "choose a new curve"
hessian=diff(vars P2,transpose diff(vars P2,C_0));
p=first kkrationalPts
coordP=transpose syz transpose jacobian p
M=sub(hessian,coordP)
det M
trim minors(2,M)
a=gens trim ideal(vars P2%p)
A=contract(transpose vars P2,gens p|a)
coordChange=map(P2,P2,vars P2*inverse A)
coordChange p
localEq=sub(coordChange C_0,{y_2=>1})
tangCone=localEq%(ideal(y_0,y_1))^3
y01=matrix{{y_0,y_1}}
hessian=diff(y01,transpose diff(y01,tangCone))
not (det hessian == 0)
-- => C has an ordinary double point at p.
--Actually points p where the jacobian ideal
--is reduced, i.e., the ideal mm_p is a primary
--component of singI,
```

```
--are always ordinary double points.
--
--To detect this for the absolute primary decomposition
--one can use the following theorem:
--
--Let C be a singular plane curve with isolated
--singularities. Then the jacobian ideal singC and its
--radical have the same degree if and only if
--C has only ordinary double points as singularities

kk=ZZ/11
P1=kk[x_0,x_1]
P2=kk[y_0..y_2]
d=5
elapsedTime tally apply(100,c->(
phi=map(P1,P2,random(P1^1,P1^{3:-d}));
C=ker phi;
singC=ideal jacobian C;
(degree C,degree singC, degree radical singC)))

kk=ZZ/5
P1=kk[x_0,x_1]
P2=kk[y_0..y_2]
d=5
elapsedTime tally apply(100,c->(
phi=map(P1,P2,random(P1^1,P1^{3:-d}));
C=ker phi;
degree C))

p=nextPrime 50
kk=ZZ/p
P1=kk[x_0,x_1]
P2=kk[y_0..y_2]
dmax=8
netList apply(toList(1..dmax),d->
     elapsedTime tally apply(100,c->(
     phi=map(P1,P2,random(P1^1,P1^{3:-d}));
     C=ker phi;
     singC=ideal jacobian C;
  (degree C,degree singC, degree radical singC))))
```

Exercise 12.2.10

```
P2=QQ[x,y,z]
f=12*x^4-44*x^3*y+20*x^2*y^2+12*x*y^3-9*x^3-30*x^2*y+
   23*x*y^2-4*y^3
```

```
fh=homogenize(f,z)
q=matrix{{y*z,x*z,x*y}}
sub(q,q)
decompose (ideal(fh)+ideal q)
fd1=factor sub(fh,q)
fd=first (toList fd1)_2
-- the coordinate change from Exercise 8.1.5
p0=matrix{{-866/1000,-1/2,1}}
p1=matrix{{866/1000,-1/2,1}}
p2=matrix{{0,1,1}}
ps={p0,p1,p2}
A=1/3*transpose (p0||p1||p2)
convexCoord=map(P2,P2,vars P2*A)
circle0=x^2+y^2-z^2
apply(ps,p->sub(circle0,p)+0.0)
circle1=sub(sub(z,vars P2*inverse A),q)
circle=sub(circle1,vars P2*transpose inverse A)
apply(ps,p->sub(circle,p))

fhc=sub(convexCoord fh,{z=>1})
fdc=sub(convexCoord fd,{z=>1})
A2floats=RR[x,y]
sub(fhc,A2floats)
sub(fhc,A2floats)
sub(sub(circle,{z=>1}),A2floats)
```

B.3 The Computation in Section 13.5

The computation below uses the package HilbertSchemeStrata which can be downloaded from https://www.math.uni-sb.de/ag-schreyer/index.php/computeralgebra.

```
loadPackage("HilbertSchemeStrata")
viewHelp HilbertSchemeStrata
S=QQ[x_0..x_3]
I=ideal(x_0^2,x_0*x_1,x_1^2)
gb I
(unf,R)=unfolding(I)
gens R
base = flatteningRelations(I, unf, R)
(J,h) = removeVariables base
dim ring J
fam1 = substituteFamily(unf,S,h)
syz fam1
```

```
use S
I=ideal(x_0^2,x_0*x_1,x_0*x_2,x_1^3)
ass(I)
primaryDecomposition I
dim I, degree I, betti res I
(unf,R)=unfolding I
SR=ring unf;
transpose unf
numgens R
betti (base=flatteningRelations(I,unf,R))
(J,h) = removeVariables base
betti base, betti J

fam=substituteFamily(unf,S,h);
Sh=ring fam
#support fam, #support unf
dim ring J
dim J
J_0
unf_{2}, support J_0
J1=trim( J+ideal ((support J_0)_0))
betti(J2=trim (J:J1))
dim J1, dim J2

(J1s,h1)=removeVariables J1
betti(fam1=substituteFamily(fam,S,h1))
SR1=ring fam1; gens SR1
m=vars S|matrix{apply(numgens SR1-4,c->random(19)-9)}
fiber1=ideal sub(fam1,m)
betti res fiber1
cFib1=primaryDecomposition fiber1
apply(cFib1,c->(dim c, degree c,betti c))
E=first cFib1
singE=saturate(minors(2,jacobian E)+E)
saturate(E+last cFib1)
-- => E is a smooth plane cubic and the fiber fiber1
--    is the union of E with a disjoint point


(J2s,h2)=removeVariables J2
betti(fam2=substituteFamily(fam,S,h2))
SR2=ring fam2; gens SR2
m=vars S|matrix{apply(numgens SR2-4,c->random(19)-9)}
fiber2=ideal sub(fam2,m)
betti res fiber2
```

```
cFib2=primaryDecomposition fiber2
apply(cFib2,c->(dim c, degree c, betti c))
NC=first cFib2
singNC=saturate(minors(2,jacobian NC)+NC)
degree NC, genus NC, betti (fNC=res NC)
-- => NC=fiber2 is a smooth curve of degree 3 and genus 0,
--    i.e., a twisted cubic.
fNC.dd_2

betti(J12=trim(J1+J2))
(J12s,h12)=removeVariables J12
betti(fam12=substituteFamily(fam,S,h12))
SR12=ring fam12; gens SR12
m=vars S|matrix{apply(numgens SR12-4,c->random(19)-9)}
fiber12=ideal sub(fam12,m)
betti res fiber12
cFib12=primaryDecomposition fiber12
apply(cFib12,c->(dim c, degree c, betti c))
ass(fiber12)
C=first cFib12
singC= saturate(minors(2,jacobian C)+C)
singC== radical last cFib12
-- => C is a nodal plane cubic and the fiber12
--    has an embedded point at the node
```

B.4 Solutions to Exercises in Chapter 14 and 16

Exercise 14.1.7
```
P2=QQ[x..z]
f = 8*x^4-2*x^2*y^2+6*y^4+60*x^2*y*z-37*y^3*z-
    16*x^2*z^2+78*y^2*z^2-60*y*z^3+8*z^4
singC=ideal jacobian ideal f
P2d=QQ[u,v,w]
PSegre=P2**P2d
m1x3=sub(transpose jacobian ideal f,PSegre)
m2x3=m1x3||sub(vars P2d,PSegre);
Igraph1=minors(2,m2x3)+ideal sub(f,PSegre);
Igraph=saturate(Igraph1,sub(singC,PSegre));
degrees source gens Igraph
betti Igraph
ICdual= sub(Igraph_15,P2d)

singCdual=saturate(ideal jacobian ICdual);
```

```
betti singCdual
radSingCdual=radical singCdual
cuspsCdual=singCdual:radSingCdual
degree cuspsCdual
use P2d
g=eliminate(cuspsCdual,u)
m1x2=diff(matrix{{v,w}},gens g)
m2x2=diff(transpose matrix{{v,w}},m1x2)
eigenvalues sub(m2x2,QQ)
g_0-44548*(v+31348/44548/2*w)^2
-- => all cusps are complex
nodesCd=saturate(singCdual,cuspsCdual)
-- => two real nodes on the line u=0

hessianMat=diff(transpose vars P2,diff(vars P2,f))
hessianf=det hessianMat;
flexC=saturate(ideal(f,hessianf),singC)
degree flexC == degree cuspsCdual

nodesCd_1
quadEq=-nodesCd_1/47
disc=sub(sub(quadEq-(w+236/47/2*v)^2,v=>1),QQ)
sub(quadEq,{v=>1})==(w+236/47/2)^2+disc
factor (27648*2209)
sqrt (-disc)
wValues={-236/47/2-sqrt(-disc),-236/47/2+sqrt(-disc)}
nodesCdual=apply(wValues,c->matrix{{0,1,c}})
fdualAff=sub(ICdual,{v=>1})
use P2d
gradfd=diff(matrix{{u,w}},fdualAff)
apply(nodesCdual,n->sub(gradfd,n))
hessMatsfd=diff(matrix{{u,w}},transpose gradfd)
apply(nodesCdual,n->sub(hessMatsfd,n))
--=>
nodesCdual_0 -- is an isolated point
nodesCdual_1 -- has two real tangents
```

Exercise 16.3.28

```
P2=QQ[x..z]
f=random(4,P2)
gradf = diff(vars P2,f)
radical ideal gradf
hessf=det diff(vars P2,transpose gradf)
flex=ideal(f,hessf)
degree flex==degree radical flex
```

B.4 Solutions to Exercises in Chapter 14 and 16   287

```
--Macaulay2 random elements in QQ are rather special
apply(100,c->random QQ)
max apply(1000,c->random(ZZ,Height=>199)-99)
min apply(1000,c->random(ZZ,Height=>199)-99)
apply(10,c->random(ZZ,Height=>199)-99)
tal=tally apply(1000,c->random(ZZ,Height=>19)-9)
-- => roughly the equal distribution on the 19 values

mons4=basis(4,P2)
coeffs=matrix{apply(15,c->random(ZZ,Height=>199)-99)}
f=mons4*transpose coeffs
gradf = diff(vars P2,f)
radical ideal gradf
hessf=det diff(vars P2,transpose gradf)
flex=ideal(f,hessf)
degree flex==degree radical flex

mons4a=gens intersect(ideal(y,x^4),ideal mons4)
coeffs=matrix{apply(11,c->random(ZZ,Height=>199)-99)}
f=mons4a*transpose coeffs
gradf =transpose jacobian f
radical ideal gradf
--=> f defines a smooth curve
P2d=QQ[u,v,w]
PSegre=P2**P2d
m2x3=sub(gradf,PSegre)||sub(vars P2d,PSegre);
Igraph1=minors(2,m2x3)+ideal sub(f,PSegre);
Igraph=saturate(Igraph1,ideal(sub(vars P2,PSegre)));
degrees source gens Igraph
betti Igraph
fCdual= sub(Igraph_22,P2d)
singCd=saturate ideal jacobian fCdual;
--singCred=radical singCd
degree singCd
betti singCd
use P2d
p=ideal(u,w)
otherSings=saturate(singCd,p);
degree otherSings
28-1+(24-2)*2==degree otherSings
-- => p counts for 2 cusps (flex)
--       and one node (bitangent)
singCp=singCd:otherSings;
singCp
```

288   B Code for Macaulay2 computations

```
degree singCp
singCpLoc=sub(singCp,{v=>1})
twoMonomials=transpose  matrix{{w^2,u^2*w}}
m2x4=sub(contract(twoMonomials,gens singCpLoc),QQ)
m2x1=m2x4*syz transpose syz m2x4^{1}
parabel=(w+m2x1_(1,0)/m2x1_(0,0)/2*u^2)
trim(singCpLoc+ideal parabel^2)==singCpLoc
g=sub(fCdual,{v=>1})
thirdPower=sub(contract(w^3,g),QQ)*(parabel)^3
g1=g-thirdPower
fourthPower=sub(contract(u^4,g1),QQ)*u^4
E6sing=(thirdPower+fourthPower)
jacE6sing=trim ideal jacobian E6sing
g2=g-E6sing
g2%jacE6sing
g2%singCpLoc
-- => (Cd,p) is analytically equivalent to w^3+u^4
```

Exercise 16.4.9

```
P2=QQ[x_0..x_2]
sixPts=apply(6,i->matrix{apply(3,j->random(19)-9)})
sixIdeals=apply(sixPts,c->ideal(vars P2*syz c))
IGamma=intersect sixIdeals
IG6=gens intersect(IGamma,ideal basis(4,P2));
coeffs=transpose matrix{apply(9,c->random(19)-9)}
C=ideal(IG6*coeffs)
genus C==3
radical ideal jacobian C
-- (C, IGamma) is a pair comprising a smooth plane curve
-- of genus 3 and  6 points
betti(fGamma=res IGamma)
H=fGamma.dd_1*random(P2^{4:-3},P2^{-3})
E=ideal H+C:IGamma;
betti E, degree E
P3=QQ[y_0..y_3]
PC=P2/C;
phi=map(PC,P3,gens E);
spaceC=ker phi;
degree spaceC,genus spaceC
betti res spaceC
```

Exercise 16.5.21

```
P2t=QQ[x,y,z,t,Degrees=>{1,1,1,0}]
ft = x*y*(x+y)-3*(x^5+y^5)-2*(x^6+y^6)+
    t*(-2*x*y+12*(x^4+y^4)+x^3*y+x*y^3-20*x^2*y^2+
```

B.4 Solutions to Exercises in Chapter 14 and 16 289

```
    8*(x^5+y^5)-12*(x^4*y+x*y^4)+6*(x^3*y^2+x^2*y^3)-
    5*(x^5*y+x*y^5)-2*(x^4*y^2+x^2*y^4)+14*x^3*y^3)+
        t^2*(-12*(x^3+y^3)-2*(x^2*y+x*y^2)-8*(x^4+y^4)+
    24*(x^3*y+x*y^3)-44*x^2*y^2+
    10*(x^4*y+x*y^4)+20*(x^3*y^2+x^2*y^3)+
    10*(x^4*y^2+x^2*y^4)+24*x^3*y^3)
    Ct=homogenize(ft,z)
    singCt=ideal diff(matrix{{x,y,z}},Ct)
    p3=ideal(x+z,y+z)
    p12=ideal(x*y,x+y-2*t*z)
    p0=ideal(x,y)
    pencil=intersect(p0,p12,p3)
    m=gens intersect(pencil,(ideal(x,y,z))^(6-3))
    P5t=QQ[w_0..w_5,t,Degrees=>{6:1,0}]
    PCt=P2t/ideal Ct;
    phi=map(PCt,P5t,m|matrix{{t}})
    Jt=ker phi;

    P5=QQ[w_0..w_5]
    m0=vars P5|matrix{{0}}
    J0=ideal map(P5^1,,gens sub(Jt3,m0));
    betti res J0
    m1=vars P5|matrix{{1/10}}
    J1=ideal map(P5^1,,gens sub(Jt3,m1));
    betti res J1
```

Exercise 16.5.22

```
    p=nextPrime 10^3
    kk=ZZ/p
    R=kk[x_0..x_2]
    g=10
    r=2
    d=ceiling(g+r-g/(r+1))
    rho=g-(r+1)*(g-d+r)
    rho>=0
    delta=binomial(d-1,2)-g
    -- => g,r,d,rho and delta have
    --     the expected values as
    --     in the Exercise
    pts=apply(delta,j->random(R^1,R^3))
    Ipts=apply(pts,q->ideal(vars R*syz q));
    IGamma=intersect Ipts;
    -- The random choice in M2 has lead
    -- to delta distinct points, since
    degree IGamma==delta
```

```
J=intersect apply(Ipts,I->I^2);
betti res J
-- The expected value for the number of forms of
-- degree d which double at delta points is
binomial(d+2,2)-3*delta
J'=intersect(J,(ideal gens R)^d);
tal=tally flatten degrees source gens J'
tal#d==binomial(d+2,2)-3*delta
f=gens J*random(source gens J,R^{-d});
IC=ideal f
degree IC==d
singC=saturate ideal jacobian IC;
singC==IGamma
-- => C has the expected delta ordinary double
--    as singularities and is smooth otherwise.

-- Since the eighteen points do not lie on a cubic,
-- no 14 of them lie on a conic, and
-- no 8 of them lie on a line,
-- the curve C is absolutely irreducible by Bezout.
betti(fIGamma= res IGamma)
-- => there are no linear relations between the forms of
--    degree d-4 of IGamma, which proves that the
--    Petri map is injective at the given point.
S=kk[y_0..y_(g-1)] -- the coordinate ring of PP^(g-1).
betti(LW=gens intersect(IGamma,(ideal gens R)^(d-3)))
RC=R/IC
phi=map(RC,S,LW)
betti(Ican=trim ker phi)
elapsedTime minimalBetti Ican

-- Replacing g by a number g' < 10
-- in the code above
-- proves the analogous results
-- for general curves of genus g'.
```

References

1. Marian Aprodu and Gavril Farkas. Green's conjecture for curves on arbitrary $K3$ surfaces. *Compos. Math.*, 147(3):839–851, 2011.
2. Marian Aprodu, Gavril Farkas, Ştefan Papadima, Claudiu Raicu, and Jerzy Weyman. Koszul modules and Green's conjecture. *Invent. Math.*, 218(3):657–720, 2019.
3. E. Arbarello, M. Cornalba, P. A. Griffiths, and J. Harris. *Geometry of algebraic curves. Volume I*, volume 267 of *Grundlehren Math. Wiss.* Springer, Cham, 1985.
4. Enrico Arbarello and Maurizio Cornalba. A few remarks about the variety of irreducible plane curves of given degree and genus. *Ann. Sci. Éc. Norm. Supér. (4)*, 16:467–488, 1983.
5. Enrico Arbarello, Maurizio Cornalba, and Phillip A. Griffiths. *Geometry of algebraic curves. Volume II. With a contribution by Joseph Daniel Harris*, volume 268 of *Grundlehren Math. Wiss.* Berlin: Springer, 2011.
6. Michael F. Atiyah and I. G. Macdonald. *Introduction to commutative algebra.* Reading, Mass.-Menlo Park, Calif.-London-Don Mills, Ont.: Addison-Wesley, 1969.
7. Christine Berkesch and Frank-Olaf Schreyer. Syzygies, finite length modules, and random curves. In *Commutative algebra and noncommutative algebraic geometry. Volume I: Expository articles*, pages 25–52. Cambridge: Cambridge University Press, 2015.
8. Christian Bopp. Syzygies of 5-gonal canonical curves. *Doc. Math.*, 20:1055–1069, 2015.
9. M. Boratyński and S. Greco. Hilbert functions and Betti numbers in a flat family. *Ann. Mat. Pura Appl. (4)*, 142:277–292, 1985.
10. Joel Briancon. Weierstrass prepare à la Hironaka. *Astérisque* 7-8(1973), 67-73 (1974), 1974.
11. Egbert Brieskorn and Horst Knörrer. *Plane algebraic curves. Transl. from the German by John Stillwell.* Basel-Boston-Stuttgart: Birkhäuser Verlag, 1986.
12. Alexander Brill and M. Nöther. Über die algebraischen Functionen und ihre Anwedung in der Geometrie. *Math. Ann.*, 7:269–310, 1874.
13. David A. Buchsbaum and David Eisenbud. Algebra structures for finite free resolutions, and some structure theorems for ideals of codimension 3. *Am. J. Math.*, 99:447–485, 1977.
14. R.-O. Buchweitz. Über Deformation monomialer Kurvensingularitäten und Weierstraßpunkte auf Riemannischen Flächen. Dissertation an der Universität Hannover, 1980.
15. Ciro Ciliberto and Rick Miranda. Degenerations of planar linear systems. *J. Reine Angew. Math.*, 501:191–220, 1998.
16. Marc Coppens and Gerriet Martens. Secant spaces and Clifford's theorem. *Compos. Math.*, 78(2):193–212, 1991.
17. Pierre Deligne and D. Mumford. The irreducibility of the space of curves of a given genus. *Publ. Math., Inst. Hautes Étud. Sci.*, 36:75–109, 1969.
18. David Eisenbud. *Commutative algebra. With a view toward algebraic geometry*, volume 150 of *Grad. Texts Math.* Berlin: Springer-Verlag, 1995.
19. David Eisenbud and Joe Harris. The monodromy of Weierstrass points. *Invent. Math.*, 90:333–341, 1987.

20. David Eisenbud and Joe Harris. On varieties of minimal degree. (A centennial account). Algebraic geometry, Proc. Summer Res. Inst., Brunswick/Maine 1985, part 1, Proc. Symp. Pure Math. 46, 3-13 (1987), 1987.
21. David Eisenbud, Herbert Lange, Gerriet Martens, and Frank-Olaf Schreyer. The Clifford dimension of a projective curve. *Compos. Math.*, 72(2):173–204, 1989.
22. Shalom Eliahou and Michel Kervaire. Minimal resolutions of some monomial ideals. *J. Algebra*, 129(1):1–25, 1990.
23. Burçin Eröcal, Oleksandr Motsak, Frank-Olaf Schreyer, and Andreas Steenpaß. Refined algorithms to compute syzygies. *J. Symb. Comput.*, 74:308–327, 2016.
24. Hubert Flenner, Liam O'Carroll, and Wolfgang Vogel. *Joins and intersections.* Springer Monogr. Math. Berlin: Springer, 1999.
25. Otto Forster. *Lectures on Riemann surfaces. Transl. from the German by Bruce Gilligan*, volume 81 of *Grad. Texts Math.* Springer, Cham, 1981.
26. W. Fulton. *Algebraic curves.* New York-Amsterdam: W.A. Benjamin, Inc., 1969.
27. William Fulton. *Intersection theory*, volume 2 of *Ergeb. Math. Grenzgeb., 3. Folge.* Springer, Cham, 1984.
28. I. M. Gelfand, M. M. Kapranov, and A. V. Zelevinsky. *Discriminants, resultants, and multidimensional determinants.* Boston, MA: Birkhäuser, 1994.
29. D. Gieseker. Stable curves and special divisors: Petri's conjecture. *Invent. Math.*, 66:251–275, 1982.
30. Roger Godement. Topologie algébrique et théorie des faisceaux. Actualités Scientifiques et Industrielles. 1252. Publications de l'Institut de Mathématique de l'Université de Strasbourg. XIII. Paris: Hermann & Cie. viii, 283 p. (1958)., 1958.
31. Hans Grauert. Über Modifikationen und exzeptionelle analytische Mengen. *Math. Ann.*, 146:331–368, 1962.
32. Hans Grauert. Über die Deformation isolierter Singularitäten analytischer Mengen. *Invent. Math.*, 15:171–198, 1972.
33. Hans Grauert and Reinhold Remmert. *Coherent analytic sheaves*, volume 265 of *Grundlehren Math. Wiss.* Springer, Cham, 1984.
34. Daniel R. Grayson and Michael E. Stillman. Macaulay2, a software system for research in algebraic geometry. Available at http://www2.macaulay2.com.
35. Mark L. Green. Koszul cohomology and the geometry of projective varieties. Appendix: The nonvanishing of certain Koszul cohomology groups (by Mark Green and Robert Lazarsfeld). *J. Differ. Geom.*, 19:125–167, 168–171, 1984.
36. Gert-Martin Greuel and Gerhard Pfister. *A Singular introduction to commutative algebra. With contributions by Olaf Bachmann, Christoph Lossen and Hans Schönemann.* Berlin: Springer, 2nd extended ed. edition, 2007.
37. Phillip Griffiths and Joseph Harris. On the variety of special linear systems on a general algebraic curve. *Duke Math. J.*, 47:233–272, 1980.
38. Wolfgang Gröbner. Über den Multiplizitätsbegriff in der algebraischen Geometrie. *Math. Nachr.*, 4:193–201, 1951.
39. Alexandre Grothendieck. Techniques de construction et théoremes d'existence en géométrie algébrique. IV: Les schemas de Hilbert. Sem. Bourbaki 13(1960/61), No. 221, 28 p. (1961), 1961.
40. Mark Haiman and Bernd Sturmfels. Multigraded Hilbert schemes. *J. Algebr. Geom.*, 13(4):725–769, 2004.
41. Robin Hartshorne. Connectedness of the Hilbert scheme. *Publ. Math., Inst. Hautes Étud. Sci.*, 29:5–48, 1966.
42. Robin Hartshorne. *Algebraic geometry*, volume 52 of *Grad. Texts Math.* Springer, Cham, 1977.
43. Grete Hermann. On the question of finitely many steps in the theory of polynomial ideals. (With the use of posthumous theorems by K. Hentzelt). *Math. Ann.*, 95:736–788, 1926.
44. D. Hilbert. Über die Theorie der algebraischen Formen. *Math. Ann.*, 36:473–534, 1890.
45. A. Hirschowitz and S. Ramanan. New evidence for Green's conjecture on syzygies of canonical curves. *Ann. Sci. Éc. Norm. Supér. (4)*, 31(2):145–152, 1998.

References

46. Melvin Hochster. Cohen-Macaulay rings, combinatorics, and simplicial complexes. Ring Theory II, Proc. 2nd Okla. Conf. 1975, 171-223 (1977)., 1977.
47. A. Hurwitz. On Riemann surfaces with monogenic transformations into themselves. *Math. Ann.*, 41:403–442, 1893.
48. Dale Husemöller. *Elliptic curves. With appendices by Otto Forster, Ruth Lawrence, and Stefan Theisen*, volume 111 of *Grad. Texts Math.* New York, NY: Springer, 2nd ed. edition, 2004.
49. Birger Iversen. *Cohomology of sheaves*. Universitext. Berlin: Springer, 1986.
50. Nathan Kaplan and Lynnelle Ye. The proportion of Weierstrass semigroups. *J. Algebra*, 373:377–391, 2013.
51. Michael Kemeny. Betti numbers of curves and multiple-point loci. *J. Pure Appl. Algebra*, 226(11):40, 2022. Id/No 107090.
52. Frances Kirwan. *Complex algebraic curves*, volume 23 of *Lond. Math. Soc. Stud. Texts.* Cambridge etc.: Cambridge University Press, 1992.
53. János Kollár. Sharp effective Nullstellensatz. *J. Am. Math. Soc.*, 1(4):963–975, 1988.
54. Emanuel Lasker. On the theory of modules and ideals. *Math. Ann.*, 60:20–115, 1905.
55. Robert Lazarsfeld. Brill-Noether-Petri without degenerations. *J. Differ. Geom.*, 23:299–307, 1986.
56. Frank Loose. On the graded Betti numbers of plane algebraic curves. *Manuscr. Math.*, 64(4):503–514, 1989.
57. J. Lüroth. Note on branch cuts and crosscuts on a Riemann surface. *Math. Ann.*, 3:181–184, 1871.
58. Yuri V. Matiyasevich. *Hilbert's tenth problem. With a foreword by Martin Davis*. Cambridge, MA: MIT Press, 1993.
59. Ernst W. Mayr and Albert R. Meyer. The complexity of the word problems for commutative semigroups and polynomial ideals. *Adv. Math.*, 46:305–329, 1982.
60. Rick Miranda. *Algebraic curves and Riemann surfaces*, volume 5 of *Grad. Stud. Math.* Providence, RI: AMS, American Mathematical Society, 1995.
61. Ferdinando Mora. An algorithm to compute the equations of tangent cones. Computer algebra, EUROCAM '82, Conf. Marseille/France 1982, Lect. Notes Comput. Sci. 144, 158-165 (1982), 1982.
62. Ferdinando Mora. An algorithmic approach to local rings. Computer algebra, EUROCAL '85, Proc. Eur. Conf., Linz/Austria 1985, Vol. 2, Lect. Notes Comput. Sci. 204, 518-525 (1985), 1985.
63. L. J. Mordell. On the rational solutions of the indeterminate equations of the third and fourth degrees. *Proc. Camb. Philos. Soc.*, 21:179–192, 1922.
64. Shigeru Mukai. Curves, K3 surfaces and Fano 3-folds of genus ≤ 10. Algebraic geometry and commutative algebra, in Honor of Masayoshi Nagata, Vol. I, 357-377 (1988), 1988.
65. D. Mumford. The topology of normal singularities of an algebraic surface and a criterion for simplicity. *Publ. Math., Inst. Hautes Étud. Sci.*, 9:5–22, 1961.
66. D. Mumford, J. Fogarty, and F. Kirwan. *Geometric invariant theory*, volume 34 of *Ergeb. Math. Grenzgeb.* Berlin: Springer-Verlag, 3rd enl. ed. edition, 1994.
67. David Mumford. *Lectures on curves on an algebraic surface*, volume 59 of *Ann. Math. Stud.* Princeton University Press, Princeton, NJ, 1966.
68. David Mumford. Curves and their Jacobians. Ann. Arbor: The University of Michigan Press. 104 p. (1975), 1975.
69. Max Nöther. On the invariant representation of algebraic functions. Erl. Ber. 1880, Clebsch Ann. XVII. 263-284 (1880), 1880.
70. Karl Petri. Über die invariante Darstellung algebraischer Funktionen einer Veränderlichen. *Math. Ann.*, 88:242–289, 1923.
71. Ragni Piene and Michael Schlessinger. On the Hilbert scheme compactification of the space of twisted cubics. *Am. J. Math.*, 107:761–774, 1985.
72. Bjorn Poonen. Undecidability in number theory. *Notices Am. Math. Soc.*, 55(3):344–350, 2008.
73. J. A. Robinson. *Logic: Form and function. The mechanization of deductive reasoning*. Edinburgh: Edinburgh University Press, 1979.

74. Michael Sagraloff. Special linear series and syzygies of canonical curves of genus 9. Dissertation an der Universität des Saarlandes, 2006.
75. Frank-Olaf Schreyer. Die Berechnung von Syzygien mit dem verallgemeinerten Weierstraßchen Divisionssatz und eine Anwendung auf analytische Cohen-Macaulay Stellenalgebren minimaler Multiplizität. Diplomarbeit an der Universität Hamburg, 1980.
76. Frank-Olaf Schreyer. Syzygies of canonical curves and special linear series. *Math. Ann.*, 275:105–137, 1986.
77. Frank-Olaf Schreyer. A standard basis approach to syzygies of canonical curves. *J. Reine Angew. Math.*, 421:83–123, 1991.
78. Frank-Olaf Schreyer. Some topics in computational algebraic geometry. In *Advances in algebra and geometry. Proceedings of the international conference on algebra and geometry, Hyderabad, India, December 7–12, 2001*, pages 263–278. New Delhi: Hindustan Book Agency, 2003.
79. Jean-Pierre Serre. Faisceaux algébriques cohérents. *Ann. Math. (2)*, 61:197–278, 1955.
80. Jean-Pierre Serre. *Algèbre locale. Multiplicités. Cours au Collège de France, 1957-1958, rédigé par Pierre Gabriel. 2e ed.*, volume 11 of *Lect. Notes Math.* Springer, Cham, 1965.
81. F. Severi. Sulla classificazione delle curve algebriche e sul teorema d'esistenza di *Riemann. Rom. Acc. L. Rend. (5)*, 24(1):877–888, 1011–1020, 1915.
82. B. L. van der Waerden. *Moderne Algebra. Unter Benutzung von Vorlesungen von E. Artin und E. Noether, Tl. 1. 3. verb. Aufl*, volume 33 of *Grundlehren Math. Wiss.* Springer, Cham, 1950.
83. Claire Voisin. Courbes tétragonales et cohomologie de Koszul. (Tetragonal curves and Koszul cohomology). *J. Reine Angew. Math.*, 387:111–121, 1988.
84. Claire Voisin. Green's generic syzygy conjecture for curves of even genus lying on a $K3$ surface. *J. Eur. Math. Soc. (JEMS)*, 4(4):363–404, 2002.
85. Claire Voisin. Green's canonical syzygy conjecture for generic curves of odd genus. *Compos. Math.*, 141(5):1163–1190, 2005.
86. Joachim von zur Gathen and Jürgen Gerhard. *Modern computer algebra*. Cambridge: Cambridge University Press, 3rd ed. edition, 2013.
87. O. Zariski. *Algebraic surfaces. With appendices by S.S. Abhyankar, J. Lipman, and D. Mumford. 2nd suplemented ed*, volume 61 of *Ergeb. Math. Grenzgeb.* Springer-Verlag, Berlin, 1971.
88. Oscar Zariski and Pierre Samuel. *Commutative algebra. Vol. 1. With the cooperation of I. S. Cohen.* Princeton, N.J.-Toronto-London-New York: D. Van Nostrand, 1958.

Glossary

$>$ monomial order

$>_{lex}$ lexicographic monomial order

$>_{rdlex}$ degree reverse lexicographic monomial order

$>_{ldrlex}$ local degree reverse lexicographic monomial order

$\mathbb{A}^n$ affine n-space over an algebraically closed field

$D \sim E$ two linearly equivalent divisors

$\mathcal{F}_p$ stalk of a (pre)-sheaf $\mathcal{F}$ at a point p

$\mathcal{G}_d^r$ relative Brill-Noether variety

$H^i(X, \mathcal{F})$ the i-th cohomology group of the (coherent) sheaf $\mathcal{F}$ on the variety X

$I : J$ the colon ideal of I and J

$\mathbf{I}(A)$ the vanishing ideal of an subset $A \subset \mathbb{A}^n$, or the homogeneous ideal of $A \subset \mathbb{P}^n$

$K(A)$ the function field if a variety A

$K[A]$ the coordinate ring of a affine variety A

$L(D)$ the Riemann-Roch space of a divisor D

$L(d; r_1 p_1, \ldots, r_s p_s)$ the vector space of homogeneous forms of degree d with assigned base points p_i of muliplicity r_i

$\mathrm{Lt}(f)$ the leading term of a polynomial f with respect to a monomial order

$\mathrm{Lt}(I)$ the ideal of leading terms of an ideal I with respect to a monomial order

$\mathfrak{M}_g$ the moduli space of curve of genus g

$\mathcal{O}(d)$ twisted sheaf on $\mathbb{P}^n$

$\mathcal{O}_X$ sheaf of regular functions on X

$\mathcal{O}_X(D)$ sheaf of rational functions on X with poles bounded by D

$\mathcal{O}_{A,p}$ the local ring of A at a point p

$\Omega(C)$ the $K(C)$ vector space of rational differential forms on a curve C

Ω^1_X the sheaf of Kähler differentials on a variety X

$\mathbb{P}^n$ projective space of dimension n

$T_p(A)$ tangent space of a variety A at a point p

$V(I)$ the vanishing locus of the ideal I

$X(k)$ the set of k-rational points of a variety X

$\chi(X,\mathcal{F})$ Euler characteristic of a sheaf $\mathcal{F}$ on X

g^r_d an r-dimensional linear system of divisors of degree d

$h^i(X,\mathcal{F})$ the dimension of $H^i(X,\mathcal{F})$

$\ell(D)$ the dimension of $L(D)$

$p_a(C)$ the arithmetic genus of a curve C

$\pi_*\mathcal{F}$ direct image sheaf of $\mathcal{F}$

v_p valuation at a point p of a curve

ω_X dualizing sheaf on X

Index

$>$, monomial order, 8
$>_{\text{lex}}$, lexicographic order, 9
$>_{\text{rdlex}}$, reverse degree lexicographic order, 9
$>_{\text{ldrlex}}$, local degree reverse lexicographic order, 133
$D \sim E$, linear equivalence of divisors, 210
$H^i(X, \mathcal{F})$, i-th cohomology group of X with values in $\mathcal{F}$, 255
$I : J$, colon ideal, 15
$K(A)$, function field of A, 62
$K[A]$, coordinate ring of A, 36
$L(D)$, Riemann-Roch space of the divisor D, 209, 261
$L(d; r_1 p_1, \ldots, r_s p_s)$, 177
$V(I)$, vanishing loci of the ideal I, 24
$X(k)$, set of k-rational points of an algebraic set X, 6
$\mathbb{A}^n$, affine n-space, 6
$\Gamma(U, \mathcal{F})$, space of sections of the sheaf $\mathcal{F}$ over U, 253
$\Gamma_{\geq 0}(\mathcal{F})$, module of global sections of twists of $\mathcal{F}$, 254
$I(A)$, vanishing ideal of A, 31
$Lt(I)$, ideal of leading terms of I, 12
$Lt(f)$, leading term of f, 8
$\Omega(C)$, the vector space of rational differential forms on C, 216
Ω^1_X, sheaf of Kähler differentials on X, 261
$\mathbb{P}^n$, projective space of dimension n, 92
$\chi(X, \mathcal{F})$, Euler characteristic of the sheaf $\mathcal{F}$ on X, 259
$\ell(D)$, dimension of $L(D)$, 210

$\mathcal{G}^r_{g,d}$, relative Brill-Noether variety, 247
$\mathfrak{M}_g$, moduli space of curves of genus g, 234, 247
ω_X, dualizing sheaf on X, 262
$\pi_* \mathcal{F}$, direct image sheaf of $\mathcal{F}$, 259
$\mathcal{F}_p$, stalk of the (pre)sheaf $\mathcal{F}$ at the point p, 250
$\mathcal{O}(d)$, twisted sheaf on $\mathbb{P}^n$, 253
$\mathcal{O}_X$, sheaf of regular functions on X, 250
$\mathcal{O}_X(D)$, sheaf of rational function on X with poles bounded by D, 261
$\mathcal{O}_{A,p}$, local ring of A at p, 63, 251
g^r_d, an r-dimensional linear system of divisors of degree d, 211
$h^i(X, \mathcal{F})$, dimension of $H^i(X, \mathcal{F})$, 259
$p_a(C)$, arithmetic genus of C, 109, 204
v_p, valuation at the point p of a curve, 145

absolutely
　primary, 59
　prime, 59
adjunction
　formula, 266
　sequence, 262
affine
　k-algebra, 5
　n-space, 6
algebraic
　over k, 66
　set, 6
　　projective, 93
　subset, 6, 32
algebraically independent, 66
algorithm
　Buchberger, 14

298   Index

colon ideals, 81
computation of syzygies, 80
division with remainder, 23
elimination, 83
finite free resolution, 103
Hom(M,N), 89
homogenization, 99
homology of a R-module complex, 88
intersection of ideals, 79
kernel
   of a ring homomorphism, 83
   of an R-module homomorphism, 86
lifting, 85
Mora division, 135
Riemann-Roch space
   on a smooth projective curve, 230
   via a plane model, 224
submodule membership, 23
tangent cone, 141
analytically isomorphic, 142
annihilator, 42
arithmetic genus, 109, 204
assigned base points, 177
associated
   graded ring, 141
   prime, 46
   primes, 45
     of a module, 47
     of an ideal, 48
associated sheaf
   of a module, 253

base point, 177
   free, 211
Bertini's Theorem, 192
Bézout's theorem
   for plane curves, 116
   for the intersection with a hypersurface, 120
bidegree, 148
bihomogeneous
   coordinate ring, 150
   polynomial, 148
   vanishing ideal, 150
binomial ideal, 24
birational map, 65
blow-up, 165
boundaries, 104
bounded complex, 104
branch
   point, 196
Briancon's Theorem, 195
Buchberger's criterion, 14
   second version, 20

canonical
   curve, 229
   divisor, 219, 265
Čech cohomology, 256
chain
   complex, 104
   of prime ideals, 75
chart
   of the standard atlas of $\mathbb{P}^n$, 93
Clifford index
   of a curve, 238
   of a divisor, 238
Clifford's theorem, 237
closed
   embedding, 160
coherent
   sheaf, 252
cohomology
   groups, 255
cokernel, 18
colon ideal, 15
complete linear system, 210
complex, 104
component decomposition, 39
contracted ideal, 56
convex coordinates, 96
coordinate points of $\mathbb{P}^n$, 96
coordinate ring, 36
coprime, 49
Cremona resolution, 172
cubic
   scroll, 168
curve, 70
cusp, 115, 142
cycles, 104

d-uple embedding, 154
degree
   of a divisor, 208
   of a polynomial, 9
   of a projective algebraic set, 107
   of an ideal, 107
degree shifted module, 102
dimension
   of a ring, 75
   of an algebraic set, 66
dimension bound, 153
direct image sheaf, 259
discrete valuation, 143
   ring, 144
division with remainder
   in free modules, 19
   in one variable, 4
   multivariate, 11

Index   299

divisor, 208
　cut out by a hypersurface, 208
　of multiple points, 223
　principle, 209
Dixon's Lemma, 10
domain of definition
　of a function, 63
　of a rational map, 64
dominant rational map, 64
double lines, 176
dualizing sheaf, 261

Eagon-Northcott complex, 242
effective divisor, 208
elimination ideal, 22
embedded prime, 45
Euler characteristic, 259
exceptional curve, 166
extended ideal, 56

fiber, 160
field
　algebraically closed, 6
　of definition, 6
　of rational functions, 62
finite
　free resolution, 100, 103
　presentation, 18
　ring extension, 71
finitely generated
　module, 18
flex, 194
formal Laurent series, 144
fundamental
　points of the quadratic transformation, 171
　theorem
　　of algebra, 6
　　of elimination, 159

genus
　arithmetic, 109, 204
　geometric, 203
　topological, 196, 204
geometric genus, 203, 268
germ, 250
　of a section, 250
going-up theorem, 76
gonality, 238
graded
　module, 102
　ring, 97
graded Betti numbers
　of a module, 127
　of a resolution, 103

Grassmannian, 179
Grauert division, 130
Gröbner basis, 12
　reduced, 14
group law
　on a cubic, 231

height
　of a prime ideal, 76
Hessian, 205
higher secant variety, 164
Hilbert
　function, 105
　polynomial, 105
　scheme, 185
Hilbert's
　basis theorem, 12
　Nullstellensatz
　　strong version, 33
　　weak version, 5, 24
　syzygy theorem, 100
　　graded case, 103
homogeneous
　coordinate ring, 98
　coordinates, 92
　ideal, 98
　polynomial, 92
homogenization
　of a polynomial, 94
　of an ideal, 99
homology of a complex, 87, 104
homomorphism
　of K-algebras, 37
　of graded modules, 102
　theorem, 17
homotopy, 88
　of complexes, 104
Hopf fibration, 94
hyperelliptic curve, 229
hyperplane
　at infinity, 94
hypersurface
　of bidegree (d, e), 150

ideal, 3
　binomial, 24
　colon, 15
　generated by, 3, 10
　maximal, 35
　monomial, 10
　of denominators of a rational function, 63
　of initial forms, 141
　of leading terms, 12
　primary, 43

prime, 34
radical, 34
ideal membership problem, 5
infinitesimal near point
  of first order, 203
  of seconf order, 203
integral
  over an ideal, 71
  ring extension, 71
intersection
  divisor, 209
  multiplicity
    along Z, 120
    of plane curves, 112
  number, 265
irreducible
  algebraic set, 35
  ideal, 44
irredundant
  component decomposition, 39
irregularity, 268
irrelevant ideal, 98
isomorphism
  of affine algebraic sets, 37

join, 154

k-rational point
  of X, 6
Key property of $>_{\text{lex}}$, 23
Krull dimension, 75
Krull's
  intersection theorem, 126
  prime existence lemma, 74
  principal ideal theorem, 139

leading term, 8
  module, 20
length
  of a chain of prime ideals, 75
Leray's theorem, 256
lifting
  of A along B, 85
linear
  eqivalent, 210
  Noether normalization, 153
  projection from L, 153
  system
    of hypersurfaces, 176
linearly
  equivalent divisors, 210
local ring, 125
  of A at p, 63
localization

at a prime, 54
in f, 54
in a multiplicative subset U, 52
lying-over theorem, 73

$\mathfrak{m}$-adic
  topology, 129
Macaulay's theorem, 12
map of complexes, 104
maximal ideal, 35
  of a local ring, 125
minimal
  free resolution, 127
  generator
    of a monomial ideal, 10
  polynomial, 66
  primary decomposition, 45
  prime, 45
  set of generators, 126
minimum
  of divisors, 211
module, 16
  graded, 102
  homomorphism, 17
  of leading terms, 20
monomial, 7
  ideal, 10
  in a free module, 18
monomial order, 8
  global, 9
  induced, 21
  lexicographic, 9
  local, 9
  product order, 82
  reversed degree lexicographic, 9
  weight order, 10
Mora division, 135
morphism
  between affine algebraic sets, 36
  between quasi-projective algebraic sets, 151
  between sheaves, 251
multiplicative subset, 51
multiplicity
  of a plane curve at a point, 114

Nakayama's lemma, 126
net, 177
node, 115, 142
Noether's
  AF+BG Theorem, 257
  reduction lemma, 227
Noether's isomorphism theorems, 49
Noetherian
  module, 43

Index

ring, 40
non-reduced subscheme, 184
Nullstellensatz
   projective, 98
   strong version, 33
   weak version, 5
numerical polynomial, 109

ordinary m-fold point, 114
ordinary double point, 142

$\mathfrak{P}$ lies over $\mathfrak{p}$, 73
pairs of lines, 176
Pascal's Theorem, 179
pencil, 177
Plücker
   coordinates, 180
   embedding, 180
   formulas, 204
   quadric, 181
plane conics, 176
polar, 205
power series ring
   formal, 128
presheaf, 249
primary decomposition, 43
   minimal, 45
primary ideal, 43
prime
   embedded, 45
   ideal, 34
   isolated, 45
prime divisor, 260
principal divisor, 209
projection
   from L, 153
   theorem, 25
projective
   algebraic set, 93
   morphism, 160
   space, 92

quadratic transformation, 171
quasi-affine, 150
quasi-projetive, 150
quotient module, 17

radical, 33
ramification
   divisor, 221
   index, 221
   point, 196
rational
   differential form, 216

function field, 62
   of a projective variety, 111
map, 64
   dominant, 64
regular
   differential form, 219
   local ring, 140
residue
   class, 3
   field of a local ring, 125
   of a meromorphic differential form, 227
   ring, 3
   theorem, 226
resolution of plane curve singularities, 168
resultant, 118
Riemann's
   existence theorem, 232
   inequality, 214
Riemann-Hurwitz formula, 197
   second version, 222
Riemann-Roch
   on surfaces, 267
   space, 209
   theorem, 222
ring
   of regular functions, 151
Roman surface, 97

S-polynomial, 14
saturated ideal, 184
scaling point, 96
Schreyer's corollary, 22
Schubert varieties, 182
secant variety, 163
section
   of a sheaf, 250
Segre product, 149
separable
   morphism, 221
sheaf, 250
short exact sequence
   of modules, 47
singular point
   of a plane curve, 114
   of an algebraic set, 140
smooth point
   of a plane curve, 114
special divisor, 237
stalk
   of a sheaf, 250
stalk of a sheaf, 250
standard atlas of $\mathbb{P}^n$, 93
standard graded polynomial ring, 97
Stanley-Reisner ring, 128

Steiner surface, 97
Steinitz's theorem, 6
strict transform, 166
structure sheaf, 250
submodule, 17
subscheme
   of $\mathbb{A}^n$, 183
   of $\mathbb{P}^n$, 184
surface, 70
Sylvester matrix, 118
syzygies
   computation of, 80
syzygy, 17
   module, 18
   theorem, 100

tacnode, 115
tame ramification, 221
tangent
   cone, 141
   lines, 114
   space, 138, 140
term, 7
3-fold, 70
topological
   Euler characteristic, 196
   genus, 196, 204
topology
   $\mathfrak{m}$-adic, 129
total
   ramification number, 196
   transform, 166
transcendence
   basis, 67
   degree, 67
   transcendental over k, 66
transversal intersection, 116
triple point, 115
twisted cubic
   curve, 31
   projective closure, 99

unmixed, 70

valuation ring, 144
vanishing
   ideal, 31
   locus, 2, 24
   order, 208
variety, 35
Veronese
   embedding, 154
   surface, 155, 239

web, 177
Weierstrass
   $\wp$-function, 232
   points, 231
   preparation theorem, 132
Weil divisor, 261
wild ramification, 221
Wronski determinant, 231

Zariski
   open, 32
   tangent space, 140
   topology, 32
   of $\mathbb{P}^n$, 93

The manufacturer's authorised representative in the EU is Springer Nature Customer Service Centre GmbH, Europaplatz 3, 69115 Heidelberg, Germany. If you have any concerns regarding our products, please contact ProductSafety@springernature.com

Printed and bound by CPI Group (UK) Ltd, Croydon, CR0 4YY

26/03/2026

02078953-0008